Human Exposure to Pollutants via Dermal Absorption and Inhalation

Editors
Mihalis Lazaridis
Technical University of Crete
Department of Environmental Engineering
Polytechneioupolis
73100 Chania
Greece
lazaridi@mred.tuc.gr

Ian Colbeck
University of Essex
Interdisciplinary Centre for Environment and Society
Department of Biological Sciences
Colchester, CO4 3SQ
United Kingdom
colbi@essex.ac.uk

ISBN 978-90-481-8662-4 e-ISBN 978-90-481-8663-1
DOI 10.1007/978-90-481-8663-1
Springer Dordrecht Heidelberg London New York

Library of Congress Control Number: 2010924437

Printed on acid-free paper

Springer is part of Springer Science+Business Media (www.springer.com)

Preface

Estimates of the air pollution health impact play a crucial role in environmental protection. These estimates require accurate data on the pollutant exposure and dose to the population as well as the dose–response relationships to calculate the health impact. From an air quality manager's perspective there is concern about the validity and accuracy of these calculations. There is a need for information and possible ways to adjust the assessment.

One important topic for air quality managers is to understand the relative contribution of sources to the total exposure. These sources may be coming from both different outdoor sources from sectors such as transport, industry and energy industries, and from a number of indoor sources, such as heating, ventilation and indoor activities as well as out-gassing from building material and furniture.

Indoor air quality is now drawing the attention of policy makers. The basic right to, and importance of, healthy indoor air was emphasized by the World Health Organization as early as 2000 and several countries have described target concentrations for various pollutants. The WHO Air Quality Guidelines 2005 recommended the development of specific guidelines for indoor air quality and these are expected to be published soon. Indoor air pollutants have not been as extensively monitored as outdoor air pollutants and the evidence base for contributions to health effects needs to be strengthened.

This book reviews information necessary to address the steps in the exposure assessment relevant to air pollution. The aim is to identify available information including data sources and models, and show that an integrated multi-route exposure model can be built, validated and used as part of air quality management process.

Environmental levels of air pollutants are reviewed based on monitoring information from background and urban areas, and compared to the current EU air quality legislation. Further, available information about particle speciation is also assessed. This is one of major gaps of knowledge currently, as most information is based on background measurement sites, and it can not be linked to health effect information due to expected differences in composition.

Many epidemiological studies have focused on inhalation exposure. Whilst this is appropriate for many substances failure to consider the importance of exposure and uptake of material deposited on the skin may lead to an over/under estimation of the risk. Hence dermal exposure is also considered. This involves an assessment of the quality of drinking water across Europe, providing information about differences between water originating from surface and ground water sources, and providing approximate models for disinfection by-product formation.

Indoor exposures are discussed on the basis of literature information, but more importantly, based on recent measurement campaigns that were targeted towards a description of particulate matter indoors. Other elements of exposure models such as aspects related to human behaviour are also reviewed.

A short review of health aspects connected to ambient levels of air pollution and to drinking water contamination by disinfection by products is also included.

The book also gives significant attention to micro-environmental modelling. This is one of the bases of exposure assessment that has often been identified as an area with significant knowledge gaps. A number of models and other information are identified that addresses the most important elements such as ventilation and infiltration rates, sources and sinks, or general concepts of compartmental modelling.

To further advance all the elements necessary to build and implement a comprehensive exposure model, the review also gives information about ongoing or recently published studies relevant to the topic, that are not yet available in open literature.

The book also includes a review on internal dose modelling both from inhalation and from dermal absorption and show how these can be incorporated into an air quality management system.

Chapter 1 summarizes the environmental levels of air pollutants and quality of drinking water, while Chapters 2 and 3 describe indoor–outdoor relationships of air pollutants and key chemical processes occurring in the indoor environment. In indoor environments, chemistry can significantly alter the composition of the air we breathe and chemical transformations reduce our exposure to reactants and increase our exposure to products. Personal exposure measurements are analyzed in Chapter 4 and this topic is of great importance for accurate exposure evaluation since people move, commute, and frequently change their place they can be exposed every day to various kinds and mixtures of gases and airborne particles. Health effects from air pollutants are studied in Chapter 5 as well as a risk analysis and health impact assessments. Chapter 6 gives an overview of particulate matter deposition in the human respiratory tract including model developments and experimental measurements. Chapter 7 focuses on dermal absorption modelling of chemical pollutants into and across human skin since the skin is a primary route of systemic exposure to a number of environmental pollutants. Micro-environmental modelling is studied in Chapter 8. Indoor air models can be utilized to predict the indoor air quality and are also useful from the engineering point of view to maintain an acceptable indoor air quality in a building. Finally, Chapter 9 studies human exposure to air pollutants using an air quality management computer platform. An Environmental Management System incorporating a human exposure module enables local authorities and scientists to reduce the environmental impacts of air pollution to population and to increase the efficiency with which environmental policies are implemented.

Greece Mihalis Lazaridis
Colchester, UK Ian Colbeck

Contents

Contributors

Trond Bøhler
Norwegian Institute for Air Research, P.O. Box 100, N-2007 Kjeller, Norway
trond.bohler@nilu.no

Martin Braniš
Faculty of Science, Institute for Environmental Studies,
Charles University in Prague, Albertov 6, 128 43 Prague 2, Czech Republic
branis@natur.cuni.cz

Ian Colbeck
Department of Biological Sciences, Interdisciplinary Centre for Environment and Society, University of Essex, Colchester, CO4 3SQ, UK
colbi@essex.ac.uk

Christos Housiadas
"Demokritos" National Centre for Scientific Research, Institute of Nuclear Technology-Radiation Protection, 15310 Agia Paraskevi, Athens, Greece
christos@ipta.demokritos.gr

Tareq Hussein
Department of Physics, University of Helsinki, P.O. Box 64,
FI–00014, UHEL, Finland

Markku Kulmala
Department of Physics, University of Helsinki, P.O. Box 64,
FI–00014, UHEL, Finland
markku.kulmala@helsinki.fi

Mihalis Lazaridis
Department of Environmental Engineering, Technical University of Crete,
Polytechneioupolis, 73100 Chania, Greece
lazaridi@mred.tuc.gr

Glenn Morrison
Missouri University of Science and Technology, 221 Butler-Carlton Hall, Rolla MO 65409, USA
gcm@mst.edu

Zaheer Ahmad Nasir
Department of Biological Sciences, Interdisclpinary Centre for Environment and Society, University of Essex, Colchester, CO4 3SQ, UK

Caroline Ochieng
Public & Environmental Health Research Unit, London School of Hygiene & Tropical Medicine, Keppel Street, London WC1E 7HT, UK

Xiyu Phoon
Public & Environmental Health Research Unit, London School of Hygiene & Tropical Medicine, Keppel Street, London WC1E 7HT, UK

Jim E. Riviere
Center for Chemical Toxicology Research and Pharmacokinetics, College of Veterinary Medicine, North Carolina State University, Raleigh, NC 27606, USA
jim_riviere@ncsu.edu

Sotiris Vardoulakis
Public & Environmental Health Research Unit, London School of Hygiene & Tropical Medicine, Keppel Street, London WC1E 7HT, UK
sotiris.vardoulakis@lshtm.ac.uk

Chapter 1
Environmental Levels

Mihalis Lazaridis and Ian Colbeck

Abstract One of the most important environmental concerns of today is the negative impact of pollution on human health. The air we breathe and the water we drink are essential ingredients for a healthy life. Unfortunately polluted water and air are common throughout the world. Over a day a healthy adult will consume between 2 and 3 l of fluid and inhale around 11 m^3 of air. While exposure to pollutants in air is via inhalation that for water may occur via the ingestion, dermal absorption and inhalation routes. In this chapter we review the sources and concentrations of various air pollutants before considering drinking water quality. For the latter we concentrate on the potentially harmful disinfection by-products.

1.1 Introduction

The atmosphere contains thousands of chemical species in trace quantities (ppm to ppt levels) (Finlayson Pitts and Pitts 1986, 2000). The troposphere can be viewed as a huge container which includes gaseous and particulate matter pollutants. The atmosphere is a dynamic system with continuous exchange of its gaseous components between the atmosphere and the earth surface including the vegetation and the oceans. Emissions of pollutants are transported into the atmosphere at long distances from their sources. The dynamic of the atmosphere and the chemical

M. Lazaridis (✉)
Department of Environmental Engineering, Technical University of Crete, Polytechneioupolis, 73100 Chania, Greece
e-mail: lazaridi@mred.tuc.gr

I. Colbeck
Department of Biological Sciences, Interdisciplinary Centre for Environment and Society, University of Essex, Colchester CO4 3SQ, UK
e-mail: colbi@essex.ac.uk

M. Lazaridis and I. Colbeck (eds.), *Human Exposure to Pollutants via Dermal Absorption and Inhalation*, Environmental Pollution 17, DOI 10.1007/978-90-481-8663-1_1,

reactivity of the pollutants, as well as, the size of particulate matter determine their residence time and their effects to humans and ecosystems (Seinfeld and Pandis 2006). Table 1.1 presents the different spatial scales of pollutant transport in the atmosphere and related physico-chemical processes.

Figure 1.1 presents the different time and length scales related to atmospheric processes ranging from molecular diffusion to climatic impacts. In the atmosphere the chemical composition of atmospheric species can be divided into four main groups, namely sulfur, nitrogen, carbon and halogen containing compounds (Finlayson-Pitts and Pitts 1986; Seinfeld and Pandis 2006). Of course there are chemical compounds in the above groups which include atoms from other groups such as compounds which include both sulfur and carbon atoms. The chemical compounds which are emitted into the atmosphere eventually are removed and there exists a cycle for these compounds which is called the biogeochemical cycle.

Table 1.1 Spatial scales of pollutant transport in the atmosphere and related phenomena

Scale	Dimension	Examples of physical and chemical processes in the atmosphere
Molecular scale	<<2 mm	Molecular diffusion
Microscale	2 mm–2 km	Industry emissions, clouds
Mesoscale	2–2,000 km	Cloud coalesce, storms, air pollution at urban centers
Synoptic scale	500–10,000 km	Low and high pressure systems, ozone hole in the Antarctica
Global scale	>10,000 km	Decrease of stratospheric ozone, planetary wind systems

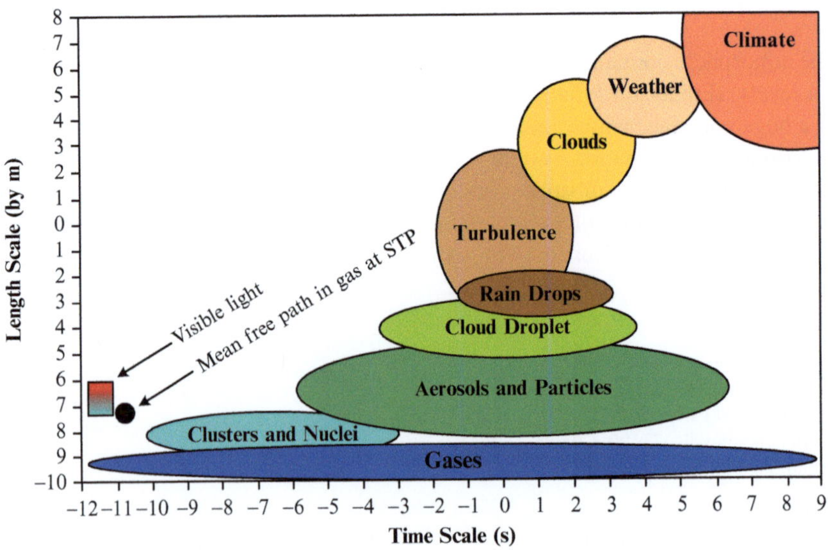

Fig. 1.1 Time (s) and length (m) scales in the atmosphere and related phenomena. The numbers in the x and y axes are the power factor in the scale of 10. For example the number –8 is equivalent to 10^{-8} s

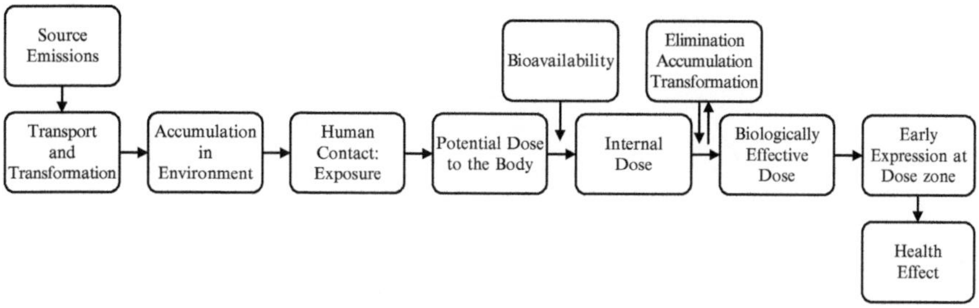

Fig. 1.2 Schematic representation of the complex sequence from emissions of air pollutants to health effects (Adapted from Lioy 1990)

The term "air pollution" is used when chemical compounds, emitted from mainly anthropogenic activities, are at concentrations above their normal ambient levels and have measurable effects on humans and ecosystems.

The understanding of the complex sequence of events starting from the emissions of air pollutants to the atmosphere with the human health effects as a final event is necessary for the prognosis of potential risk to humans from specific chemical compounds and mixtures of them (see Fig. 1.2). Furthermore, the understanding of the chemical composition/size distribution characteristics of particulate matter (PM) and the chemical reactivity of gaseous pollutants, together with their indoor-outdoor characteristics and their relation to human exposure and internal dose, are necessary steps for the quantification of human exposure to air pollutants.

Specifically, concerning particulate matter, there are recent cohort studies (see for a summary HEI 2001) which have followed the successful implementation of the Harvard Six Cities Study (Dockery et al. 1993) and the American Cancer Society (ACS) study (Pope et al. 1995) and they show significant effects of particulate matter levels on mortality and morbidity. Long-term average mortality rates ranging from 17% to 26% higher for people living in communities with higher levels of $PM_{2.5}$ were observed (USEPA 2004).

In addition to the above findings other studies re-examined the previous epidemiological findings to ensure the reproduction and the consistency of previous results. The HEI US-wide National Morbidity, Mortality and Air Pollution Study (NMMAPS) (Samet et al. 2000) is a notable example of these efforts. The NMMAPS study shows a 0.5% increase in total non accidental mortality associated with a 10 $\mu g/m^3$ increase in the 90 largest US cities where daily average PM_{10} ranged from 15 to 53 $\mu g/m^3$. Similar results had been previously obtained from the European APHEA study (0.6% per 10 $\mu g/m^3$) (Katsouyanni 1996) and with meta-analysis in Europe and North and South America (0.7% per 10 $\mu g/m^3$) (Levy et al. 2000).

Even though the epidemiological studies have shown an important relationship between PM levels and health effects, many important questions remain to be addressed. There is a need for a better understanding of the factors responsible for the variability in health effects due to PM among geographical regions. Also, there is a clear need to identify the mechanisms by which short and long term exposure to PM affects the cardiovascular and respiratory system (HEI 2001). Additionally the

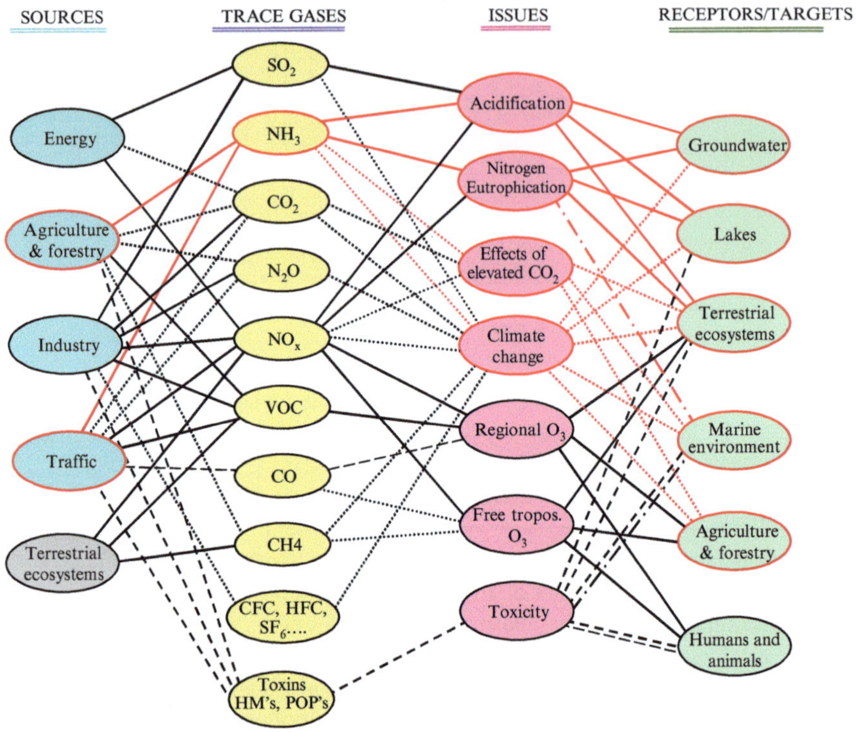

Fig. 1.3 Conceptual linkages between the sources of trace gas emissions, environmental issues and receptors of concern. The line styles represent the main international policy structure dealing with each link (Adapted from Sutton et al. 2003; Erisman et al. 2001; Grennfelt and Schjoldager 1984)

physico-chemical features of PM responsible for respiratory allergies, asthma and health effects, in general, are not still clearly identified.

A more general conceptual picture of the effects that air pollutants have to ecosystems and humans is shown in Fig. 1.3. This picture relates the different air pollutants together with their sources and their effects including the different international protocols related to specific compounds. The transport of pollutants and emissions from marine and atmospheric sources outside Europe influence the air pollution climate and policies within Europe. The importance of this transport varies between the different compounds due to differences in their atmospheric residence times. The multiplicity of linkages demonstrate the complexity of the interactions between the different pollutants at different receptors.

1.2 Primary Emissions of Air Pollutants

Air pollution and its effect on air quality and the environment is one of the paramount environmental problems internationally. The most important atmospheric pollutants are carbon dioxide and carbon monoxide (CO_2, CO), Non Methane

Volatile Organic Compounds (NMVOCs), methane (CH_4), nitrogen oxides (NO_x, N_2O), ammonia (NH_3), sulfur dioxide (SO_2), Particulate Matter (PM), Heavy Metals (HM) and Persistent Organic Pollutants (POPs). The emissions of the above air pollutants originate both from anthropogenic and natural sources (Finlayson-Pitts and Pitts 1986; EMEP/CORINAIR 2002).

As an example, primary particles can be divided as anthropogenic or natural depending on their origin. Secondary formed particles in the atmosphere are of both natural and anthropogenic origin and originate from chemical transformations of gaseous precursors such as sulfur dioxide, nitrogen oxides and VOCs. Recent research studies highlight the importance of biogenic hydrocarbons (such as terpenes) in the formation of organic aerosols (Seinfeld and Pandis 2006).

In Europe anthropogenic sources are dominant because of the urbanization of many countries and the large number of vehicle and combustion sources (industrial and residential) (CAFE 2004). Natural sources of primary aerosols in Europe include sea spray, fugitive dust (e.g. soil resuspension by the wind), long range transport of Sahara dust, volcanic and biogenic emissions. There is a consistent pattern of geographical variability in Europe with lower concentrations of PM in the far north and higher concentrations in the southern countries) (CAFE 2004). This is possible due to natural emissions of unsaturated hydrocarbons (including isoprene) which are highly reactive. However, increased concentration of PM is found also in central Europe due to high anthropogenic emissions. Many research studies have been performed in north-western Europe, where aerosol concentrations show little variation between urban and non-urban areas (CAFE 2004). This can be attributed to high emissions from a dense population area, the small weather variability between the measurement sites and the importance of long-range transport of air pollutants.

The available emission data were used by the Co-operative Programme for Monitoring and Evaluation of the Long-Range Transmission of Air Pollutants in

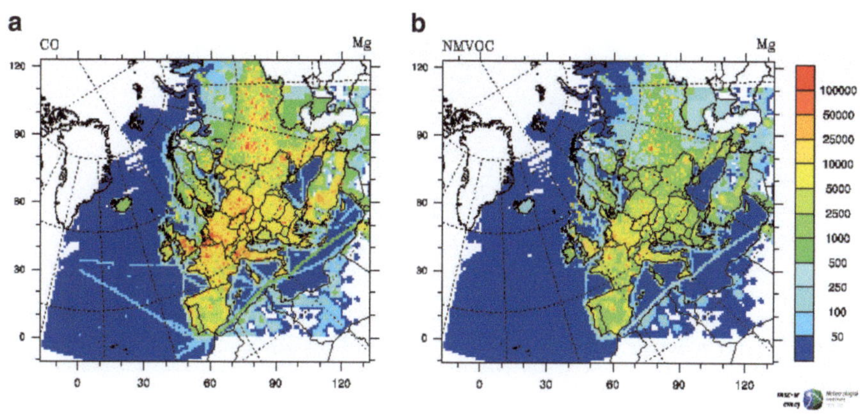

Fig. 1.4 Emissions of (**a**) CO (**b**) NMVOC

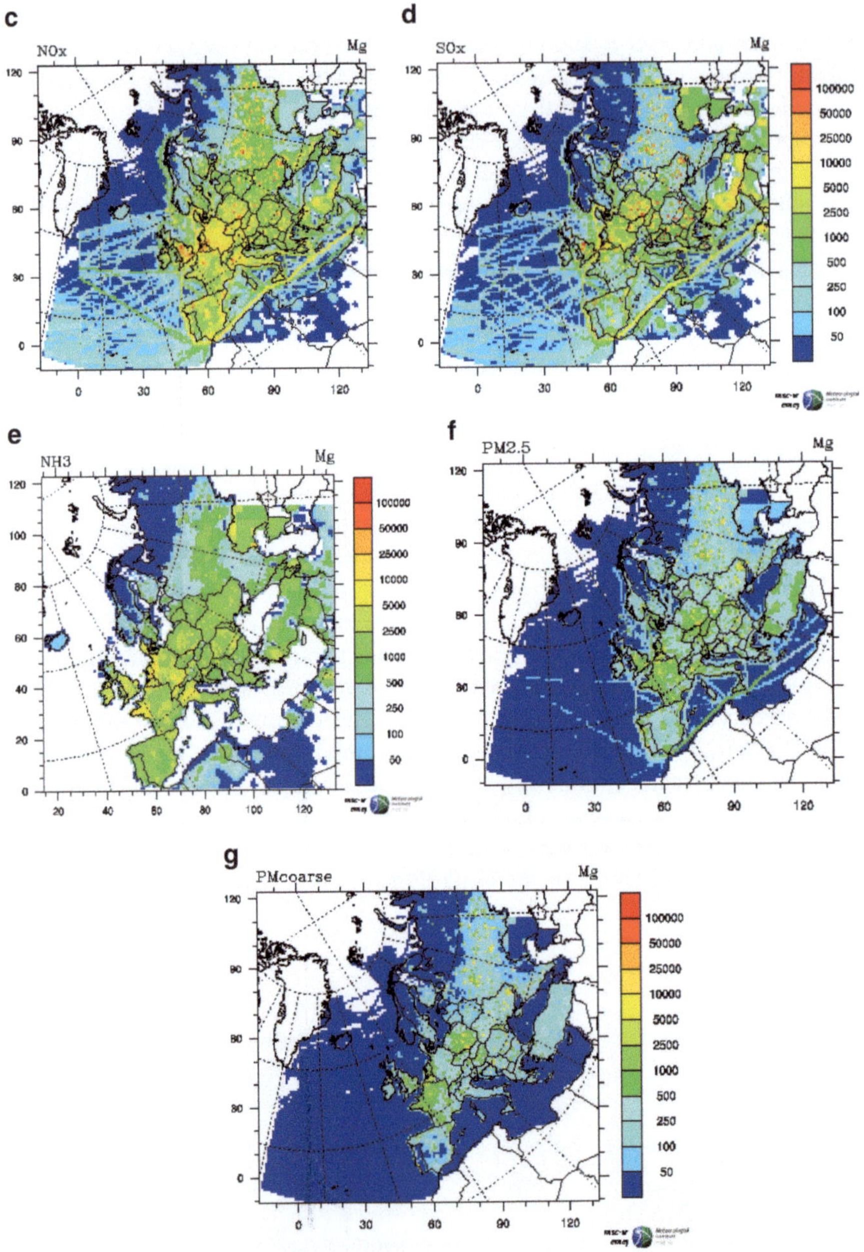

Fig. 1.4 (continued) (**c**) NO_x (**d**) SO_x (**e**) NH_3 (**f**) $PM_{2.5}$ and (**g**) PM_{coarse} during 2004 at 50 km resolution (Mg) in Europe (MSC-W [Adapted from EMEP 2006])

Europe (EMEP) (EMEP 2006) for determination of the spatial resolution of PM_{10} and $PM_{2.5}$ emissions in Europe as the base for the modeling of dispersion and the PM concentrations. Figure 1.4 shows the results of grided emissions for the main

Table 1.2 Air quality guidelines for particulate matter, ozone, nitrogen dioxide and sulfur dioxide proposed by the World Health Organization (WHO 2006)

Compound	Annual mean (μg/m³)	24-h mean (μg/m³)	8-h mean (μg/m³)	1-h mean (μg/m³)	10-min mean (μg/m³)
PM_{10}	20	50			
$PM_{2.5}$	10	25			
O_3			100		
NO_2	40			200	
SO_2		20			500

air pollutants and particles by EMEP for the year 2004 in Europe. The hot spots with highest PM emissions are located in the Netherlands, Germany, Poland, Ukraine, Russia, Romania, Turkey, Bulgaria, Italy and France.

1.3 Ambient Air Quality Standards

Ambient air quality standards have been introduced at national and international levels with the aim to protect human health and the environment. The threshold pollutant concentrations described in the legislation are based on a detailed review of the scientific information related to their effects on human health.

Table 1.2 presents the air quality guidelines for particulate matter, ozone, nitrogen dioxide and sulfur dioxide proposed by the World Health Organization (WHO) (WHO 2000). The proposed annual mean concentration together with the 24-h, 8-h, 1-h and 10 min mean values are also presented.

Legislation for PM_{10} concentrations has recently been implemented by the European Union (EU). The roles, goals and methods of air quality management are determined EU directive 96/62/EC. The later daughters directives describe the objectives for air protection policy and standards addressed to EU countries as well as candidates countries as the platform for air quality assessment. For PM_{10} the obligatory standards have been established at levels as shown at Table 1.3. In addition in Table 1.4 the proposed ambient concentration levels for ozone and particulate matter proposed by the US Environmental Protection Agency (EPA) are shown. Ambient concentration levels of exposure to specific air pollutants proposed by the World Health Organization (WHO) are shown in Tables 1.5 and 1.6 (WHO 2006).

1.4 Ambient Air Pollution Concentrations Over Europe

1.4.1 Ambient Levels of Gaseous Pollutants

Measurements of air pollutants in Europe are concentrated in urban areas and, in addition, there is no sufficient reporting of the data to international organizations,

Table 1.3 Limit values for particulate matter (PM_{10})

	Averaging period	Limit value	Margin of tolerance	Date by which limit value is to be met
24-h limit value for the protection of human health	24 h	50 μg/m³ PM_{10}, not to be exceeded more than 35 times a calendar year	50% on the entry into force of this Directive, reducing on 1 January 2001 and every 12 months thereafter by equal annual percentages to reach 0% by 1 January 2005	1 January 2005
Annual limit value for the protection of human health	Calendar year	40 μg/m³ PM_{10}	20% on the entry into force of this Directive, reducing on 1 January 2001 and every 12 months thereafter by equal annual percentages to reach 0% by 1 January 2005	1 January 2005

which could allow a comprehensive geographical overview of their concentrations. The ambient concentration of air pollutants at a defined spatial and temporal point is a complex functional of emissions, dispersion, physico-chemical processes and removal. In the present chapter an overview of ambient levels of specific gaseous pollutants and particulate matter in Europe is presented. The data were obtained from the Co-operative Programme for Monitoring and Evaluation of the Long-Range Transmission of Air Pollutants in Europe (EMEP) (2006).

Geographical distributions based on annual means of SO_2 and NO_2 are shown in Figs. 1.5 and 1.6. Higher concentrations are found in central Europe due to anthropogenic emissions.

Ozone is a natural constituent of the atmosphere and plays a vital role in many atmospheric processes. However, man-made emissions of volatile organic compounds and nitrogen oxides have increased the photochemical formation of ozone in the troposphere. The formation of ozone is due to a large number of photochemical reactions taking place in the atmosphere and depends on the temperature, humidity and solar radiation as well as the primary emissions of nitrogen oxides and volatile organic compounds.

The critical ozone levels defined by ECE for protection of vegetation are 150 μg/m³ for hourly mean, 60 μg/m³ for 8-h mean and 50 μg/m³ for 7-h mean (9 a.m.–4 p.m.) averaged over the growing season (April–September) (EMEP/CCC 2006). In EU the ozone directive (Directive 2002/3/EC) has defined a number of target values

Table 1.4 National ambient air quality standards for "criteria" pollutants (USEPA 2004)

Pollutant	Primary standard[a]	Averaging times	Secondary standard[b]
Carbon monoxide	9 ppm (10 mg/m³)	8-h[c]	None
	35 ppm (40 mg/m³)	1-h[c]	None
Lead	1.5 μg/m³	Quarterly average	Same as primary
Nitrogen dioxide	0.053 ppm (100 μg/m³)	Annual (arithmetic mean)	Same as primary
Particulate matter (PM_{10})	Revoked[d]	Annual[d] (arithmetic mean)	
	150 μg/m³	24-h[e]	
Particulate matter ($PM_{2.5}$)	15.0 μg/m³	Annual[f] (arithmetic mean)	Same as primary
	35 μg/m³	24-h[g]	
Ozone	0.08 ppm	8-h[h]	Same as primary
	0.12 ppm	1-h[i] (applies only in limited areas)	Same as primary
Sulfur oxides	0.03 ppm	Annual (arithmetic mean)	–
	0.14 ppm	24-h[c]	–
	–	3-h[c]	0.5 ppm (1,300 μg/m³)

[a]Primary standards set limits to protect public health, including the health of "sensitive" populations such as asthmatics, children, and the elderly.
[b]Secondary standards set limits to protect public welfare, including protection against decreased visibility, damage to animals, crops, vegetation, and buildings.
[c]Not to be exceeded more than once per year.
[d]Due to a lack of evidence linking health problems to long-term exposure to coarse particle pollution, the agency revoked the annual PM_{10} standard in 2006 (effective December 17, 2006).
[e]Not to be exceeded more than once per year on average over 3 years.
[f]To attain this standard, the 3-year average of the weighted annual mean $PM_{2.5}$ concentrations from single or multiple community-oriented monitors must not exceed 15.0 μg/m³.
[g]To attain this standard, the 3-year average of the 98th percentile of 24-h concentrations at each population-oriented monitor within an area must not exceed 35 μg/m³ (effective December 17, 2006).
[h]To attain this standard, the 3-year average of the fourth-highest daily maximum 8-h average ozone concentrations measured at each monitor within an area over each year must not exceed 0.08 ppm.
[i](a) The standard is attained when the expected number of days per calendar year with maximum hourly average concentrations above 0.12 ppm is < 1; (b) As of June 15, 2005 EPA revoked the 1-h ozone standard in all areas except the fourteen 8-h ozone non attainment Early Action Compact (EAC) Areas.

and long-term objectives for the protection of vegetation and human health. The target value for human health for 2010 is 120 g/m³ (8 h mean) which is not to be exceeded on more than 25 days per year averaged over 3 years. Information should be given to the population when hourly means exceed 180 μg/m³ and an alert warning should be issued if hourly means exceed 240 μg/m³ (EMEP/CCC 2006). However, the critical ozone level formulated by WHO for the protection of health is 120 μg/m³ for 8-h mean.

Table 1.5 Ambient annual average concentrations and guideline values for carbon monoxide, lead, nitrogen dioxide, ozone and sulfur dioxide

Compound	Annual average concentration (µg/m³)	Health effect	Critical level (µg/m³)	Guideline value (µg/m³)	Averaging time
Carbon monoxide	500–7,000	CO critical blood level <2.5%	NA	100,000	15 min
				60,000	30 min
				30,000	1 h
				10,000	8 h
Lead	0.01–2	Pb critical blood level <100–150 µg Pb/l	NA	0.5	1 year
Nitrogen dioxide	10–150	Small change in lung function among asthmatics	365–565	200	1 h
				40	1 year
Ozone and other photochemical oxidants	10–100	Change in lung function	NA	120	8 h
Sulfur dioxide	5–400	Change in lung function among asthmatics	1,000	500	10 min
		Increase in symptom exacerbations among adults or asthmatics	250	125	24 h
			100	50	1 year

Table 1.6 Guideline values for pollutants based on effects other than cancer or odor/annoyance

Pollutants	Time-weighted average	Averaging time	Lifetime risk (unit concentration)
Organic pollutants			
Acrylonitrile	–	–	2×10^{-5} (μg/m³)
Benzene	–	–	6×10^{-6} (μg/m³)
Butadiene			
Carbon disulfide[a]	100 μg/m³	24 h	
Carbon monoxide	100 mg/m³[b]	15 min	
	60 mg/m³[b]	30 min	
	30 mg/m³[b]	1 h	
	10 mg/m³	8 h	
1,2-Dichloroethane[c]	0.7 mg/m³	24 h	
Dichloromethane	3 mg/m³	24 h	
	0.45 mg/m³	1 week	
Formaldehyde	0.1 mg/m³	30 min	
Polycyclic aromatic hydrocarbons (PAHs as BaP)	–	–	8.7×10^{-5} (ng/m³)
Polychlorinated biphenyls (PCBs[c])	–	–	
Polychlorinated dibenzodioxins and dibenzofurans (PCDDs/PCDFs)[d]	–	–	
Styrene	0.26 mg/m³	1 week	
Tetrachloroethylene	0.25 mg/m³	annual	
Toluene	0.25 mg/m³	1 week	
Trichloroethylene			
Vinyl chloride			4.3×10^{-7} (μg/m³)
Inorganic pollutants			
Arsenic			1.5×10^{-3} (μg/m³)
Asbestos			
Cadmium[e]	5 ng/m³[g]	Annual	
Chromium ($Cr^{(VI)}$)	–	–	4×10^{-2} (μg/m³)
Fluoride[f]	–	–	
Hydrogen sulfide[b]	15 μg/m³	24 h	
Lead	0.5 μg/m³	annual	
Manganese	0.15 μg/m³	annual	
Mercury	1 μg/m³	annual	
Nickel	–	–	3.8×10^{-4} (μg/m³)
Platinum[g]	–	–	
Vanadium[a]	1 μg/m³	24 h	
Classical pollutants[h]			
Nitrogen dioxide	200 μg/m³	1 h	
	40 μg/m³	annual	

(continued)

Table 1.6 (continued)

Pollutants	Time-weighted average	Averaging time	Lifetime risk (unit concentration)
Ozone and other photochemical oxidants	120 (100) μg/m³	8 h	
Particulate matter[i]	Dose–response	–	
PM_{10}	(20/50) μg/m³	(annual/24 h)	
$PM_{2.5}$	(10/25) μg/m³	(annual/24 h)	
Sulfur dioxide	500 μg/m³	10 min	
	125 (20) μg/m³	24 h	
	50 μg/m³	annual	

[a]Not re-evaluated for the second edition of the guidelines.
[b]Exposure at these conditions should be for no longer than the indicated times and should not be repeated within 8 h.
[c]No guideline values has been recommended for PCBs because inhalation constitutes only a small proportion (about 1–2%) of the daily intake from food.
[d]No guideline values has been recommended for PCDDs/PCDFs because inhalation constitutes only a small proportion (generally less than 5%) of the daily intake from food.
[e]The guideline value is based on the prevention of a further increase of cadmium in agricultural soils, which is likely to increase the dietary intake.
[f]Because there is no evidence that atmospheric deposition of fluorides results in significant exposure through other routes than air, it was recognized that levels below 1 μg/m³, which is needed to protect plants and livestock, will also sufficiently protect human health.
[g]It is unlikely that the general population, exposed to platinum concentrations in ambient air at least three orders of magnitude below occupational levels where effects were seen, may develop similar effects. No specific guideline value has therefore been recommended.
[h]Values in brackets are the new air quality guideline values (WHO 2006).
[i]The available information for short- and long-term exposure to PM_{10} and $PM_{2.5}$ does not allow a judgement to be made regarding concentrations below which no effects would be expected. For this reason no guideline values have been recommended, but instead risk estimates have been provided.

Graphical distributions of the 99-percentile for the period April–September 2004 are shown in Fig. 1.7. The lowest values are found in the Northern parts of Central Europe, Baltics, Ireland, Scotland and Scandinavia, where the 99-percentile is below 140 μg/m³. The concentrations are higher in Central Europe, where the 99-percentile generally ranges from 140–160 μg/m³. A few sites in Central Europe had values above 160 μg/m³. The concentration levels on the Iberian Peninsula are variable, possibly due to local influence and topographical differences. The highest hourly mean values were found at K-puszta, Hungary (253 μg/m³, 12th September), Krvavec, Slovenia (218 μg/m³, 10th June) and in Sibton in Great Britain (212 μg/m³, 8th August).

Figure 1.8 shows the number of exceedances of the threshold value of 180 μg/m³ formulated by the EU for informing the public. At the station Montelibretti (Italy), the threshold value was exceeded 8 days, while five additional sites measured above 180 μg/m³ for at least 4 days. In total values above 180 μg/m³ were measured at 33 sites.

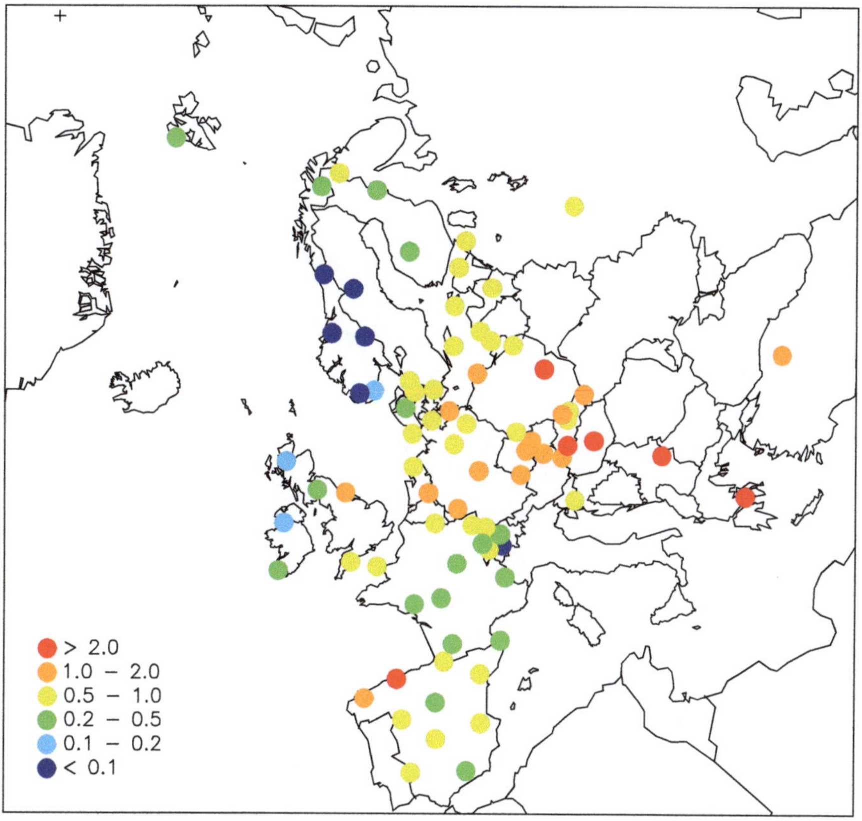

Fig. 1.5 Geographical distribution of sulfur dioxide 2003. Unit: μg S/m^3. (Adapted from EMEP 2005b)

In addition to the measurements performed in Europe by the EMEP network model simulations were also performed by the EMEP Unified Model (EMEP 2001). Figure 1.9 shows the average surface ozone concentration during the year 2004. The model results show elevated ozone concentrations in the central Europe but also higher levels in the southern part of Europe during to the increased photochemical activity. The combination of modeled and measured results is an area of future research in the EMEP framework (EMEP 2005a).

1.4.2 Ambient Levels of Particulate Matter

Airborne particulate matter is a complex mixture of many different chemical species originating from a variety of sources. Composition, morphology, physical and thermodynamic properties of PM vary with respect to time and location and quite

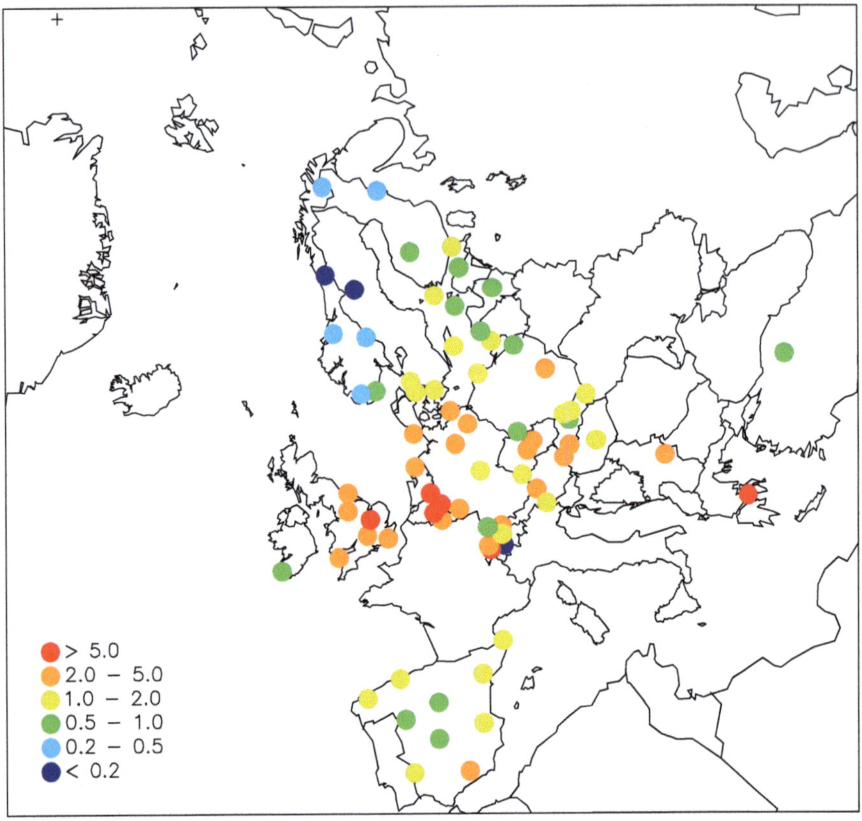

Fig. 1.6 Geographical distribution of nitrogen dioxide 2003. Unit: μg N/m^3 (Adapted from EMEP 2005b)

typically have a seasonal variability (IPCC 2001; Finlayson-Pitts and Pitts 1986; Seinfeld and Pandis 2006; EPA 2003; CAFE 2004).

The dynamics of the particulate matter in the atmosphere involve various physical and chemical processes under different time scales. Furthermore, aerosol dynamics involve a wide range of particle sizes ranging from a few nanometers to several 100 μm, different compositions and chemical reactivity and undergo complex physical transformations (nucleation, condensation, coagulation and deposition processes). Aerosols arise from natural and anthropogenic sources and are mixture of primary emissions and secondary species. Crustal material, biogenic matter and sea-salt comprise the majority of natural aerosols. Anthropogenic aerosols are composed of primary emitted soot (elemental carbon) and secondary formed carbonaceous material (organic carbon) and inorganic matter (nitrates, sulfates, ammonium and water). Therefore modelling or measuring atmospheric aerosols involves many challenging tasks and is a fast evolving scientific area (Seinfeld and Pandis 2006).

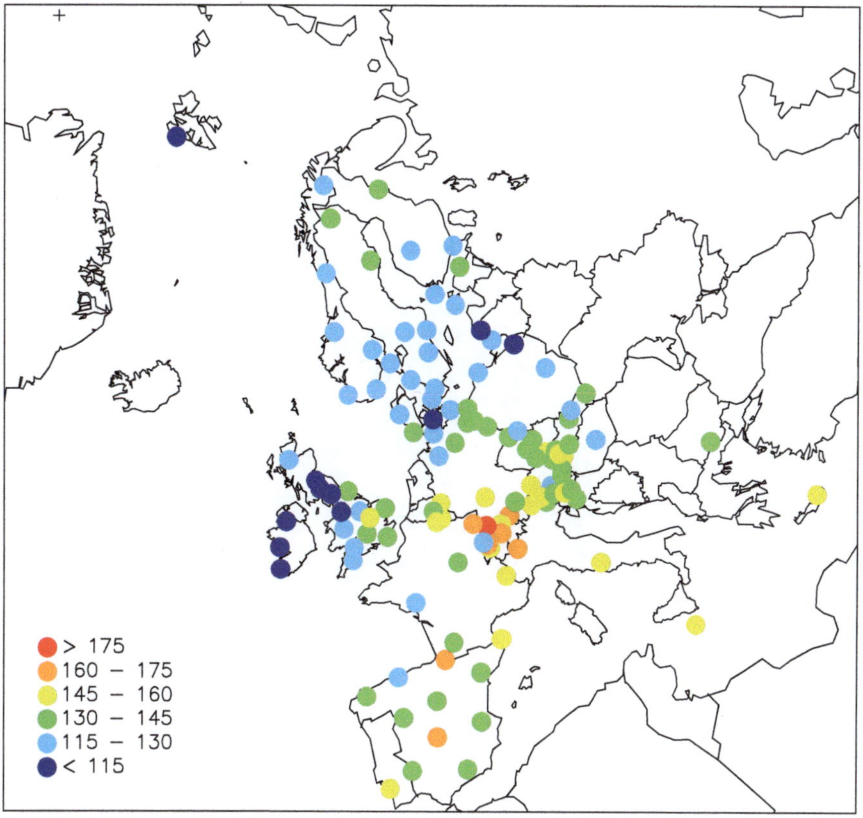

Fig. 1.7 Ozone April–September 2004. 99-percentiles (μg/m³) (Adapted from EMEP/CCC 2006)

There is a considerable literature concerning patterns and trends of ambient aerosols and their concentration characteristics. In Europe, PM measurements are performed mainly in urban areas for the purpose of monitoring human PM exposure. Monitoring of PM in rural areas is not as extensive as seen for urban areas. Despite this, results presented in the CAFE report (2004) for the period 1997–2001 show that the gap between concentrations of PM_{10} at rural and urban sites in Europe can be quite narrow. In addition, the annual variation of the PM concentrations for these two site categories is similar for the period 1997–2001. Hence, the regional contribution seems to be highly important for the concentrations of PM_{10} observed in urban areas. Also the level of PM_{10} at traffic-influenced sites in Europe follows the annual variation observed for urban and rural areas, although its concentration level is considerably higher.

Particulate matter has negative effects on the environment (Andreae and Crutzen 1997; Charlson et al. 1992) and there is an association of exposure to particles through inhalation to adverse effects on human health as many epidemiological and

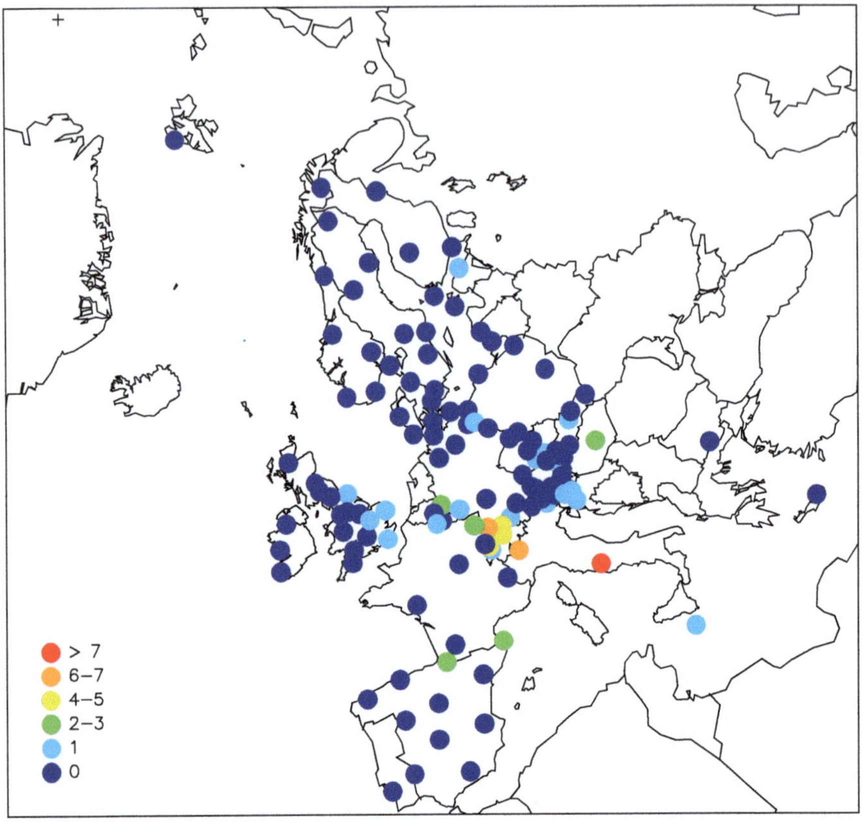

Fig. 1.8 Number of exceedances of the threshold value of 180 μg/m³ for the year 2004 (Unit: number of days) (Adapted from EMEP/CCC 2006)

toxicological studies have shown (e.g. Pope et al. 1995; Schlesinger 1995; Neuberger et al. 2004). More specifically, health effects caused by inhaled particles are pulmonary function decrements, respiratory symptoms, neurological dysfunction, cardiovascular dysfunction, morbidity and mortality (Rombout et al. 2000; EPA 2003; Pope et al. 1995; Dockery et al. 1993; Schwartz 2000; Pope and Dockery 2006).

There various monitoring networks active for the determination of air pollution in Europe. These include the Co-operative Programme for Monitoring and Evaluation of the Long-Range Transmission of Air Pollutants in Europe (EMEP) (www.emep.int), Global Atmosphere Watch (GAW) programme (http://www.empa.ch/gaw/gawsis/) and the European Air Quality monitoring Network (EuroAirnet) which delivers the data to the AIRBASE database (http://etc-acc.eionet.eu.int/databases/airbase.html) managed by the European Topic Centre on Air Quality and climate change under contract to the European Environmental Agency (EEA).

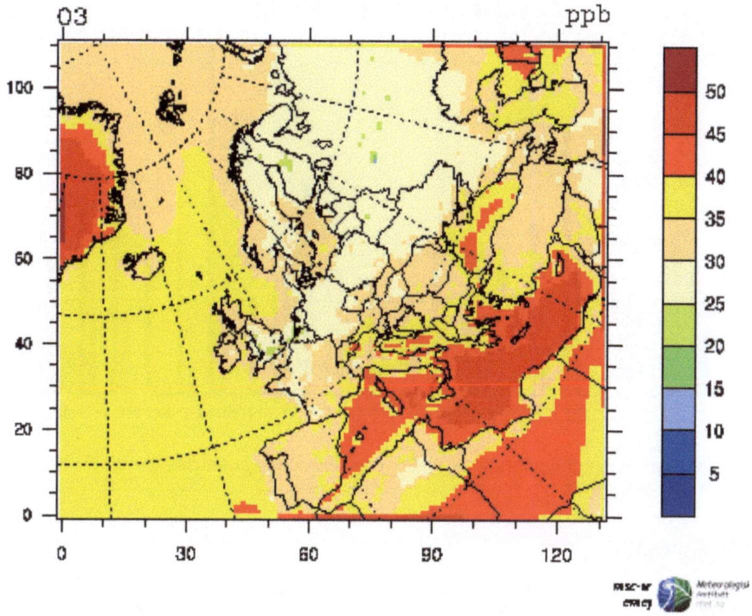

Fig. 1.9 Average modeled surface ozone concentration during the year 2004 (Adapted from EMEP 2005a)

Data from the Co-operative Programme for Monitoring and Evaluation of the Long-Range Transmission of Air Pollutants in Europe (EMEP) and the AIRBASE database system of the European Topic Centre on Air Quality (ETC/AQ) were used in this chapter for the evaluation of the ambient particulate matter concentration over Europe. Both the EMEP and the AIRBASE systems include data on gaseous pollutants and particulate matter concentrations from air quality stations in Europe. The stations included in the AIRBASE database are classified into three main station types, traffic, background and industrial, and according to the type of the area where they are situated, into urban, suburban and rural (Larssen et al. 1999).

In Fig. 1.10, the annual mean PM_{10} concentrations at AIRBASE and EMEP stations during 2004 are depicted. It is observed that the PM_{10} concentrations were higher at traffic stations and stations situated in urban areas. The annual limit value was exceeded at 208 of 1,801 stations during 2004. Most of these stations were background (52) and traffic (92) sites situated in urban areas, specifically in Spain (51) and Italy (49). Moreover, PM_{10} concentrations at seven background rural stations were above the limit value. Concentrations measured at EMEP sites in each country were on an average lower than concentrations at stations in rural and urban areas or traffic stations. However the mean annual PM_{10} concentration of all the EMEP stations in Spain and Italy was higher than the averaged traffic and background rural stations. This is probably associated with Saharan dust intrusions for Spain whereas in Italy it is due to the high concentrations measured at the Ispra station

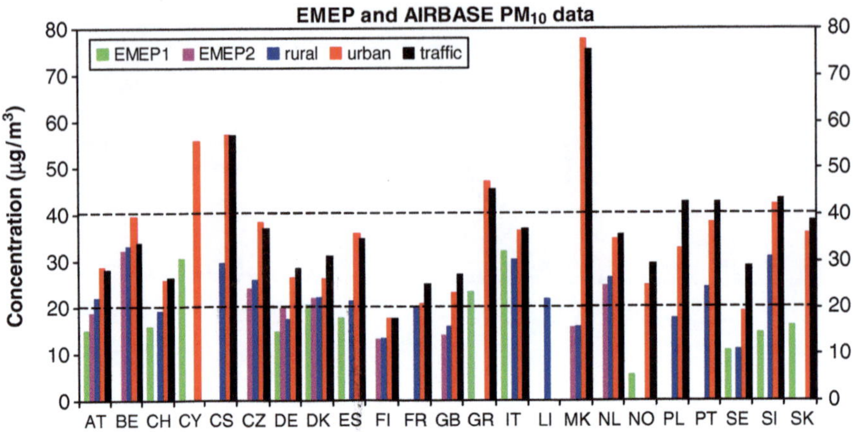

Fig. 1.10 Annual mean concentrations of PM_{10} in European countries for 2004, based on data from EMEP and AIRBASE

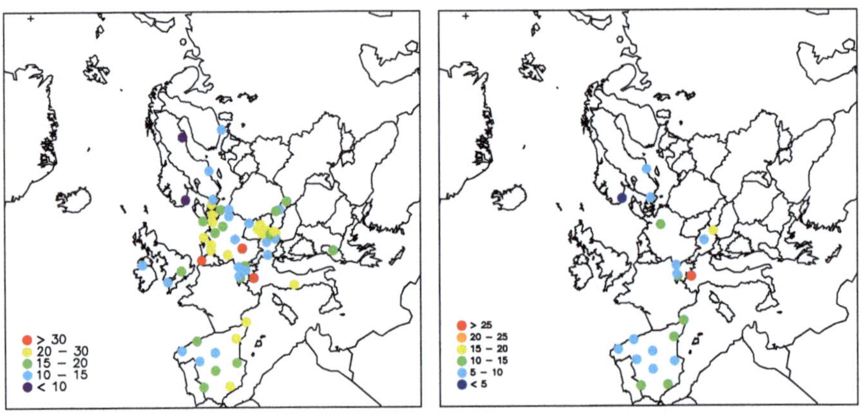

Fig. 1.11 Annual mean concentrations of PM_{10} (*left*) and $PM_{2.5}$ for 2004 (µg/m³) (Adapted from EMEP/CCC 2006)

located at the northern part of the country. The term EMEP2 refers to EMEP stations for which data have been obtained from the AIRBASE database.

It is also observed that EMEP2 stations experience higher concentrations than stations reporting PM_{10} data directly to EMEP and rural stations in each country. This is because some of these background stations are situated in urban and suburban areas (e.g. stations in Belgium) rather than rural. The concentrations at these stations were in fact higher than the annual mean concentration at industrial and background suburban stations in Finland and the Netherlands.

From Fig. 1.11 it can be seen that stations in the central and the continental part of western Europe have higher annual mean concentrations of PM_{10} than stations in

Great Britain and in the Scandinavian countries during 2004. Country annual mean PM_{10} concentrations ranged from 5.3 μg/m^3 for Norway to 32.1 μg/m^3 for Belgium. The high levels can be attributed to the local anthropogenic emissions (traffic, industries) in western and central Europe whereas for southern Europe this can rather attributed to biogenic emissions, resuspended dust from local sources and also Saharan dust events. A similar spatial distribution of PM_{10} concentrations for Europe was observed during 2003. During 2004, the annual limit value of 40 μg/m^3 of PM_{10} was not exceeded in any of the EMEP stations.

The majority of the sites reporting $PM_{2.5}$ concentrations during 2004 were concentrated in Spain, central and western Europe. From Fig. 1.11 it can be seen that the annual mean concentrations of $PM_{2.5}$ increased from northern to southern parts of Europe, and that the highest concentrations were reported for the two central European sites Ispra (Italy) (28.3 μg m^{-3}) and Illmitz (Austria) (19.1 μg m^{-3}). Hence, the US EPA annual limit value of 15 μg m^{-3} was exceeded at both these sites. During 2003, four stations exceeded the US EPA annual limit value, and the maximum concentration of 28.6 μg m^{-3} was reported for Ispra, as for 2004. The annual mean concentration of PM_1 did not exceed 15 μg m^{-3} at any of the sites stations reporting this parameter in 2004.

1.5 Composition of PM_{10}

Atmospheric chemistry involves a large number of reactive species which are in ppm and ppb levels. For example the formation of ozone and nitrogen dioxide involves a large number of nonlinear chemical reactions (e.g. Seinfeld and Pandis 2006). Furthermore, the chemical reaction rates depend also on the background concentration of the various chemical species that is determined from the emission and meteorological characteristics. The modeling of gaseous chemical reactions in the atmosphere is a difficult task because of the complex chemical reactions and the stochastic mixing processes due to turbulence. Several simplifications are adopted in describing the gaseous phase chemical reactions in air quality models.

Airborne particulate matter is a complex mixture of many different chemical species originating from a variety of sources (see Fig. 1.12). At EMEP stations, speciation of the ambient particulate matter is mainly focused on the determination of sulfate, nitrate and ammonium constituents and for certain time periods on the organic mass (EMEP 2005a). The mixture of air pollutants which is potential harmful to humans is presented in Fig. 1.13.

Previous studies have shown that total particulate sulfate and nitrate are the dominant components, with ammonium nitrate prevailing in Western Europe and particulate sulfate (mainly as ammonium sulfate) prevailing in Eastern Europe (Lazaridis et al. 2002; Van der Zee et al. 1998; Zappoli et al. 1999). In addition, the fraction of gaseous nitric acid is typically 20–30%, relative to the sum of nitrate and nitric acid (Lazaridis et al. 2002).

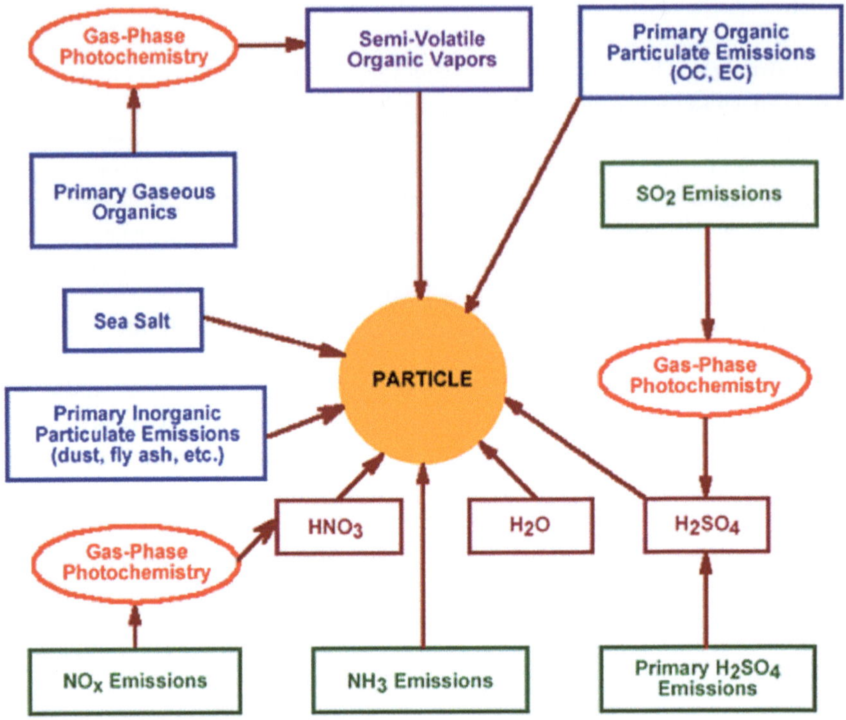

Fig. 1.12 Schematic representation of the chemical reactions and processes associated with particulate matter (Adapted from Meng and Seinfeld 1996)

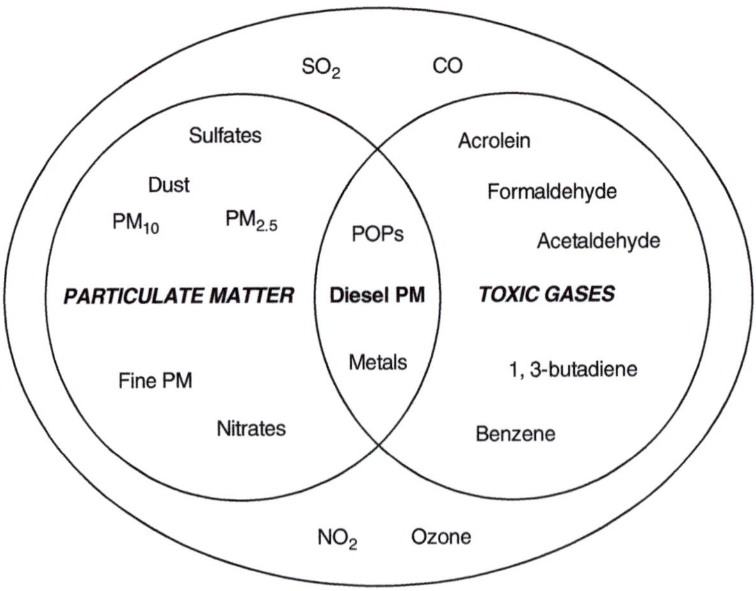

Fig. 1.13 The air pollution mixture (Adapted from HEI 2001)

The importance of spatial and temporal distribution mapping of the nitrate aerosol over Europe inside the EMEP framework due to the importance of nitrates to radiative forcing, acidification and eutrophication has been already pointed out (Lazaridis et al. 2002).

A significant fraction of the ambient particulate matter is secondary particulate matter in the form of sulfate, nitrate, ammonium and organic aerosol particles formed by the oxidation of sulfur dioxide, nitrogen oxides and organic gaseous species. It is apparent that particulate matter is not a single pollutant and its mass includes a mixture of many pollutants distributed differently at different sizes. Daily measurements of inorganic constituents in aerosols are performed within the EMEP framework. The EMEP database (http://www.nilu.no/projects/ccc/emepdata.html) contains concentration data for a large number of inorganic chemical species in Europe. Measurements of the organic constituents (Elementary Carbon/Organic Carbon) of the ambient particulate matter at the EMEP stations have been also performed during specific measurement campaigns the last years (EMEP 2005a).

Previous studies have shown that there is relatively good correlation between observed particulate matter concentrations and concentrations of aerosol sulfate (and nitrate) at many of the EMEP sites, even if the sulfate and nitrate concentrations account only for 20–40% of the aerosol mass (Lazaridis et al. 2002; EMEP 2006).

The chemical mass closure of ambient PM in Europe is a challenging task. Figure 1.14 shows as an example the chemical mass closure of PM_{10} aerosols at the

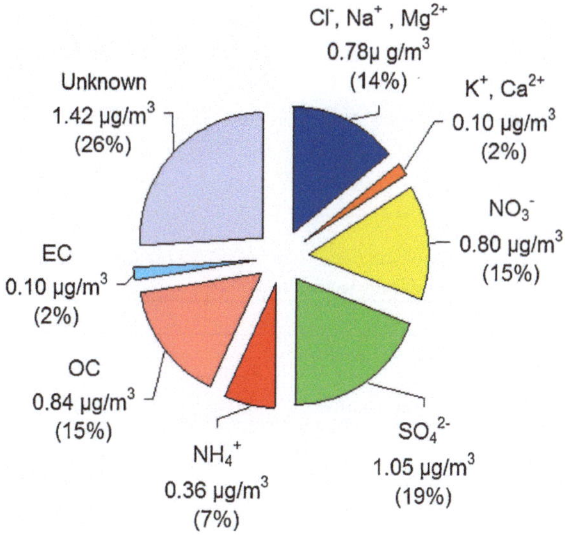

Fig. 1.14 Chemical mass closure of PM_{10} aerosols at the Birkenes station in Norway during 2004 (Adapted from EMEP/CCC 2006)

Birkenes station in Norway during 2004. The inorganic part is the most important part of the aerosol mass.

1.6 Indoor Air Pollution

People spend about 85% of their time indoors and therefore are exposed to PM from both outdoor (ambient) sources through infiltration of outdoor air and indoor sources (cigarette smoking, cooking, personal activities, dusting of indoor areas, and various other indoor activities), which may have different composition and possibly different toxicities.

There are numerous recent papers on indoor air quality measurements and modelling studying the different aspects of the indoor air quality ranging from the characterization of the indoor/outdoor ratio (I/O) of specific pollutants, indoor source strength, infiltration, deposition of pollutants on surfaces, chemical reactions and human exposure (e.g. Wallace 2000; Lachenmyer and Hidy 2000; Thatcher and Layton 1995; Lebowitz et al. 1995; Morawska et al. 2003; Abt et al. 2000). A presentation of recently performed microenvironmental PM measurement studies is summarized by the United States Environmental Protection Agency (USEPA 2004). Table 1.7 presents a comparison of workplace air quality standards for selected gasses (Jacobson 2002).

There are several indoor sources of air pollutants. These include combustion sources due to heating, cooking and smoking. Additional sources are from building material, carpets, cleaning materials and wooden furniture. The outdoor environment is also a source for air pollutants indoors as well as the soil (radon). The volatile organic compounds (VOCs) indoors arise from human activities, building materials and from the outdoor air. Indoor sources of VOCs include paintings, solvents, air fresheners as shown in Table 1.8. Examples of measurements of indoor concentrations of various pollutants are given in Chapter 2.

Table 1.7 Comparison of Indoor Workplace Standards with outdoor Federal and California State standards for selected gases. Data from Jacobson (2002)

	Indoor			Outdoor	
Gas	8-h PEL and TWA-TLV (ppmv)	15 min STEL (ppmv)	Ceiling (ppmv)	NAAQS (ppmv)	California Standard (ppmv)
Carbon monoxide	35	–	200	9.5 (8 h)	9 (8 h)
Nitrogen dioxide	–	1	–	0.053 (annual)	0.25 (1 h)
Ozone	0.1	0.3	–	0.08 (8 h)	0.09 (1 h)
Sulfur dioxide	2	5	–	0.14 (24 h)	0.05 (24 h)

PEL: Permissible exposure limits; TWA-TLV: Time-Weighted average threshold limit values; STEL: Short-time exposure limits; NAAQS: National Ambient Air Quality Standard.

Table 1.8 Indoor pollutants and sources (Su [1996] and Nero [1988])

Pollutants	Sources
Radon and radioactive daughters (^{222}Rn)	Soil, ground water, building materials
Nitrogen oxides (NO_x)	Combustion
Volatile Organic Compounds (including HCHO)	Building materials, carpets, solvents, paints, personal care products, house cleaning products, room fresheners, pesticides, mothball, humans
Carbon Monoxide (CO)	Combustion
Ozone (O_3)	Outdoor air, photocopying machines, electrostatic air cleaners
Sulfur Dioxide (SO_2)	Combustion
Particulate matter	Combustion
Asbestos	Building materials
Bioaerosols	Air conditioners, cold water spray humidifiers

1.7 Drinking Water Quality

1.7.1 Standards

In 1995 the world withdrew 3,906 cubic kilometers (km^3) of water for use by humans. By 2025 water withdrawal is projected to increase by at least 50%. Figures vary from region to region (Table 1.9). North America, on a per capita basis, withdraws seven times more freshwater than Africa (Gleick 2006).

Access to safe drinking water and sanitation is critical to maintain health, particularly for children. But more than one billion people across the globe lack enough safe water to meet minimum levels of health and income. Although the domestic and industrial sectors use far less water than agriculture, the growth in water consumption in these sectors has been rapid. Globally, withdrawals for domestic and industrial uses quadrupled between 1950 and 1995, compared with agricultural uses, for which withdrawals slightly more than doubled.

The World Health Organisation has as one its priorities that "all people, whatever their stage of development and their social and economic conditions, have the right to have access to an adequate supply of safe drinking water". To help achieve this they publish 'Guidelines on Drinking Water Quality' which countries should meet to ensure the health of their population.

The Guidelines describe reasonable minimum requirements of safe practice to protect the health of consumers and/or derive numerical "guideline values" for constituents of water or indicators of water quality (Table 1.10). In order to define mandatory limits, it is preferable to consider the guidelines in the context of local or national environmental, social, economic and cultural conditions. The approach

Table 1.9 Freshwater withdrawal by country and sector. Data from The World's Water 2006–2007 (P.H. Gleick, Island Press, Washington, 2006)

Country	Total freshwater withdrawal (km^3/year)	Domestic use (%)	Industrial Use (%)	Agricultural use (%)	2005 Population (millions)
Australia	24.06	15	10	75	20.16
Brazil	59.30	20	18	62	186.41
Canada	44.72	20	69	12	32.27
Chad	0.23	17	0	83	9.75
China	549.76	7	26	68	1,323.35
Egypt	68.30	8	6	86	74.03
Germany	38.01	12	68	20	82.69
India	654.84	8	5	86	1,103.37
Pakistan	169.39	2	2	96	157.94
Somalia	3.29	0	0	100	8.23
UK	11.75	22	75	3	59.67
USA	477.00	13	46	41	298.21
Zambia	1.74	17	7	76	11.67

Table 1.10 Guidelines on drinking water quality, 3rd Edition, Geneva. Data from WHO (2006)

Chemical	Guideline value (mg/l)	Remarks
Acrylamide	0.0005	
Alachlor	0.02	
Aldicarb	0.01	Applies to aldicarb sulfoxide and aldicarb sulfone
Aldrin and dieldrin	0.00003	For combined aldrin plus dieldrin
Antimony	0.02	
Arsenic	0.01 (P)	
Atrazine	0.002	
Barium	0.7	
Benzene	0.01	
Benzo[*a*]pyrene	0.0007	
Boron	0.5 (T)	
Bromate	0.01 (A, T)	
Bromodichloromethane	0.06	
Bromoform	0.1	
Cadmium	0.003	
Carbofuran	0.007	
Carbon tetrachloride	0.004	
Chlorate	0.7 (D)	
Chlordane	0.0002	
Chlorine	5 (C)	For effective disinfection, there should be a residual concentration of free chlorine of ≥0.5 mg/l after at least 30 min contact time at pH <8.0

(continued)

Table 1.10 (continued)

Chemical	Guideline value (mg/l)	Remarks
Chlorite	0.7 (D)	
Chloroform	0.3	
Chlorotoluron	0.03	
Chlorpyrifos	0.03	
Chromium	0.05 (P)	For total chromium
Copper	2	
Cyanazine	0.0006	
Cyanide	0.07	
Cyanogen chloride	0.07	For cyanide as total cyanogenic compounds
2,4-D (2,4-dichlorophenoxyacetic acid)	0.03	Applies to free acid
2,4-DB	0.09	
DDT and metabolites	0.001	
Di(2-ethylhexyl)phthalate	0.008	
Dibromoacetonitrile	0.07	
Dibromochloromethane	0.1	
Dibromo-3-chloropropane, 1,2-	0.001b	
Dibromoethane, 1,2-	0.0004 (P)	
Dichloroacetate	0.05 (T, D)	
Dichloroacetonitrile	0.02 (P)	
Dichlorobenzene, 1,2-	1 (C)	
Dichlorobenzene, 1,4-	0.3 (C)	
Dichloroethane, 1,2-	0.03	
Dichloroethene, 1,2-	0.05	
Dichloromethane	0.02	
1,2-Dichloropropane (1,2-DCP)	0.04 (P)	
1,3-Dichloropropene	0.02	
Dichlorprop	0.1	
Dimethoate	0.006	
Dioxane, 1,4-	0.05	
Edetic acid (EDTA)	0.6	Applies to the free acid
Endrin	0.0006	
Epichlorohydrin	0.0004 (P)	
Ethylbenzene	0.3 (C)	
Fenoprop	0.009	
Fluoride	1.5	Volume of water consumed and intake from other sources should be considered when setting national standards
Hexachlorobutadiene	0.0006	
Isoproturon	0.009	
Lead	0.01	
Lindane	0.002	
Manganese	0.4 (C)	

(continued)

Table 1.10 (continued)

Chemical	Guideline value (mg/l)	Remarks
MCPA	0.002	
Mecoprop	0.01	
Mercury	0.006	For inorganic mercury
Methoxychlor	0.02	
Metolachlor	0.01	
Microcystin-LR	0.001 (P)	For total microcystin-LR (free plus cell- bound)
Molinate	0.006	
Molybdenum	0.07	
Monochloramine	3	
Monochloroacetate	0.02	
Nickel	0.07	
Nitrate (as NO3 -)	50	Short-term exposure
Nitrilotriacetic acid (NTA)	0.2	
Nitrite (as NO2 -)	3	Short-term exposure
	0.2 (P)	Long-term exposure
Pendimethalin	0.02	
Pentachlorophenol	0.009 (P)	
Permethrin	0.3	Only when used as a larvicide for public health purposes
Pyriproxyfen	0.3	
Selenium	0.01	
Simazine	0.002	
Styrene	0.02 (C)	
2,4,5-T	0.009	
Terbuthylazine	0.007	
Tetrachloroethene	0.04	
Toluene	0.7 (C)	
Trichloroacetate	0.2	
Trichloroethene	0.02 (P)	
Trichlorophenol, 2,4,6-	0.2 (C)	
Trifluralin	0.02	
Trihalomethanes		The sum of the ratio of the concentration of each to its respective guideline value should not exceed 1
Uranium	0.015 (P, T)	Only chemical aspects of uranium addressed
Vinyl chloride	0.0003	
Xylenes	0.5 (C)	

P = provisional guideline value, as there is evidence of a hazard, but the available information on health effects is limited; T = provisional guideline value because calculated guideline value is below the level that can be achieved through practical treatment methods, source protection, etc.; A = provisional guideline value because calculated guideline value is below the achievable quantification level; D = provisional guideline value because disinfection is likely to result in the guideline value being exceeded; C = concentrations of the substance at or below the health based guideline value may affect the appearance, taste or odour of the water, leading to consumer complaints.

followed in the Guidelines is intended to lead to national standards and regulations that can be readily implemented and enforced and are protective of public health. Approaches that may work in one country or region will not necessarily transfer to other countries or regions. It is essential that each country review its needs and capacities in developing a regulatory framework.

Drinking water quality is of particular concern. In the United States it is regulated by the Safe Drinking Water Act (SDWA) and applies to every public water system. The SDWA authorizes the United States Environmental Protection Agency (US EPA) to set national health-based standards for drinking water to protect against both naturally-occurring and man-made contaminants that may be found in drinking water. US EPA, states, and water systems then work together to make sure that these standards are met.

In Europe Directive 98/83/EC, the Drinking Water Directive (DWD), covers the quality of water intended for human consumption. The objective of the Drinking Water Directive is to protect the health of the consumers in the European Union and to make sure the water is wholesome and clean. The Drinking Water Directive sets standards for the most common substances (a total of 48 microbiological and chemical parameters) that can be found in drinking water. The Directive focuses on those parameters of importance to human health but it also includes others that relate to the control of water treatment processes and the aesthetic quality of drinking water. While translating the Directive into their own national legislation, the Member States can include additional requirements, for example regulate additional substances that are relevant within their territory or set higher standards. But Member States are not allowed to set lower standards as the level of protection of human health should be the same within the whole EU.

1.7.2 Disinfection By-products

A full discussion of pollutants in drinking water is beyond the scope of this chapter. Instead we shall only consider disinfection by-products. Disinfection is vital to most drinking water but can lead to contamination.

It is of paramount importance to ensure the microbiological quality of drinking water. Disinfection can be defined as a process that removes or renders inactive pathogenic micro-organisms. Disinfection of drinking water is achieved through a number of treatment processes that are used in various combinations depending on source type and quality. Whilst each of these processes will have an effect on microbial loading, disinfection is only effective when they are used together, to provide multiple barriers to the passage of micro-organisms. Source protection constitutes the initial barrier. Chemical disinfection, using chlorine, or other oxidant, provides the final barrier. Groundwaters commonly require little pre-treatment before chemical disinfection whereas surface water treatment generally includes processes such as coagulation, clarification and filtration before the final chemical disinfection step.

Chemical disinfection, using principally chlorine but also ozone, chlorine dioxide, or other oxidants has been practised for many years and has played a major role in the virtual eradication of waterborne disease in Europe. However, all chemical disinfectants produce either or both organic and inorganic by-products that may have potential adverse health effects through ingestion. For example, chlorination results in the formation of trihalomethanes (THMs) from naturally occurring precursors; ozone oxidises bromide to form bromate; and chlorine dioxide reacts to form chlorite as an inevitable decomposition product.

Whilst chemical disinfection has been practised since the early twentieth century, it was not until the 1970s that it was discovered that chlorination generates THMs (Rook 1974). Since then considerable efforts have been made, world-wide, to identify, study and control the formation of disinfection by-products (DBPs). Many DBPs have been identified, some of which are potentially hazardous, but the associated health risks are considered to be small compared to the risks associated with inadequate disinfection (Galal-Gorchev and Bartram 1996).

The principal DBPs formed during chlorination are trihalomethanes (THMs) – chloroform, bromodichloromethane, dibromochloromethane and bromoform – as a result of chlorination of naturally occurring organic precursors such as humic substances. Chloroform is normally the predominant THM; the brominated THMs are formed as a result of the oxidation of naturally present bromide ion to form hypobromous acid, which then reacts with precursors. Some THMs, including chloroform, are classified as possible human carcinogens. The total THM (TTHM) concentrations found in European drinking waters vary quite widely (from a few μg/l to hundreds of μg/l) and, because of differences in sampling, sample points and so on, it was reported to be not possible, in 1997, to assess accurately trends in exposure of the population (Premazzi et al. 1997). Other DBPs found following chlorination include chlorinated acetic acids, chlorinated ketones and halogenated acetonitriles. Chloramination produces lower THM concentrations than chlorination but does produce other DBPs including cyanogen chloride.

Ozone can form brominated THMs through the oxidation of bromide to form hypobromous acid which brominates precursors. A range of other DBPs including aldehydes and carboxylic acids may also be formed.

Chlorine dioxide does not form THMs by reaction with humic substances. As an example, even under exhaustive reaction conditions, using abnormally high (≥20 mg/l) doses of pure chlorine dioxide on raw River Meuse water, TTHMs were undetectable and the non-purgeable organic halide (NPOX) concentration was much lower than with chlorine under equivalent conditions (Savoir et al. 1987).

The concentrations of THMs formed depend on several factors including:

- Concentrations of precursors
- Disinfectant, applied dose and contact time and
- Process variables, such as pH

After the water leaves the treatment plant the THM concentrations can increase or decrease. This is due to either continued reactions forming THMs or processes within the distribution system that consume THMs. The evolution of THM concentrations during distribution appears to depend upon several factors including residence time, water flow patterns, condition and management of the system and the practice of maintaining a disinfectant residual (Premazzi et al. 1997).

In general, lower THM levels are found in groundwaters because both precursor concentrations and disinfectant doses are lower than with surface waters. The practice in some countries of using multiple-point application of chlorine, including dosing to the raw water, can lead to much higher THM levels. Some countries, for example Germany, strictly regulate the use and dosage of disinfection chemicals as well as imposing tight limits on THM concentrations. There are also differences in the disinfectant used – for example chlorine dioxide, which does not form THMs, is quite widely used in France, Germany and Italy but hardly at all in the UK and some other countries. Whilst differences exist it is clear that the presence of DBPs in drinking water is an important public health issue with a European dimension.

The 1998 Drinking Water Directive (EC 1998) includes limits on total THMs (100 mg/l by 2008) and bromate (10 mg/l by 2008). It is important to note that these values refer to water taken at the tap, not ex-works.

The basic strategies that can be adopted for reducing the concentrations of disinfection by-products are:

- Changing process conditions (including removal of precursor compounds)
- Using a different chemical disinfectant or
- Using non-chemical disinfection (e.g. UV irradiation or membrane processes)

The formation of THMs during chlorination can be reduced by removing precursors, for example by installing or enhancing coagulation (this may involve using higher coagulant doses and/or lower coagulation pH than are applied conventionally). THM formation can also be reduced by lowering the applied chlorine dose but if this is done it has to be ensured that disinfection is still effective.

It may be feasible to change disinfectant in order to comply with limits for DBPs. The extent to which this is possible will be dependent on raw water quality and installed treatment, for example for precursor removal. It may be effective to change from chlorine to chloramine, at least to provide a residual disinfectant within distribution, in order to reduce THM formation and subsequent development within distribution. Whilst chloramine provides a more stable residual within distribution, it is generally considered to be a less powerful disinfectant.

In some Northern European countries the use of final disinfection using oxidants has been terminated on some supplies. For example, in the Netherlands chlorine or other oxidant is now only used for about 20% of the total volume of drinking water. This approach relies either on aquifers which act as filters for the removal of micro-organisms or on dune infiltration to utilise the natural geology as a very large filter; in general treatment is intensive with many barriers to micro-organisms. The success of this strategy is also dependent on having clean distribution systems with reasonably short residence times.

1.7.3 THM Concentrations

THMs should only be present at very low concentrations in source waters and in the majority of sources the levels are below the normal analytical detection limits (~0.1 μg/l). In some cases the sources may contain chloroform from industrial effluents and spillages and THMs in general because of the use of chlorine to disinfect effluents and cooling waters.

Literature from the 1970s generally reports higher concentrations than are likely to be found more recently because treatment processes have been introduced or modified in order to comply with the various regulations in different countries. Premazzi et al. (1997) reviewed the exposure of the European population to THMs in drinking water. A summary is given in Table 1.11.

A review of THMs and other organohalogens in chlorinated drinking water in EU countries based on data from several European countries and covering the period 1980–2000 has been published (Palacios et al. 2000). Published data from different utilities using chlorination in Belgium, France, Germany, Spain, Netherlands and Italy were analysed.

The statistical results for THMs identified and quantified in drinking waters originating from chlorinated treated surface water and groundwaters are given in Tables 1.12 and 1.13, respectively. The range of concentrations for THMs was explained by different treatment conditions. THM concentrations were generally lower in groundwaters compared to surface waters.

The occurrence of THMs in public water supplies in Greece was studied in the years 1993 to 1998 (Golfinopoulos 2000). Samples were collected from three of the four water works of Athens, which treat 940,000 m^3/day and supply water to nearly four million people. The raw water sources are lakes supplemented by borehole water in times of drought and treatment consists of prechlorination, coagulation, sedimentation, filtration and final chlorination. Samples were also taken from the distribution systems of Mytilene and Chalkida which are supplied with chlorinated water from wells and boreholes.

The concentrations of THMs ranged from 5 to 106 μg/l in the distribution system of Athens, from 4 to 27 μg/l in the distribution system of Mytilene and from 5 to 96 μg/l in Chalkida. The wide ranges of concentrations in the three cities were attributable to differences in the concentrations of organic substances in the raw water. In all cases the THM concentrations were higher during the summer because of the high temperature and increased chlorine dose during this period. In almost all samples the concentrations were lower than the maximum allowable level of 100 μg/l for THMs in the 1998 drinking water Directive (EC 1998). However, in the distribution system of Athens (Piraeus) the maximum level was exceeded during the warm months of the year. The groundwater sources showed little THM production, while the surface waters and well sources showed high THM generation.

DBPs were measured in the final water of 35 Finnish major municipal water works, which utilised different treatment processes and raw water sources. DBPs were measured also from the distribution systems supplied by three water works

Table 1.11 Typical THM concentrations (μg/l) in European drinking water (Data from Premazzi et al. 1997)

	$CHCl_3$		$CHBrCl_2$		$CHBr_2Cl$		$CHBr_3$	
Country	Max.	Mean	Max.	Mean	Max.	Mean	Max.	Mean
Austria		9–18		1–2		0.1		<0.1
Belgium	100	10–50		2–37		8–37		2–33
Finland	80	<0.5–80	4	<1–4	<2	<2	<3	<3
France	60	3–60	40	1–40	35	2–35		<2.5
Germany[a]	75	1–20						
Ireland[a]	125	<100						
Italy	60	<1–15	15	<1–15	15	<1–15	15	<1–15
Luxembourg	8	3–8	5.5	1.5–5.5				
Netherlands		34[a]		<1–15		<1–15		<1–25
Portugal	>80	<1≥80	>50	<1≥50	>50	<1≥50	>50	<1≥50
Spain	60	<1–60	>50	<1–50	>50	<1–50	>50	<1–50
Sweden	225	10	26	2	7	<1	3	<2
UK	718	1–52	102	1–21				

[a]TTHM concentration.

Table 1.12 Descriptive statistics for treated surface waters (Data from Palacios et al. 2000)

	N	Mean μg/l	Median	Min.	Max.	Standard deviation	Lower quartile	Upper quartile
TTHM	19	77.70	47.0	n.d.	390.0	99.93	11.0	99.0
$CHCl_3$	26	13.38	4.95	0.07	68.5	20.32	0.30	19.0
$CHBrCl_2$	12	26.72	9.95	n.d.	98.4	35.33	3.05	44.65
$CH\ Br_2Cl$	17	28.49	3.0	n.d.	160.0	50.66	1.27	17.3
$CHBr_3$	9	37.38	13.5	0.6	156.1	55.99	4.0	110.1

n.d. = not detected.
N = number of observations.

Table 1.13 Descriptive statistics for treated groundwaters (Data from Palacios et al. 2000)

	N	Mean μg/l	Median	Min.	Max.	Standard deviation	Lower quartile	Upper quartile
TTHM	3	7.61	8.90	0.34	13.6	6.72	–	–
$CHCl_3$	5	11.73	7.93	0.07	31.0	14.26	1.63	14.0
$CHBrCl_2$	3	6.26	5.90	1.87	11.0	4.58	–	–
$CH\ Br_2Cl$	5	2.13	2.05	0.0	4.4	1.88	1.4	3.31
$CHBr_3$	1	3.83	–	3.83	3.83	–	–	–

n.d. = not detected.
N = number of observations.

(Nissinen et al. 2002). The concentration of THMs found in Finnish drinking waters was similar to that measured in other European countries. The EU limit value for THMs (100 μg/l) was exceeded only in one waterworks, where it was attributed to the elevated concentration of bromide in the raw water, leading to the increased formation of brominated THMs. Since there were very low concentrations of bromide in most of the waters studied, chlorinated derivatives predominated.

THM concentrations in samples of Irish tap water were reported by Stack et al. (2000); the results are summarised in Table 1.14. In most of these samples the concentrations of brominated THMs exceeded that of chloroform. Villanueva et al. (2003) reported on THM concentrations in tap water from four regions of Spain: Asturias (good quality surface water), Alicante and Barcelona (poorer quality surface water) and Tenerife (good quality groundwater). The results are summarised in Table 1.15. Results for total THMs for other areas of Spain were reported in a separate paper (Villanueva et al. 2003). For Sabadell the minimum, maximum and mean TTHM concentrations were 64.1, 100.9 and 82.5 μg/l based on nine samples. For Manresa the minimum, maximum and mean were 28.3, 101.8 and 56.1 μg/l respectively based on nine samples.

Studies on total THM concentrations in the tap water of Prague, Czech Republic showed seasonal variations from 5 to 30 μg/l approximately. The TTHM concentration depends mainly on the amount of chlorine used for disinfection concentration of humic substances in water, and time elapsed after the disinfection process during transport to the customers. According to the Prague Water Works the

Table 1.14 Concentration of THMs (μg/l) in Irish tap water (Data from Stack et al. 2000)

Site number	$CHCl_3$	$CHBrCl_2$	$CHBr_2Cl$	$CHBr_3$	Total THM
1	21.3	20.4	17.3	2.8	61.8
2	8.4	20.4	18.2	<1.2	47.0
3	<2.8	9.3	20.0	<1.2	29.3
4	16.1	11.7	7.3	<1.2	35.1
5	<2.8	3.7	12.4	<1.2	16.1
6	4.2	6.6	7.8	<1.2	18.6
7	5.5	18.1	17.2	2.1	42.9
8	9.3	7.5	4.5	<1.2	21.3
9	4.1	12.5	13.7	<1.2	30.3
10	6.7	6.5	4.9	<1.2	18.1

Sites 1–5: regional water supply; Sites 6–10: rural water supplies.

average concentration of TTHM in tap water in Prague was 20.7 μg/l in 2001 (average based on 871 analyses). Of this it is estimated that the chloroform concentration was 15 μg/l.

Other studies on THM formation in the water supply of South Bohemia, Czech Republic (Abd El-Shafy and Grünwald 2000) showed that the quantity of THMs formed inside the distribution system was more than 120% of that formed in the water treatment plant. A very good correlation was found between the formation of THMs and the cumulative chlorine decay and the residence time in the pipelines. This study concentrated on pipelines that take their drinking water from Plav water works in the city of Ceske Budejovice. Raw water from a surface water reservoir is treated by ferric sulfate coagulation, sedimentation, rapid gravity filtration and chlorination. The distribution pipes are about 80 km long and they carry drinking water to the towns and villages situated between the plant and the city of Tabor. The concentrations of chloroform and THM in water leaving the water works were on average 4.61 and 5.2 μg/l respectively, whereas at the end of the pipelines they reached 8.50 and 10.3 μg/l. It was noted that the formation of THMs increased with an increase in distance from Plav water works. The formation of THMs in the treatment plant represented only about 45% of the THMs found at the end of the pipelines, while the rest was formed in the pipelines. The dominant THM was chloroform, which represented about 82% of the THMs formed.

In Poland water samples were collected for analysis from consumer water pipes in ten different municipalities: Katowice, Siemianowice, Sosnowiec, Zabrze, Gliwice, Knurów, Sławków, Olkusz, Kozłowa Góra and Wojkowice. Concentrations of trihalomethanes: chloroform, dichlorobromomethane, chlorodibromomethane and bromoform found in finished drinking water of Upper Silesia are given in Table 1.16.

The highest value of THMs, 37 μg/l, was in sample 3 from the city of Siemianowice. Among all the investigated samples, sample No 10 – from Gliwice – did not contain THMs. Gliwice is supplied almost exclusively with water from deep boreholes. In samples containing THMs, chloroform was the largest fraction,

Table 1.15 Concentration of THMs (μg/l) in Spanish tap water (Data from Villanueva et al. 2003)

	Alicante (19 samples)			Barcelona (25 samples)			Asturias (34 samples)			Tenerife (ten samples)		
	Min	Max	Mean	Min	Max	Mean	Min	Max	Mean	Min	Max	Mean
$CHCl_3$	4.93	24	13.6	8.26	35.3	20.0	2.56	29.5	14.6	<0.1	0.68	0.39
$CHCl_2Br$	11.2	30.8	24.7	7.00	31.5	22.7	1.1	12.4	5.0	<0.5	1.95	0.85
$CHClBr_2$	6.6	41.0	25.7	2.86	36.1	10.8	0.07	6.7	2.23	0.5	3.2	1.18
$CHBr_3$	4.5	44.3	21.9	0.02	40.1	10.2	0.05	1.84	0.56	2.8	11.8	5.55
TTHM	35.2	125	85.9	34.6	121.7	63.6	6.36	44.5	22.3	5.08	16.3	8.00

Table 1.16 Concentration of halomethanes (μg/l) in finished drinking waters of Upper Silesia (Data from Luks-Betlej and Bodzek 2002)

Number and city	$CHCl_3$	$CHBrCl_2$	$CHBr_2Cl$	$CHBr_3$	Total THM
1. Sławków	12.84	n.d.	n.d.	0.29	13.13
2. Katowice	19.54	2.07	0.08	0.14	21.83
3. Siemianowice	32.86	3.84	0.21	n.d.	36.91
4. Olkusz	10.43	n.d.	n.d.	0.097	19.53
5. Sosnowiec	5.76	3.10	0.86	<0.05	9.72
6. Kozłowa Góra	6.93	0.82	0.19	n.d.	7.94
7. Wojkowice	1.35	0.84	n.d.	n.d.	2.19
8. Zabrze	2.98	2.76	1.51	0.36	7.61
9. Knurów	8.87	2.35	0.5	n.d.	11.72
10. Gliwice	n.d.	n.d.	n.d.	n.d.	n.d.

Number of determinations = 5–7.
n.d. – not detected.

followed by bromodichloromethane (sample numbers 3, 5, 6, 7, 8, and 9), dibromochloromethane and bromoform. Such a distribution is representative for household water in the investigated region and thus provides information about the realistic range of environmental concentrations. Concentrations of brominated THMs in water were low, although in most water intakes in Upper Silesia bromide occurrence is found to range from 0.4 to 0.7 mg/l as Br. As these investigations were carried out in late November, it can be expected that the annual average contents of THMs will be higher.

In the UK samples of drinking water are routinely analysed for four trihalomethanes: chloroform ($CHCl_3$), bromodichloromethane ($CHBrCl_2$), dibromochloromethane ($CHBr_2Cl$), and bromoform ($CHBr_3$). Whitaker et al. (2003) report values for three water companies. They report that chloroform was the predominant THM in all water types (mean concentration 17.9–38.1 μg/l). The supplier that distributed more ground and lowland surface water had higher dibromochloromethane and bromoform levels and lower chloroform levels than the other two suppliers. In the water of two suppliers, seasonal fluctuations in bromodichloromethane and dibromochloromethane levels were found with levels peaking in the summer and autumn. For all three water suppliers, chloroform levels declined throughout 1995 when there was a drought period.

In 1974, it was discovered that some DBPs are carcinogenic in laboratory animals (Rook 1974; Bellar et al. 1974). This raised public concern about the possible adverse health effects from exposure to DBP. To date, several hundreds of DBPs have been identified. The regulation is being updated continuously as more research results on occurrence of DBPs and potential health effects are emerging, all the time attempting to balance the control of health risks from DBPs against the risks from pathogenic microbial organisms. Ongoing studies are providing toxicological and occurrence data to improve the regulatory basis.

Nieuwenhuijsen et al. (2000) have reviewed the relevant issues in the exposure assessment of disinfection by-products of chlorination for epidemiological and health risk assessment. Various DBPs can be detected in drinking water, but the values are below the current health standards. Chloroform is generally, but not always, the most predominant DBP. Both Boorman et al. (1999) and Woo et al. (2002) have reviewed the possible carcinogenic properties of DBPs. Studies suggest potential developmental, reproductive or carcinogenic health effects in humans exposed to DBPs. As a result of the association with adverse health effects numerous models have been developed to estimate the formation and the fate of DBPs (Sadiq and Rodriguez 2004).

References

Abd El-Shafy, M., & Grünwald, A. (2000). THM formation in water supply in south Bohemia, Czech Republic. *Water Research, 34*(13), 3453–3459.

Abt, E., Suh, H. H., Catalano, P., & Koutrakis, P. (2000). Relative contribution of outdoor and indoor particle sources to indoor concentrations. *Environmental Science & Technology, 34*, 3579–3587.

Andreae, M. O., & Crutzen, P. J. (1997). Atmospheric aerosols: Biochemical sources and role in atmospheric chemistry. *Science, 276*, 1052–1058.

Bellar, T. A., Lichtenberg, J. J., & Kraner, R. C. (1974). The occurrence of organohalides in chlorinated drinking water. *Journal of the American Water Works Association, 66*, 703–706.

Boorman, G. A., Dellarco, V., Dunnick, J. K., Chapin, R. E., Hunter, S., Hauchman, F., et al. (1999). Drinking water disinfection by-products: Review and approach to toxicity evaluation. *Environmental Health Perspectives, 107*, 207–217.

CAFE. (2004). Second position paper on particulate matter. CAFE working group on particulate matter.

Charlson, R. J., Langner, J., Rodhe, H., Leovy, C. B., & Warren, S. G. (1992). Climate forcing by anthropogenic aerosols. *Science, 255*, 423–430.

Dockery, D. W., Pope, C. A., Xu, X., Spengler, J. D., Ware, J. H., Fay, M. E., et al. (1993). An association between air pollution and mortality in six US cities. *New England Journal of Medicine, 329*, 1753–1808.

EC. (1998). The Water Drinking Directive, 98/83/EC, European Commission.

EMEP. 2002 *Transboundary particulate matter in Europe: Status report 2002* (Tørseth, K., Tarrason, L., Markus Amman, Eds., EMEP Report 5/2002). Norwegian Institute for Air Research, Kjeller.

EMEP. (2006). *Transboundary particulate matter in Europe: Status report 2006* (EMEP Report 4/2006). Norwegian Institute for Air Research, Kjeller.

EMEP/CCC. (2005a). *Measurements of particulate matter: Status report 2005* (Karl Espen Yttri, Ed., EMEP/CCC Report 5/2005). Norwegian Institute for Air Research, Kjeller.

EMEP/CCC. (2005b). *Data report 2003. Acidifying and eutrophying compounds* (EMEP/CCC Report 3/2005). Norwegian Institute for Air Research, Kjeller.

EMEP/CCC. (2006). *Ozone measurements 2004* (EMEP/CCC Report 2/2006). Norwegian Institute for Air Research, Kjeller.

EMEP/CORINAIR. (2002). *Atmospheric emission inventory guidebook* (3rd ed.). Prepared by the EMEP Task force on Emission Inventories, EEA Technical Report No 30.

EPA. (2003). *Air quality criteria for particulate matter* (EPA/600/P-99/002aD). U.S. Environmental Protection Agency, Research Triangle Park, N.C.

Erisman, J. W., Otjes, R., Hensen, A., Jongejan, P., van den Bulk, P., Khlystov, A., et al. (2001). Instrument development and application in studies and monitoring of ambient ammonia. *Atmospheric Environment, 35*(11), 1913–1922.

Finlayson Pitts, B. J., & Pitts, J. N. (2000). *Chemistry of the upper and lower atmosphere*. San Diego: Academic.

Finlayson-Pitts, B. J., & Pitts, J. N. (1986). *Atmospheric chemistry: Fundamentals and experimental techniques*. New York: Wiley.

Galal-Gorchev, H., & Bartram, J. (1996). Health benefits of drinking-water disinfection. In Chlorine dioxide and disinfection, *Proceedings of the first European symposium on chlorine dioxide and disinfection* (pp. 41–49). Rome: C.I.P.A. ISSN 1121-8215.

Gleick, P. H. (2006). *The world's water 2006–2007*. Washington, DC: Island Press.

Golfinopoulos, S. K. (2000). The occurrence of trihalomethanes in drinking water in Greece. *Chemosphere, 41*, 1761–1767.

Grennfelt, P., & Schjoldager, J. (1984). Photochemical oxidants in the troposphere – A mounting menace. *Ambio, 13*(2), 61–67.

HEI, Health Effects Institute. (2001). *Airborne particles and health: HEI Epidemiologic evidence*. Cambridge, MA: HEI.

IPCC. (2001). *Climate change 2001; the scientific basis*. Contribution of Working group I to the third assessment report of the Intergovernmental Panel on Climate Change (J. T. Houghton et al., Eds.). Cambridge: Cambridge University Press.

Jacobson, M. Z. (2002). *Atmospheric pollution: History, science, and regulation*. Cambridge University Press, Cambridge.

Katsouyanni, K. (Ed.). (1996). The APHEA project. Short term effects of air pollution on health: A European approach using epidemiological time series data. *Journal of Epidemiology and Community Health, 50*(suppl. 1), S1–S80.

Lachenmyer, C., & Hidy, G. M. (2000). Urban measurements of outdoor-indoor PM2.5 concentrations and personal exposure in the deep south. Part I. Pilot study of mass concentrations for nonsmoking subjects. *Aerosol Science and Technology, 32*, 34–51.

Larssen, S., Sluyter, R., & Helmis, C. (1999). *Criteria for EUROAIRNET – The EEA Air Quality Monitoring and Information Network* (EEA Topic cal report No 12). European Environment Agency, Copenhagen, Denmark.

Lazaridis, M., Semb, A., Larssen, S., Hjellbrekke, A. G., Hov, Ø., Hanssen, J. E., et al. (2002). Measurements of particulate matter within the framework of the European Monitoring and Evaluation Programme EMEP I. *First results The Science of the Total Environment, 285*, 209–235.

Lebowitz, M. D., O'Rourle, M. K., Gordon, S., Moschandreas, D., & Nishioka, M. (1995). Population based exposure measurements in Arizona: A phase I field study in support of the national human exposure assessment survey. *Journal of Exposure Analysis and Environmental Epidemiology, 5*, 297–235.

Levy, J. I., Hammitt, J. K., & Spengler, J. D. (2000). Estimating the mortality impacts of particulate matter: What can be learned from between study variability? *Environmental Health Perspectives, 108*, 109–117.

Lioy, P. J. (1990). Assessing total human exposure to contaminants. *Environmental Science and Technology, 24*, 938–945.

Luks-Betlej, K., & Bodzek, D. (2002). Occurrence of trihalomethanes, particularly those containing bromine, in Polish drinking water. *Polish Journal of Environmental Studies, 11*, 255–260.

Meng, Z. Y., & Seinfeld, J. H. (1996). Time scales to achieve atmospheric gas-aerosol equilibrium for volatile species. *Atmospheric Environment, 30*(16), 2889–2900.

Morawska, L., He, C., Hitchins, J., Mengersen, K., & Gilbert, D. (2003). Characteristics of particle number and mass concentrations in Brisbane, Australia. *Atmospheric Environment, 37*, 4195–4203.

Nero, A. V., Jr. (1988). Controlling indoor air pollution. *Scientific American, 258*, 42–48.

Neuberger, M., Schimekc, M. G., Horak, F., Jr., Moshammerb, H., Kundib, M., Frischerd, T., et al. (2004). Acute effects of particulate matter on respiratory diseases, symptoms and functions: Epidemiological results of the Austrian Project on Health Effects of Particulate Matter (AUPHEP). *Atmospheric Environment, 38*, 3971–3981.

Nieuwenhuijsen, M. J., Toledano, M. B., Eaton, N., Fawell, J., & Elliott, P. (2000). Chlorination disinfection by-products in water and their association with adverse reproductive outcomes: A review. *Occupational and Environmental Medicine, 57*, 73–85.

Nissinen, T. K., Miettinen, I. T., Martikainen, P. J., & Vartiainen, T. (2002). Disinfection by-products in Finnish drinking water. *Chemosphere, 48*, 9–20.

Palacios, M., Pampillón, J. F., & Rodriguez, M. E. (2000). Organohalogenated compounds levels in chlorinated drinking water and current compliance with quality standards throughout the European Union. *Water Research, 34*, 1002–1016.

Pope, C. A., III, & Dockery, D. W. (2006). Health effects of fine particulate air pollution: Lines that connect. *Journal of the Air & Waste Mangaement Association, 56*, 709–742.

Pope, C. A., Thun, M. J., Namboodiri, M. M., Dockery, D. W., Evans, J. S., Speizer, F. E., et al. (1995). Particulate air pollution as a predictor of mortality in a prospective study of U.S. adults. *American Journal of Respiratory and Critical Care Medicine, 151*, 669–674.

Premazzi, G., Cardoso, C., Conio, O., Palumbo, F., Ziglio, G., Borgioli, A., et al. (1997). *Exposure of the European population to trihalomethanes (THMs) in drinking water – Volume 2* (112 pp.). European Commission Report EUR 17335. ISBN 92-828-0385-6.

Rombout, P., Bloemen, H., Bree, L., Buringh, E., Cassee, F., Fischer, P., et al. (2000). *Health risks in relation to air quality, especially particulate matter*. RIVM Report No. 650010 020.

Rook, J. J. (1974). Formation of haloforms during chlorination of natural waters. *Journal Society for Water Treatment and Examination, 23*, 234–243.

Sadiq, R., & Rodriguez, M. J. (2004). Disinfection by-products (DBPs) in drinking water and predictive models for their occurrence: A review. *Science of the Total Environment, 321*, 21–46.

Samet, J. M., Dominici, F., Zeger, S. L., Schwartz, J., & Dockery, D. W. (2000). *The national morbidity, mortality and air pollution study, part I: Methods and methodologic issues* (Research Report 94, Part I). Cambridge MA: Health Effects Institute.

Savoir, R., Romnee, L., & Masschelein, W. J. (1987). Assessment of chlorine dioxide as a means of limiting the formation of organohalogenated compounds. *Aqua, 36*, 114–117.

Schlesinger, R. B. (1995). Toxicological evidence for health effects from inhaled particulate pollution: Does it support the human experience? *Inhalation Toxicology, 7*, 99–109.

Schwartz, J. (2000). Assessing confounding, effect modification, and thresholds in the association between ambient particles and daily deaths. *Environmental Health Perspectives, 108*, 563–568.

Seinfeld, J. H., & Pandis, S. N. (2006). *Atmospheric chemistry and physics, from air pollution to climate change* (2nd ed.). New York: Wiley.

Stack, M. A., Fitzgerald, G., O'Connell, S., & James, K. J. (2000). Measurement of trihalomethanes in potable and recreational waters using solid phase micro extraction with gas chromatography-mass spectrometry. *Chemosphere, 41*, 1821–1826.

Su, W.-H. (1996). Indoor air pollution resources. *Conservation and Recycling, 16*, 77–91.

Sutton, M. A., Asman, W. A. H., Ellermann, T., Van Jaarsveld, J. A., Acker, K., Aneja, V., et al. (2003). Establishing the link between ammonia emission control and measurements of reduced nitrogen concentrations and deposition. *Environmental Monitoring and Assessment, 82*(2), 149–185.

Thatcher, T. L., & Layton, D. W. (1995). Deposition, resuspension, and penetration of particles within a residence. *Atmospheric Environment, 29*, 1487–1497.

USEPA. (2004). *Air Quality Criteria for Particulate Matter* (Report EPA/600/P-99/002aD). Washington, DC: USEPA.

Van der Zee, S. K., Hoek, G., Harssema, H., & Brunekreef, B. (1998). Characterization of particulate air pollution in urban and non-urban areas in the Netherlands. *Atmospheric Environment, 32*, 3717–3729.

Villanueva, C. M., Kogevinas, M., & Grimalt, J. O. (2003). Haloacetic acids and trihalomethanes in finished drinking waters from heterogeneous sources. *Water Research, 37*, 953–958.

Wallace, L. (2000). Correlations of personal exposure to particles with outdoor air measurements: A review of recent studies. *Aerosol Science and Technology, 32*, 15–25.

Whitaker, H., Nieuwenhuijsen, M. J., Best, C., Fawell, J., Gowers, A., & Elliot, P. (2003). Description of trihalomethane levels in three UK water suppliers. *Journal of Exposure Analysis and Environmental Epidemiology, 13*, 17–23.

Woo, Y.-T., Lai, D., McLain, J. L., Manibusan, M. K., & Dellarco, V. (2002). Use of mechanism-based structure-activity relationships analysis in carcinogenic potential ranking for drinking water disinfection by-products. *Environmental Health Perspectives, 110*, 75–87.

World Health Organization (WHO). (2000). *Air quality guidelines for Europe* (2nd ed., European Series, No. 91). Copenhagen: WHO Regional Publications.

World Health Organization (WHO). (2006). *WHO Air quality guidelines for particulate matter, ozone, nitrogen dioxide and sulfur dioxide – Global update 2005 – Summary of risk assessment*. Geneva, Switzerland: WHO Press.

Zappoli, S., Andracchio, A., Fuzzi, S., Facchini, M. C., Gelencser, A., Kiss, G., et al. (1999). Inorganic, organic and macromolecular components of fine aerosol in different areas of Europe in relation to their water solubility. *Atmospheric Environment, 33*, 2733–2743.

Chapter 2
Indoor Air Pollution

Ian Colbeck and Zaheer Ahmad Nasir

Abstract Population exposure to various air pollutants is likely to be higher in the indoor micro-environment than outdoors due to the amount of time people spend there. Consequently, indoor air quality has drawn considerable attention in recent years. There are noticeable differences in the types and strength of air pollution sources across the globe and they are closely linked to socio-economic developments. Typically higher indoor concentrations occur in developing rather than developed countries. The types, concentration, and sources of indoor air pollutants vary considerably from one micro-environment to another. Hence, an understanding of the concentration of pollutants in different micro-environments is of great importance for improving exposure estimates and, in turn, for developing efficient control strategies to reduce human exposure and health risk.

2.1 Introduction

We often assume that air pollution is a modern phenomenon, and that it has become worse in recent times. However since the dawn of history, mankind has been burning biological and fossil fuel to produce heat. The walls of caves, inhabited millennia ago, are covered with layers of soot and many of the lungs of mummified bodies from Palaeolithic times have a black tone (McNeill, 2001). Brimblecombe (1987) has suggested that the high incidence of sinusitis in Anglo-Saxon Britain was related to a build-up of smoke in their poorly ventilated huts.

Air pollution problems in ancient Rome appear in many documents (Hughes 1993; Makra and Brimblecombe 2004). As residents of what had become the largest city in the world, ancient Romans were well aware of the problem of air pollution. They called it gravioris caeli (heavy heaven) or infamis aer (infamous air).

I. Colbeck (✉) and Z.A. Nasir
Department of Biological Sciences, Interdisciplinary Centre for Environment and Society, University of Essex, Colchester, CO4 3SQ, UK
e-mail: colbi@essex.ac.uk

M. Lazaridis and I. Colbeck (eds.), *Human Exposure to Pollutants via Dermal Absorption and Inhalation*, Environmental Pollution 17, DOI 10.1007/978-90-481-8663-1_2,

"*The smoke, the wealth, the noise of Rome...*" held no charms for the Roman poet Horace (65 BC–AD 8) who described the blackening of buildings by smoke (Costa 1997).

Indoor air pollution, and in particular particulate matter, was also a significant problem. Animal and vegetable oils were burned to provide artificial light and wood, vegetal materials and animal dung was used to heat their homes. All these materials produced high quantities of soot and toxic gases. Capasso (2000) examined skeletons buried by the volcanic eruptions of Vesuvius and found evidence of inflammation of the pulmonary tract. Histological assessment of the lungs of ancient human mummies has shown that anthracosis was a regular disorder in many ancient societies, including the Egyptian, Peruvian and Aleutian.

In the Bible Leviticus 14, 34–57 indicates that people were aware that residing in damp buildings was dangerous to their health. The remedial action was rather severe:

> "and he shall break down the house, the stones of it, and the timber thereof, and all the morter of the house; and he shall carry them forth out of the city into an unclean place".

In the early seventeenth century that it was recognised that "*want of ventilation*" resulted in increased rates of infectious disease and in the mid-nineteenth century Griscom (1848) highlighted the impact of poor ventilation on health stating that "*deficient ventilation ... (is) more fatal than all other causes put together.*" In the following years a number of investigations were conducted on the effect of ventilation, carbon dioxide concentrations on disease (Sundell 2004). For example Carnelly et al. (1887) measured CO_2, mould and bacteria and total organic material in public housing, schools, factories and a hospital in Dundee. They found that concentrations of all these parameters were proportional to number of occupants per room. They went on to recommend a general level of 600 CFU m^{-3} as a limit value for human exposure to bioaerosols.

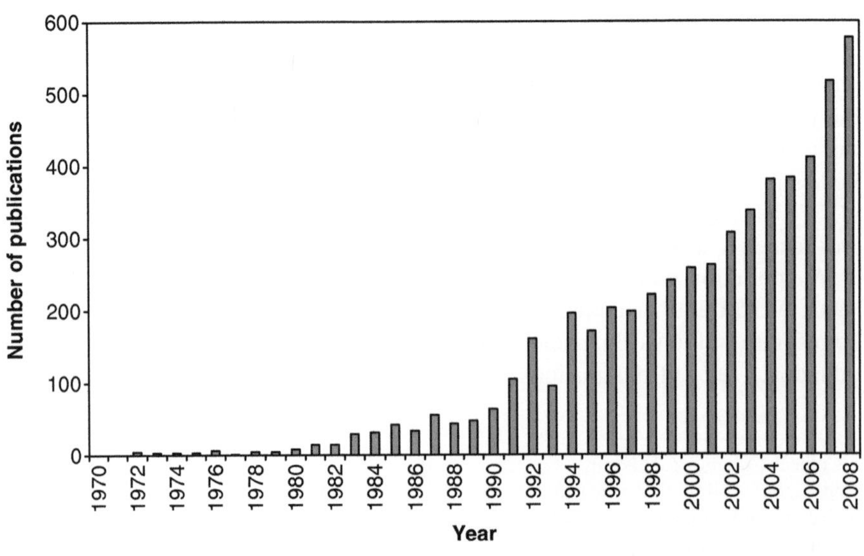

Fig. 2.1 Number of papers published per year since 1970 with 'indoor air' as the topic (Bibliographic search of Web of Knowledge [wok.mimas.ac.uk] – accessed 16 January 2009)

It wasn't until the 1960s the papers on indoor air pollution began to appear on a regular basis. The early work was related to radon and tobacco smoke before extending to formaldehyde in the early 1970s, house dust mites and sick building syndrome in the late 1970s, and allergies during the 1990s (Weschler 2009). Figure 2.1 shows the number of papers published each year from 1970 onwards based on a bibliographic search of Web of Science using indoor air as the topic. Although this will not yield every single paper it does give an overall picture of how indoor air pollution is now drawing the attention of scientists from all over the world.

It is evident that, in many circumstances, people are far more exposed to pollution indoors than outdoors. We also tend to believe that the indoor environment is cleaner, more comfortable and healthier on the obvious grounds that the building shelters us from harmful substances in the ambient environment. For this reason a number of quality indication systems in the world give warnings or advice to stay indoors during episodes of poor air quality. However, Chan (2002) pointed out that, the fundamental question is: Is indoor air really cleaner? Is it free from outdoor pollutants?

Indoor aerosol concentrations are associated with both indoor and outdoor sources. The identification of sources and the assessment of their relative contribution can be a complicated process due to the presence of a number of indoor sources, which can vary from building to building. There are also uncertainties associated with estimating the impact of outdoor sources on the indoor environment (Mitchell et al. 2007). There are numerous indoor sources and these, in residential environments, include heating, cooking, cleaning, smoking, the use of a wide variety of consumer products, building materials and furnishings, and the simple act of moving about and stirring up particles. Some of these sources emit pollutants virtually continously while others are related to specific activities (e.g. cleaning). The relative importance of any single source depends on how much of a given pollutant it emits and how hazardous those emissions are.

A wide range of pollutants have been in reported in residential environments and these are summarized in Table 2.1. There are distinct variations in the importance of the different sources in different areas of the world; closely related to level of socioeconomic development.

Outdoor pollutant concentrations may not be reliable indicators of indoor and personal pollutant sources. Lawrence et al. (2005) remarked that assessment of risk to a community resulting from exposure to air pollutants should ideally include measurements of concentration levels of pollutants in all micro-environments where people spend their time.

It is beyond the scope of this chapter to review every single pollutant in every conceivable micro-environment. The reader is referred to Ashmore and Dimitroulopoulou (2009), Fuentes-Leonarte et al. (2009), Morawska (2008), Fullerton et al. (2008), Yu et al. (2008), Mitchell et al. (2007), Ott et al. (2007); Gorny (2004), Salthammer (2009), Burroughs and Hansen (2008) and Pluschke (2004).

This chapter focuses on non-industrial buildings such as homes, schools, offices and transport micro-environments. It considers, in depth, a number of case studies. It should be remembered that the concentration of indoor particles is highly variable and house-specific.

Table 2.1 Sources of the main indoor air pollutants (Adapted from CARB 2005)

Pollutant	Main sources
Arsenic	Coal combustion
Asbestos	Building materials in older homes released during renovation, naturally occurring in some soils
Biological Agents (bacteria, fungi, viruses, house dust mites, animal dander; cockroaches, microbial VOCs)	House and floor dust; pets; bedding; poorly maintained air-conditioners, humidifiers, dehumidifiers; moist structures or furnishings; insect infestation; building occupants
Carbon monoxide	Unvented or malfunctioning gas appliances, wood stoves, fireplaces, tobacco smoke
Endocrine disruptors (phthalates; DDT, chlordane, heptachlor, *o*-phenylphenol; PBDEs)	Plastics; pesticides; flame retardants
Environmental tobacco smoke	Cigarettes, cigars and pipes
Formaldehyde, other aldehydes	Composite wood products such as plywood and particleboard; furnishings; wallpaper; paints; combustion appliances; tobacco smoke
Lead	Lead paint chips, contaminated
Nitrogen dioxide	Unvented or malfunctioning gas appliances, other combustion appliances
Organic chemicals (benzene, chloroform, paradichlorobenzene, methylene chloride, perchloroethylene, phthalates, styrene)	Solvents; glues; cleaning agents; pesticides; building materials; paints; treated water; moth repellents; dry-cleaned clothing; air fresheners
Ozone	Infiltration of outdoor air, ozone generating air "purifiers", office machines
Particulate matter	Cigarettes, wood stoves, fireplaces, cooking, candles, aerosol sprays, house dust
Polycyclic aromatic hydrocarbons	Cigarette smoke, cooking, woodburning
Radon	Uranium-bearing soil under buildings, ground-water, construction materials

2.2 Indoor-Outdoor Measurements in Oslo

As part of the Urban-Aerosol project (Characterisation of Urban Air Quality – Indoor/Outdoor Particulate Matter Chemical Characteristics and Source-to-Inhaled Dose Relationships) a series of indoor/outdoor measurements were undertaken in Athens, Oslo, London, Hannover, Prague and Milan (Lazaridis et al. 2006). In Oslo indoor-outdoor measurements, during summer and winter periods, were made at two different residential houses (one in the suburbs and the other in the city centre). Apart from the integrated indoor-outdoor particulate matter (PM_{10}, $PM_{2.5}$, size distribution) and gaseous pollutants (O_3, NO_x, VOC) measurements, infiltration rate evaluation (SF_6) and meteorological measurements were also performed along with a detailed chemical speciation and compilation of a daily diary where different indoor activities were registered. In addition continuous measurements for PM_{10}

and $PM_{2.5}$ using TEOM instruments were performed together with the use of particle number distribution measurements using a Scanning Mobility Particle Sizer (particle size range 10–450 nm), and Aerodynamic Particle Sizer (particle size range 0.7–20 μm) (Lazaridis et al. 2006, 2008).

The concentration of total particles measured outside the city centre apartment was approximately twice that measured in the suburbs. However, the concentration of total particles was higher indoors than outdoors during the winter due to the indoor activities, the operation of a heating system and reduced ventilation for energy saving. On the other hand the indoor concentration was lower than that outdoors in the summer at both sites due to the increased ventilation and the enhanced secondary aerosol formation in the outdoor environment. The concentration of particles in the outdoor environment was attributed to both anthropogenic pollution and natural sources. The results indicated the influence of marine aerosols, crustal aerosols and anthropogenic pollution from both local and remote sources to the ambient aerosol concentrations. Specifically, the concentration of nitrate, chloride, sodium, calcium and magnesium ions was higher during the winter periods due to the enhanced effect of local traffic (direct emissions and resuspension of road side dust) especially in the city centre apartment whereas sulphate and ammonium concentrations were higher during the summer periods due to the enhanced secondary aerosol formation. Furthermore, their concentrations exhibited high daily variation especially during the winter/spring periods due to the episodic strong winds blowing from marine areas (e.g. 13–17 January) due to air masses arriving at the sampling site from western locations (Fig. 2.2). Generally, when air

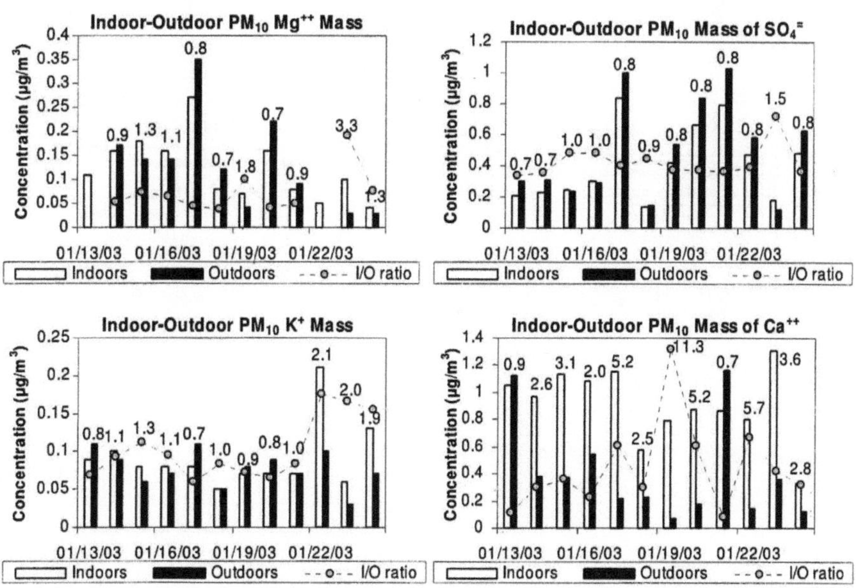

Fig. 2.2 Indoor-outdoor variability of inorganic ions, PM_{10} size faction, during the January 2003 measurement period at a city centre apartment in Oslo

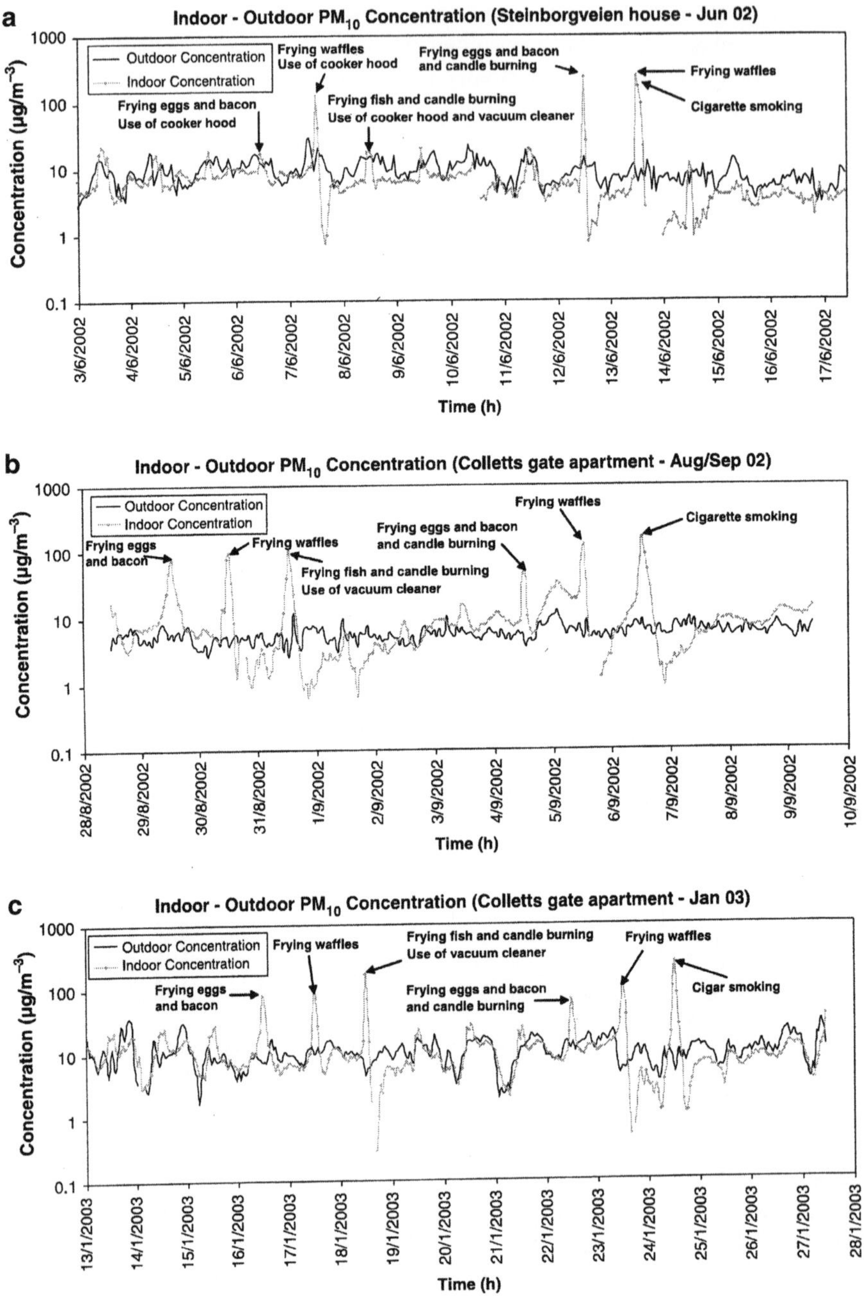

Fig. 2.3 Indoor and outdoor PM10 concentration at the: (**a**) suburban house (Steinborgveien) during June 2002; (**b**) city centre apartment (Colletts Gate) during August/September 2002; and (**c**) during January 2003. The various activities are indicated in the figure (Reproduced with permission)

masses originated from western locations the concentration of marine particles was enhanced, whereas, when the air masses originated from eastern and northern areas the concentration of the secondary and crustal component of aerosols was enhanced. When considering the indoor to outdoor relationship of specific aerosol components it was reported that the chemical concentration of the inorganic aerosol mass was higher outdoors whereas the organic mass was higher indoors, especially during cooking activities (Lazaridis et al. 2008).

Comparing simultaneous measurement of PM_{10} indoors and outdoors it is evident that specific indoor activities, such as cooking, cleaning and smoking, lead to high indoor concentrations. (Fig. 2.3). Chemical analysis indicated increased indoor concentration of organic carbon particles, sulphate, nitrate, calcium and sodium ions during cooking. The indoor concentration of fine organic carbon particles was approximately 0.65 times the outdoor concentration during days with no activity whereas on days with activities it was approximately five times higher. In addition, the indoor concentration of coarse organic carbon particles was approximately 16 times higher than that outdoors during days with indoor activities (Lazaridis et al. 2006).

2.3 Particle Emission Rates

It is evident from the above that both indoor and outdoor sources contribute to and affect the concentration and composition of particles in indoor air. While emission rates for outdoor particle sources are reasonably quantified (Mitra et al. 2002; Zhang and Morawska 2002), ignoring tobacco smoke, there was, until relatively recently, only limited data for particulate emissions from indoor sources. The most important sources include cooking, kerosene heating and wood burning (e.g. Raunemaa et al. 1989; Long et al. 2000; Dennekamp et al. 2001; Sjaastad and Svendsen 2008) while sources such as cleaning, dusting and vacuuming, showering, electric motors, movement of people and gas-to-particle conversion have also been investigated (e.g. Abt et al. 2000a, b; Waring et al. 2008). Secondary formation of ultrafines in large quantities indoors has been observed from chemical reactions of ozone and terpenes (Weschler and Shields 1999, 2003).

The method of cooking can have an impact on the emissions with significant differences between Eastern and Western cultures (He et al. 2004b; Robinson et al. 2006; See and Balasubramanian 2008). The majority of the work on particulate emissions from cooking has considered indoor PM_{10} or $PM_{2.5}$ concentrations in houses or restaurants (Lee et al. 2001; Monkkonen et al. 2005; Fortmann et al. 2001). Small particles can be high in number while contributing little to particle mass. Such particles can also penetrate more deeply into the lung and as a result have been a subject of increasing concern. Number concentration measurements have been carried out for various cooking activities (Siegmann and Sattler 1996; Dennekamp et al. 2001; Wallace et al. 2004, 2008; Afshari et al. 2005;

Table 2.2 Particle emission rates for various indoor activies

	Emission source strength (particles min^{-1})			
Activity	Gehin et al. (2008)	Afshari et al. (2005)	He et al. (2004a)	Wallace et al. (2008)
Heating electric stove	4.2×10^{11}	6.8×10^{11}	7.33×10^{11}	6×10^{11}–11×10^{12}
Grilling			7.34×10^{11}	
Frying meat/fish	5.4×10^{12}	8.27×10^{11}	4.75×10^{11}	
Cooking (oven) meat/fish	4.2×10^{11}		1.27×10^{11}	4×10^{11}–1.1×10^{12}
Candle	7.2×10^{10}	8.8×10^{10} 3.65×10^{11}		
Aerosol spray	1.8×10^{12}	2.34×10^{11}		
Vacuuming	1.2×10^{12}	3.5×10^{11}	9.7×10^{10}	
Smoking		3.76×10^{11}	1.91×10^{11}	

Hussein et al. 2005). Burning candle emissions have been reported by Wasson et al. (2002) and Afshari et al. (2005) while Lung and Hu (2003) studied incense combustion. Other sources include a clothes dryer (Wallace 2005) and office equipment (He et al. 2007; Destaillats et al. 2007). He et al. (2004a) and Gehin et al. (2008) have looked at a number of activities ranging from cooking to vacuuming. Table 2.2 summarises some of these particle emission rates. It should be remembered that the various emission rates have been based on different methods and cover slightly different size ranges. For instance the work of Gehin et al. (2008) reports emission rate for particles with diameter between 5 nm and 1 μm while that for Wallace et al. (2008) is for the size range 2–64 nm. However it is evident that various indoor activities emit a significant number of particles, the majority of which can be classified as ultrafine.

It is evident from the above that the concentration of indoor air pollution is highly dependent on the resident's activities and lifestyle. Similar houses can exhibit quite different concentrations. Even the same house can experience changes in the diurnal variation as a result of changes in lifestyle. Figure 2.4 shows the diurnal variation in PM_{10}, $PM_{2.5}$ and PM_1 for a typical house in the UK with children of school age. The solid lines represent PM concentrations on non-school days while the dashed lines are for school days. On school days the early morning peak is of short duration and there is little increase around lunch time. When the children return in the afternoon there is a sharp rise in PM concentrations which is not evident on non-school days. It is clear that there is a close relationship between the concentration and human indoor activities. Humans are responsible for their own "personal cloud", that is, exposure to airborne particles resulting from personal activities (e.g. occupation, hobbies) or physical activities (e.g. jogging, vacuum cleaning) (Rodes et al. 1991; McBride et al. 1999; Adgate et al. 2003). For $PM_{2.5}$ this cloud is around 15 $\mu g\ m^{-3}$ and can be higher for those who live an active outdoor life (Adgate et al. 2003).

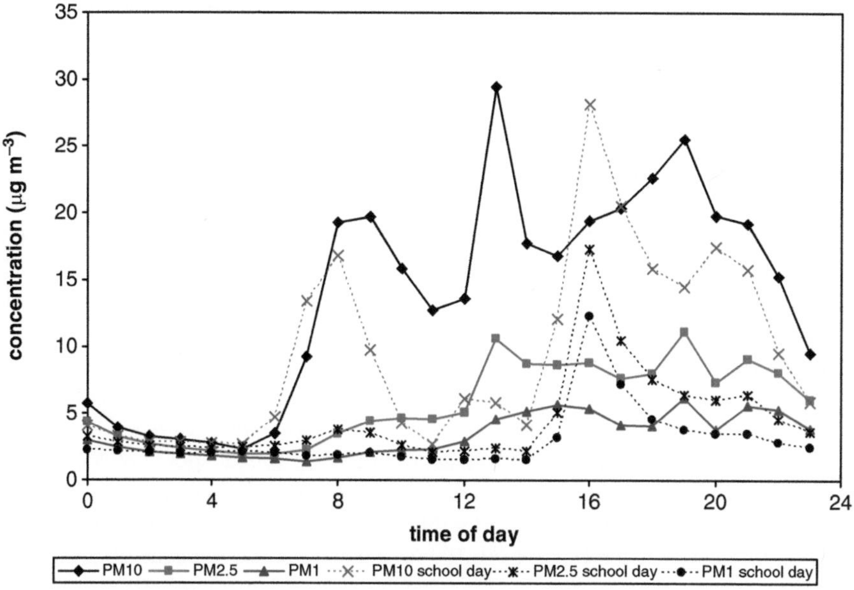

Fig. 2.4 Impact of activities on the diurnal profile of indoor air quality

2.4 Bioaerosols

Biological material is present in the atmosphere in the form of pollens, fungal spores, bacteria, viruses, and any fragments from plants and animals. The size scale ranges from about 15–400 nm for viruses, through 0.3–10 μm for bacteria to 1–100 μm for fungal spores, pollen and plant debris. There is great concern about the potential health hazards of indoor bioaerosols to humans, with a special focus on allergenic or toxigenic fungi and their association with indoor air quality. Douwes et al. (2003) concluded that the potential health effects of bioaerosol exposures are diverse including infectious diseases, acute toxic effects, allergies and cancer. There is also a growing body of scientific literature examining the relationship between dampness and mould in buildings and associated health effects (Bornehag et al. 2001; IOM 2004; Mudarri and Fisk 2007).

2.4.1 Indoor Concentrations

Bioaerosols have been studied in numerous different regions and settings: schools (Aydogdu et al. 2005), child care centres (Zuraimi and Tham 2008), markets (Narayan et al. 1982), animal feed industry (Hameed et al. 2003), animal sheds

(Rosas et al. 2001), rice mills (Savino and Caretta 1992; Desai and Ghosh 2003), saw mills (Oppliger et al. 2005; Jothish and Nayar 2004), food grain warehouse, bakery and library (Jain 2000), food processing units (Zorman and Jersek 2008).), bakeries and flour mills (Musk et al. 1989; Singh and Singh 1994; Awad 2007) hospitals kindergartens, senior care centres and nursing centres (Kim and Kim 2007), social welfare houses (Rolka et al. 2005) and offices (Kalogerakis et al. 2005).

Many measurements carried out on particulate matter (PM_{10} and $PM_{2.5}$) and their outdoor-indoor relationships automatically include biological particles in the sampling process, although the data does not differentiate between organic and inorganic matter (Conner et al. 2001; Morawska et al. 2001). Measurements taken in Eastern Europe indicated that bacterial levels may be higher indoors when high numbers of people are present, than those outdoors (Goh et al. 2000). Exposure to viable airborne bacteria has also been shown to be considerably higher in the indoor environment, with levels increasing with increased human occupancy (Ambroise et al. 1999). They found this to be in stark contrast to fungal spore concentrations, which were higher outdoors. An early review on fungal exposure was given by Miller (1992). Fungal levels have chiefly been found to be higher outdoors unless the building is ventilated (Parat et al. 1997; Goh et al. 2000; Wu et al. 2000). Living conditions have been shown to affect levels of airborne microbes (Perera et al. 2002), whereby low-income families living in poorer quality accommodation were found to suffer the highest exposure levels. The work implicates the need to comprehensively examine the effects of social class and types of housing on exposure to bioaerosols, themselves requiring accurate assessment.

An evaluation of indoor and outdoor fungal concentration in 1,717 buildings in the United States was presented by Shelton et al. (2002) who found lower concentrations than outdoors. Overall 95% of the buildings had a median indoor and outdoor fungal concentration of less than 1,300 CFU/m^3, and less than 3,200 CFU/m^3, respectively. The highest fungal levels were obtained in the fall and summer. The levels of indoor bacteria and fungi in various public places and food processing units were assessed by Zorman and Jersek (2008). The concentration of fungi was significantly higher in food processing units than in public places. In public places, the concentrations of bacteria and fungi were in the range of 0–5,860 CFU/m^3 and 5–3,579 CFU/m^3, respectively. Whereas, levels ranged from 0 to 3,506 CFU/m^3 for viable bacteria and 22–46,377 CFU/m^3 for fungi in food processing units. Lin and Li (1996) investigated indoor and outdoors fungi in six residences in Taiwan with a two-stage Anderson impactor. They revealed that the number concentrations of indoor total and respirable fungi were in the range of 420–4,200 CFU/m^3 and 250–1,000 CFU/m^3, respectively.

Monitoring of bacteria and fungi in indoor air at several schools in the city of Edrine, Turkey was carried out by Aydogdu et al. (2005). They reported a positive correlation between the concentration of bacteria and humidity and age of school. A study by Godwin and Batterman (2007) on indoor air quality in 64 Michigan schools revealed that bioaerosol concentrations were <6,500 CFU/m^3 and <4,100 CFU/m^3 for indoor and outdoor, respectively. Bioaerosol exposure in apartments located in high rise buildings in Korea was evaluated by Lee and Jo (2006).

They reported that season, room location in the apartment and floor level influenced the bioaerosol concentration.

To identify bacterial species contaminating working environments Bouillard et al. (2005) conducted sampling in 25 offices. In their study bacterial levels varied from 44–2,511 CFU/m^3 with a median of 277 CFU/m^3.They concluded that people working in offices can be exposed to large concentrations of airborne bacteria and related endotoxins. The US Environmental Protection Agency carried out a study (The Building Assessment Survey and Evaluation), in 100 large office buildings between 1994 and 1998. The summary of this study, presented by Tsai and Macher (2005), revealed that concentrations varied with a seasonal pattern. Outdoor bacterial concentrations were higher in winter than those indoors: 194 CFU/m^3 compared to 165 CFU/m^3. However indoor concentrations were higher in summer as compared to those outdoors (116 versus 87 CFU/m^3) (Tsai et al. 2007).

Haas et al. (2007) carried out a year long study to assess indoor mould in 66 apartments (29 with no visible mould growth and 37 with visible mould growth) in Austria. The median concentrations of viable fungal spores were significantly higher in apartments with mould growth (1.5×10^3 CFU/m^3) in comparison to those without mould growth (2.6×10^2 CFU/m^3). Moreover, in flats with no visible mould growth median spore concentrations were significantly higher in summer than winter and spring. On the other hand, there was no significant seasonal difference in the concentration of fungal spores in flats with mould growth, but a ten times higher indoor concentration than outdoor air was obtained. During an investigation on indoor fungal and bacterial aerosols in 60 flats of Upper Silesia, Poland Gorny et al. (1999) reported that levels were below 10^4 CFU/m^3. In another study in the same region, Pastuszka et al. (2000) showed that levels of bacterial aerosol in homes and offices were 10^3 CFU/m^3 and 10^2 CFU/m^3. The levels of fungal aerosol, during the winter, in healthy homes ranged from 10–10^2 CFU/m^3 as compared to 10–10^3 CFU/m^3 in mouldy homes. In winter, the levels increased to 10^3 and 10^4 CFU/m^3 in healthy and mouldy homes, respectively.

A study by Reponen et al. (1994) revealed the effect of a range of domestic activities. Most of the activities had a noticeable effect on the spore counts except baking, handling of house plants and vacuum cleaning, A quick decline in the concentration of large size spores was observed after the activity, clearly due to faster gravitational settling. Similar results of short term anthropogenic activity were reported by Brandl et al. (2008) during a study on bioaerosol generation in indoor air of a university hallway. Their results indicated a clear association of presence/absence of people and concentration of bioaerosols. The highest bacterial aerosol concentrations (1,200 CFU/m^3) were recorded during the presence of students while these levels fell to 200 CFU/m^3 during their absence.

The investigations discussed above demonstrate a wide variation in the concentration of bioaerosols in different microenvironments. The concentration and size distributions not only vary with geographical location but also depend on a wide range of biotic and abiotic factors. According to several studies, the moisture content of building material, relative humidity and temperature (Foarde et al. 1993; Pasanen et al. 2000; Ritschkoff et al. 2000; Viitanen et al. 2000) outdoor concentrations, air exchange rates (Kulmala et al. 1999), human activities (Buttner and Stetzenbach 1993) and number of people and pets (ACGIH 1999) significantly affect the levels of indoor bioaerosols. Moreover, housing conditions, the activities and life style of occupants considerably contribute to the varying concentrations. These factors fluctuate to a great degree between various housing types, their condition and geographic location. Hence there is need to study the indoor bioaerosols in various types of residential settings.

2.4.2 *Size Distribution*

Most studies have focused on the total concentration of bioaerosols. However, particle size is critical with regard to their fate in the air and their deposition in the human respiratory system.

Table 2.3 shows the geometric mean (GM) and geometric standard deviation (GSD) of total viable fungal and bacterial aerosol in various size fractions for three

Table 2.3 Geometric mean (GM) and geometric standard deviation (GSD) of total viable fungal and bacterial aerosol and in various size fractions for three different housing types

	Bacteria		Fungi	
Housing Type I	GM(CFU/m³)	GSD	GM(CFU/m³)	GSD
Total	1,557	1.5	925	2.9
7 μm & above	279	1.3	176	1.8
4.7–7 μm	277	1.7	241	2.3
3.3–4.7 μm	434	2.6	192	5.8
2.1–3.3 μm	240	1.3	82	9.5
1.1–2.1 μm	212	1.5	110	2.0
0.65–1.1 μm	19	2.5	15	1.9
Housing Type II				
Total	2,403	2.3	813	3.6
7 μm & above	471	3.0	63	1.6
4.7–7 μm	382	2.3	136	2.2
3.3–4.7 μm	451	2.7	232	3.7
2.1–3.3 μm	536	2.2	182	8.4
1.1–2.1 μm	355	3.5	100	2.1
0.65–1.1 μm	64	1.3	7	1.1
Housing Type III				
Total	5,036	2.5	2,124	1.38
7 μm & above	199	1.3	257	4.8
4.7–7 μm	311	1.1	294	4.2
3.3–4.7 μm	207	2.8	429	1.7
2.1–3.3 μm	774	4.2	581	2.2
1.1–2.1 μm	2,228	1.6	140	1.5
0.65–1.1 μm	659	11.1	7	1.0

different housing types. Measurement were made with an Anderson six-stage viable particle sampler in different types of residential houses: a single room in shared accommodation (Type I), single bedroom flat in three storey buildings (Type II) and two or more bedroom houses (Type III). All the accommodation was 40–50 years old and no major repairs had been undertaken over the previous year. Type I houses were occupied by one person whereas types II and III were occupied by two to three and four to six people, respectively. It is evident that the concentrations of both bacterial and fungal aerosol were almost double in type III housing accommodation as compared to types I and II. This reflects the effects of number of occupants, their activities and possibly building construction and design on indoor bioaerosol levels. The concentrations of indoor bioaerosol in this study could be due to the season as many studies have reported higher indoor levels in summer in non-mouldy houses (Pastuszka et al. 2000; Ren et al. 1999; Lee and Jo 2006; Shelton et al. 2002) due to migration of fungal spores from outdoors.

Figure 2.5 shows the size distribution of fungal and bacterial spores for the various housing types. The viable bacterial concentration exhibits different size distributions in all the housing types. However, more than 60% of viable bacteria were of <4.7 μm in housing type I and II. While in housing type III almost 88% were in the size fraction < 4.7 μm.

The differences in the size distribution of fungal spores among different housing conditions reflect the different species composition or the different ages of the spores. The aerodynamic sizes of the freshly released spores are larger than those which have been airborne for a longer time (Reponen et al. 1994). Dehydration, agglomeration and relative humidity of surrounding air (Pasanen et al. 1991; Reponen et al. 1996; Ren et al. 2001) are among other factors affecting the size of spores. In the naturally ventilated buildings hygroscopic growth of bioaerosols by condensation or water absorption, influences the kinetics of aerosols (Liao et al. 2004).

2.5 Indoor Air Quality in Developing Countries

In developing countries, population explosion along with urbanization and growing industrialization has resulted in dense urban centres with poor air quality. However, around 60% of the people in developing countries continue to live in rural areas and so the micro-environment with the greatest contribution to global person-time is the rural indoor environment. In developing countries, the most significant issue for indoor air quality is exposure to pollutants released during combustion of solid fuels, including biomass (wood, dung and crop residues) or coal (mainly in China), used for cooking and heating. Worldwide, more than three billion people, largely in developing countries, rely on biomass fuels for their domestic energy needs (WHO 2006). A number of different chemical substances are emitted when biomass is burnt including carbon monoxide, nitrogen dioxide, particulate matter, polycyclic aromatic hydrocarbons, benzene and formaldehyde (Smith 1987;

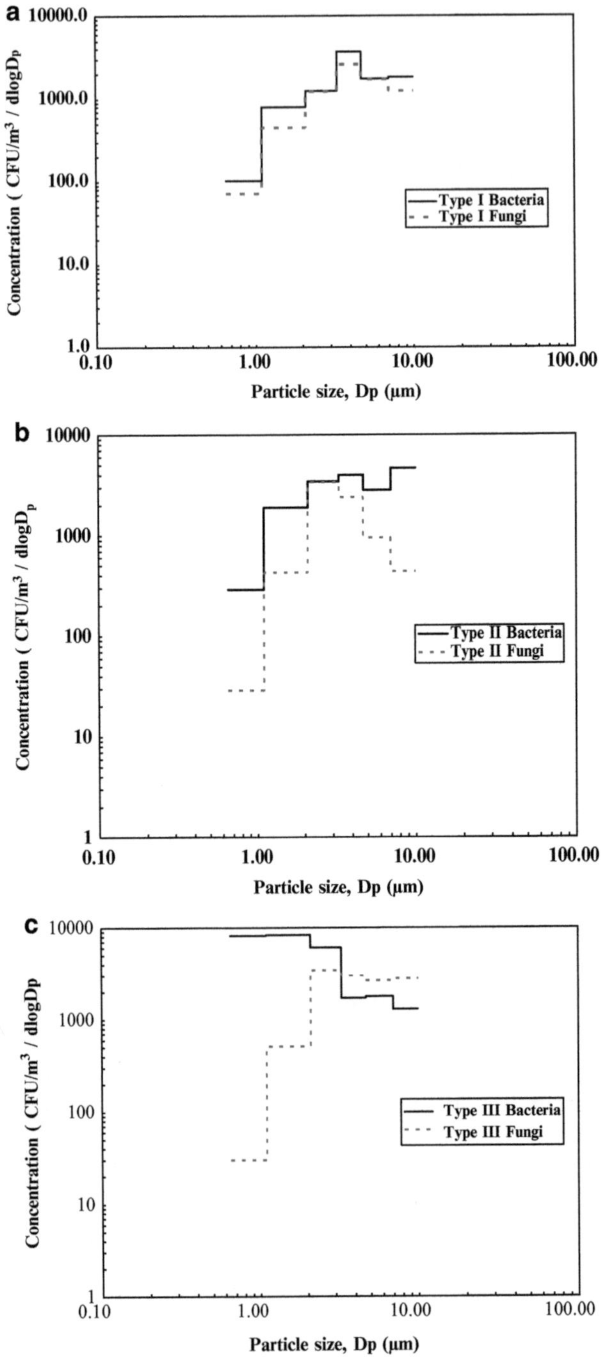

Fig. 2.5 Size distribution of bacterial and fungal spores in (**a**) type I houses, (**b**) type II houses and (**c**) type III houses

De Koning et al. 1985; Mudway et al. 2005; Naeher et al. 2007). Combustion of coal in addition to the above pollutants may release sulphur dioxide, arsenic and fluorine (Finkelman et al. 1999). Due to the variety of fuel types, together with ventilation rates and combustion temperatures, a variation in emission patterns is expected. There is strong evidence that smoke from biofuels can cause acute lower respiratory infection in childhood (WHO 2006; Smith et al. 2000; Ezzati and Kammen 2001). A recent report on national burden of diseases from indoor air pollution by the World Health Organization (2007) confirms the linkage between indoor air pollution due to solid fuels and different diseases, including acute and chronic respiratory diseases, tuberculosis, asthma, and cardiovascular disease and prenatal health outcomes. In most of cases indoor air pollution disproportionately affects women and children who spend most time near the domestic hearth. Indoor air pollution is responsible for more than 1.6 million annual deaths and 2.7% of global burden of diseases (WHO 2006) and indoor air pollution from solid fuel use is the tenth biggest threat to public health WHO (2007).

Studies on indoor air pollution from solid fuels have been conducted in various developing countries in recent years including Mexico (Zuk et al. 2007), Philippines (Saksena et al. 2007), China (Fischer and Koshland 2007; Mestl et al. 2007), Zimbabwe (Rumchev et al. 2007), Bangladesh (Dasgupta et al. 2006), India (Balakrishnan et al. 2002, 2004), Costa Rica (Park and Lee 2003), Bolivia (Albalak et al. 1999) and Kenya (Boleij et al. 1989). Indoor PM concentrations in a number of developing countries are shown in Table 2.4.

Figure 2.6 shows the mass concentration of particulate matter during cooking, using biomass fuels, at a rural location in Pakistan. A large variation in concentration is evident. Particulate levels increase rapidly during cooking and decrease quickly after cooking. Over a period of 1 week, the daily levels of PM_{10}, $PM_{2.5}$ and PM_1 during cooking ranged from 1,991 μg/m^3 to 7,881 μg/m^3, 1,531 μg/m^3 to 2,664 μg/m^3 and 1,430 μg/m^3 to 2,396 μg/m^3, respectively. Generally a wide variation in concentration of particulate matter is observed among different kitchens and even within the same kitchen during different episodes of cooking. The variation primarily depends on the quality (dryness) of biomass fuel used, duration of cooking, degree of incomplete combustion and ventilation. Ezzati and Kammen (2002) have shown that a typical 24-h average concentration of PM_{10} in homes using biofuels may range from 200 to 5,000 μg/m^3 or more throughout the year. Figure 2.7 indicates the high intensity emissions that commonly occur when using biomass fuels. The data, from a rural site in Kenya, show that the mean PM_{10} measurement near the fire was 1,250 μg/m^3 – yet levels actually peaked at over 50,000 μg/m^3 (Ezzati et al. 2000a, b). Emissions in the kitchen can vary from day to day and from season to season, due to the moisture content and density of the fuel, the amount of airflow, the type of food being cooked and any changes in the stove or fuel used.

Many improved cookstove projects currently exist worldwide, ranging from local non-governmental organization projects to nationwide initiatives (WHO 2008; Granderson et al. 2009). Methods to reduce indoor air pollution from biomass use fall into four general categories: behavioural modifications to reduce

Table 2.4 Indoor particulate concentrations in a number of developing countries (GM: geometric mean)

Location	Averaging time and size fraction	Fuel	Concentration ($\mu g\ m^{-3}$)
Nepal (Pandey et al. 1990)	Cooking period $PM_{2.5}$	Wood/crop residues	8,200 (traditional stoves)
			3,000 improved stoves
Garhwal, India (Saksena et al. 1992)	Cooking period TSP	Wood/shrubs	4,500 (GM)
	24 h exposure TSP		700–1,690 (winter)
			250–1,130 (summer)
Pune, India (Smith et al. 1994)	12–24 h PM_{10}	Wood	2,000 (area)
			1,100 (personal)
Mozambique (Ellegard 1996)	Cooking period PM_{10}	Wood	1,200
Bolivia (Albalak et al. 1999)	6 h PM_{10}	Dung	1,830 (GM, indoor kitchens)
			280 (GM, outdoor kitchens)
Kenya (Ezzati et al. 2000a, b)	Daily average exposure PM_{10}	Mixed	1,000–4,800
Tamil Nadu, India (Balakrishnan et al. 2002)	Cooking period PM_4	Wood/agricultural waste	1,307–1,535 (GM, personal)
	Daily average exposure PM_4	Wood/agricultural waste	172–226
La Victoria, Guatemala (Albalak et al. 2001)	24 h $PM_{3.5}$	Wood	1,560 (GM, traditional stove)
			250 (GM, improved stove)
			850 (GM, LPG/open fire)
La Victoria, Guatemala (Bruce et al. 2004)	24 h $PM_{3.5}$	Wood and crop residue	1,019 (GM, traditional stove)
			351 (GM, improved stove)
Andhra Pradesh, India (Balakrishnan et al. 2004)	24 h PM_4	Wood/dung/agricultural waste	297–666 (kitchen)
	Daily average exposure PM_4	Wood/dung/agricultural waste	215–357 (living area)
Bangladesh (Dasgupta et al. 2004)	24 h PM_{10}	Wood/dung/agricultural	196–264 (personal)
		waste	60–1,165 (area)
Zimbabwe (Rumchev et al. 2007)	4 h PM_4	Wood	230–7,330
China (Edwards et al. 2007)	24 h PM_4	Wood	164
		Crop residues	282–456
		Coal	142–289

(continued)

Table 2.4 (continued)

Location	Averaging time and size fraction	Fuel	Concentration (μg m^{-3})
Shenyang, China (Jiang and Bell 2008a,b)	14 h PM_{10}	Crop residue	100 mean 14–1,571 range
Gansu, China (Jin et al. 2005)	24 h PM_4	Wood and crop residue	518 spring mean 661 winter mean
Chak NO.35/2.L and Bhaun, Pakistan (Colbeck et al. 2010)	1 h PM_{10}	Wood/dung	1,581 mean 141–8,555 range
	1 h $PM_{2.5}$		1,169 mean 23–5,953 range
	1 h PM_1		913 mean 13–3,449 range
Costa Rica (Park and Lee 2003)	24 h PM_{10}	Wood	132 mean 500–18,900 (peak range)
	24 h $PM_{2.5}$		44 mean 310–8,170 (peak range)

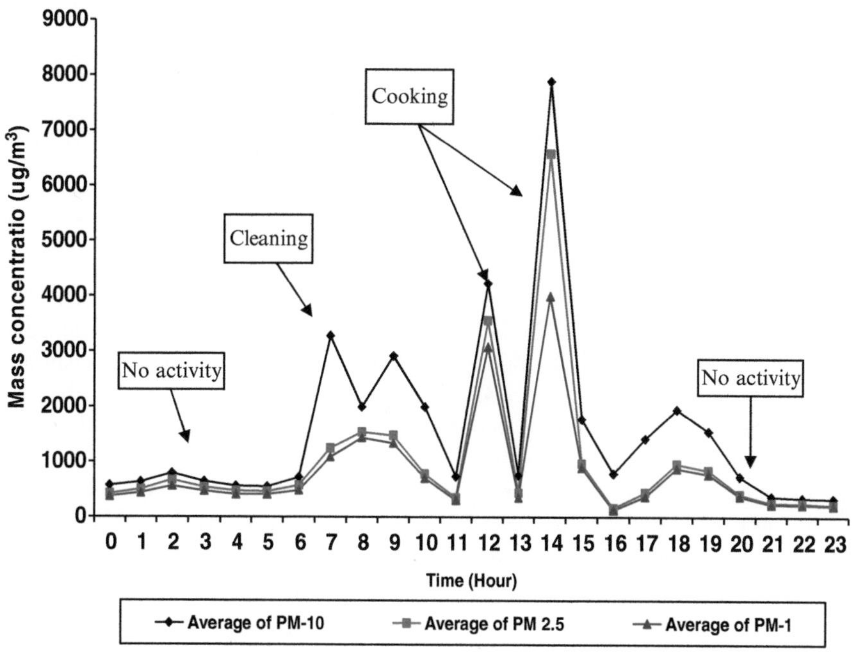

Fig. 2.6 Mass concentration of PM_{10}, $PM_{2.5}$ and PM_1 in a kitchen using biomass fuel at rural site in Pakistan

exposure; household changes to improve ventilation; improvements to cooking stoves; and interventions to enable people to use higher-quality, lower-emission liquid or gaseous fuels (Desai et al. 2004). For example switching from wood, dung

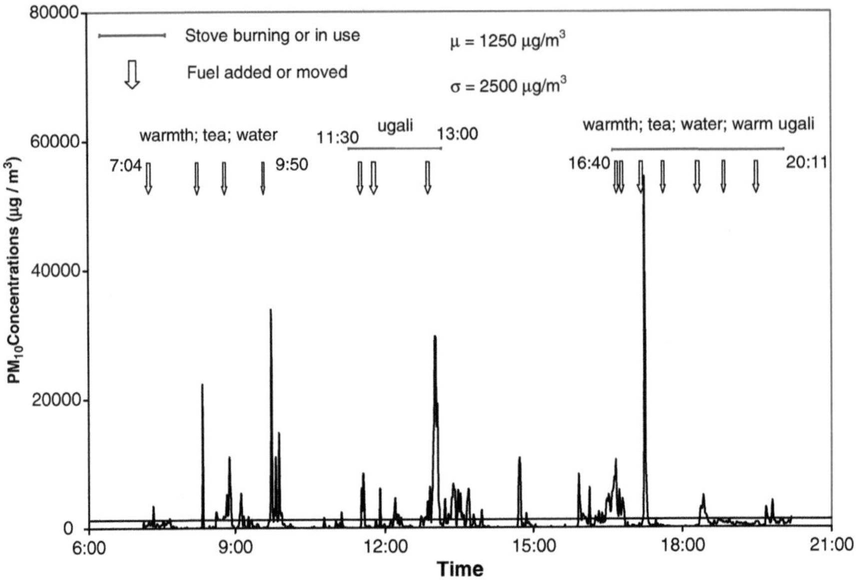

Fig. 2.7 PM_{10} concentrations, at a distance and height of 0.5 m, in a household using a three-stone stove in rural Kenya. The horizontal line represents the mean concentration for the day. As seen, mean concentration is a poor indicator of the patterns of exposure. Ugali is a common Kenyan food made from maize or sorghum flour. Reproduced with permission from Environmental Health Perspectives (Ezzati and Kammen 2002)

or charcoal to more efficient modern fuels, such as kerosene, liquefied petroleum gas and biogas, brings about the largest reductions (Parikh et al. 2001). The installation of chimneys or smoke hoods can reduce the concentration of respirable particles by up to 80% (Practical Action 2004; McCracken et al. 2007). Changing cooking behaviours are unlikely to bring about such large reductions as other interventions but are important supporting measures. Where cleaner fuels such as gas are introduced NO_2 is likely to become more important as a pollutant in kitchens.

It is not only kitchens that can experience high levels of particulate matter. For example a living room at a site in rural Pakistan experienced PM_{10} levels, up to 21,673 μg/m³ as a result of sweeping. Cleaning outside results in even higher levels as typical courtyards are dry and devoid of any grass/vegetation. In general, during most of the day indoor concentrations were higher than those outdoors (Fig. 2.8) and suggests an indoor source of fine particulates. Social gatherings take place during the evening and, in this example, up to eight smokers were in the room. One out of every two to three middle-aged men in Pakistan smoke cigarettes (Ahmad et al. 2005). During smoking there is a sharp rise in fine particulates and the indoor-outdoor ratio for PM_1 rose to a maximum of 12.95.

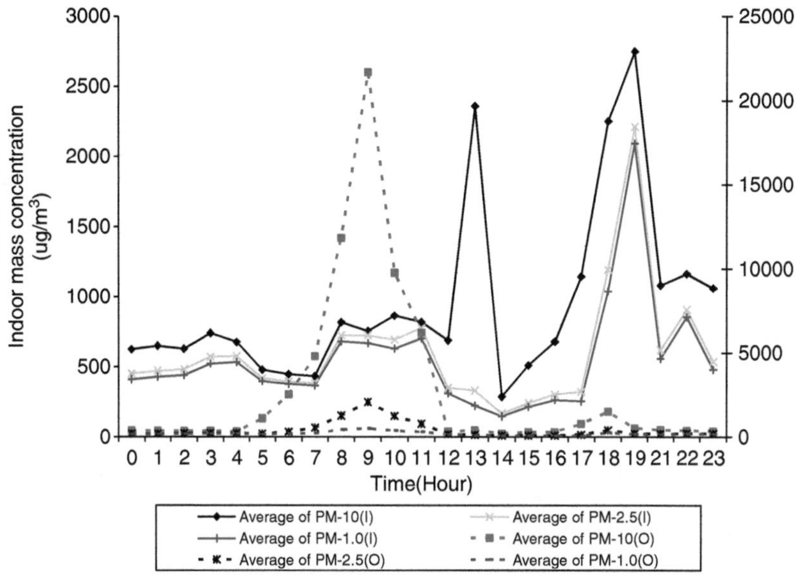

Fig. 2.8 Hourly average of indoor and outdoor mass concentration for PM_{10}, $PM_{2.5}$ and PM_1 in a living room at a rural site in Pakistan. I = Indoors; O = Outdoor

2.6 Transport Micro-environments

Exposure research frequently uses the concept of micro-environments to investigate the levels of pollutants humans are exposed to. Transport micro-environments represent are typically ones in which humans are potentially exposed to high pollutant concentrations on a regular basis.

The advancement in transport modes in the last few decades has resulted in increased mobility of people and a large proportion of the working population spend a significant time commuting. For example, in the UK average journey length by car has risen from 11.5 to 13.5 km between 1999 and 2004 and the proportion of journeys made by cars has increased from 57% to 63% (National Statistics 2006). In Great Britain, 40% of all the journeys comprise of short trips (<3.3 km) and almost 40% of these are carried out by car as compared to 55% on foot. In London alone 30% of trips are made by cars (National Statistics 2006).

A number of studies have attempted to quantify exposure to traffic related pollutants and also to relate the exposure to the travel mode. Research on the exposure of car drivers dates back to the 1960s (e.g. deBruin 1967). The majority of studies have focused on cars although, recently, results other transport modes, such as buses, trains, underground railways, bicycling or walking have been published. Table 2.5 summarises a number of studies which have investigated

Table 2.5 Comparison of particulate matter in various transport micro-environments (Adapted from Nasir and Colbeck 2009)

Reference	Location	PM size fraction	Mode of transport	Mean concentrations μg/m³
Nasir and Colbeck (2009)	Colchester, UK	PM_{10}($PM_{2.5}$) (PM_1)	In car – morning	(22)(9)(6)
			In car – evening	(21)(8)(5)
			AC train – peak	(44)(14)(12)
			AC train – off peak	(21)(6)(4)
			Non-AC train – peak	(95)(30)(19)
			Non-AC train – off peak	(95)(14)(6)
Gulliver and Briggs (2004)	Northampton, UK	PM_{10}	In car	43
		$PM_{2.5}$	Walking	15
		PM_1		7
		PM_{10}		38
		$PM_{2.5}$		15
		PM_1		7
Gulliver and Briggs (2007)	Leicester, UK	PM_{10}	In car	24
		TSP – PM_{10}	Walking	18
		PM_{10} – $PM_{2.5}$		15
		$PM_{2.5}$-PM_1		8
		PM_1		5
		PM_{10}		35
		TSP – PM_{10}		19
		PM_{10} -$PM_{2.5}$		22
		$PM_{2.5}$-PM_1		10
		PM_1		3
Pfeifer et al. (1999)	London, UK	$PM_{2.5}$	Taxi	33
Kaur et al. (2005)	London, UK	$PM_{2.5}$	Waking	27
			Cycling	33
			Bus	34
			Car	38
			Taxi	41
Briggs et al. (2006)	London, UK	PM_{10} -$PM_{2.5}$	In Car	5.87
		$PM_{2.5}$-PM_1	Walking	3.01
		PM_1		1.82
		PM_{10} -$PM_{2.5}$		27
		$PM_{2.5}$-PM_1		6
		PM_1		3
Adams et al. (2001)	London, UK	$PM_{2.5}$ Summer	Bicycle	34
		$PM_{2.5}$ Winter	Bus	39
			Car	37
			Tube – above ground	29
			Bicycle	23
			Bus	38
			Car	33
			Tube – above ground	–

(continued)

Table 2.5 (continued)

Reference	Location	PM size fraction	Mode of transport	Mean concentrations μg/m³
Gee and Raper (1999)	Manchester, UK	PM_4	Bus	338
			Bicycle	54
Seaton et al. (2005)	London, UK	$PM_{2.5}$	Tube (in cab)	170
Bevan et al. (1991)	Southampton, UK	$PM_{3.5}$	Bicycle	135
Praml and Schierl (2000)	Munich, Germany	PM_{10}	Bus	153
Fondelli et al. (2008)	Florence, Italy	$PM_{2.5}$	Bus	56
			Taxi	39
Invernizzi et al. (2004)	Italy	$PM_{2.5}$	Railway – smoking	250
			Railway – nonsmoking	15
Branis (2006)	Prague, Czech Republic	PM_{10}	Metro	114
Leutwyler et al. (2002)	Zurich, Switzerland	PM_{10}	Railway – smoking	975
			Railway – nonsmoking	209
Chillrud et al. (2004)	New York, USA	$PM_{2.5}$	Subway	62
Wohrnschimmel et al. (2008)	Mexico City	PM_{10} ($PM_{2.5}$)	Minibus	201(155)
			Bus	212(146)
			Metrobus	188(112)
Gomez-Perales et al. (2004)	Mexico City	$PM_{2.5}$	Minibus	68
			Bus	71
			Metrobus	61
Chan et al. (2002b)	Guangzhou, China	PM_{10} ($PM_{2.5}$)	Non A/C bus	184 (145)
			A/C bus	125 (101)
			Non A/C taxi	140 (106)
			A/C taxi	88(73)
Chan et al. (2002a)	Hong Kong	PM_{10} ($PM_{2.5}$)	Railway	50 (39)
			Tram	175(109)
			Non AC bus	112(93)
			AC bus	74(51)
			AC taxi	58
Park et al. (2008)	Korea	PM_{10}	High speed train	50.5
			Low speed train (A)(B)(C)	(69)(70)(83)
Kwon et al. (2008)	Korea	PM_{10}	Subway carriage	142
Lewne et al. (2006)	Stockholm, Sweden	PM_{10}	Lorry driver	57
			Bus driver	44
			Taxi driver	26
Riediker et al. (2003)	USA	$PM_{2.5}$	In car	24

(continued)

Table 2.5 (continued)

Reference	Location	PM size fraction	Mode of transport	Mean concentrations µg/m³
McNabola et al. (2008)	Dublin, Ireland	$PM_{2.5}$	In car (Route 1) (Route 2)	82(88)
			Bus (Route 1) (Route 2)	128(103)
			Cyclist (Route 1) (Route 2)	88(71)
			Pedestrian(Route 1) (Route 2)	63(46)
Han et al. (2005)	Trujillo, Peru	$PM_{2.5}$	Bus and lorry drivers	161

particulate matter concentrations for various modes of transport depends. It should also be noted that the exposure to pollutants does not only depend on travel mode but also varies significantly for different types of road layout and location, vehicle speed, vehicle design and ambient concentrations (Briggs et al. 2008). These studies give an indication of the exposure of individuals on particular journeys. However the results may only be of limited use in estimating exposure for other journeys.

As for a car, pollutant exposure on trains can vary with the time travel and type of train. Nasir and Colbeck (2009) have recently shown that most of particulate matter in trains was in the PM_{10} size fraction and this was generally derived from resuspension from the seating areas of the coaches. High concentrations of particulate matter during peak time reflected the contribution of the number of passengers. PM_{10} concentrations were well above those for $PM_{2.5}$ and PM_1 in both peak and off-peak journeys. Generally, the highest concentrations were recorded during the stoppage of train at the station (Fig. 2.9). Particulate matter concentrations during off peak journeys were typically half those during peak time.

2.7 Summary

Characterisation of indoor air pollutants is complex and requires consideration of the outdoor concentrations of the pollutants, the extent of filtering imposed by the building as air passes from outdoors to indoors, ventilation level of the building, indoor pollution sources, adsorption/desorption and chemical reactions.

There are numerous indoor sources of air pollutants. Many relate to combustion processes and others relate to human activities. Less obvious are those relating to building materials (e.g. composite woods) and products used indoors (e.g. carpets). Wallace (2009) has shown how emissions from these sources have changed over the past 50 years with a decrease in levels of "known" carcinogens and an increase in exposures to suspected endocrine disruptors. The variability of pollutants indoors is high and may be house specific. What is evident is that many

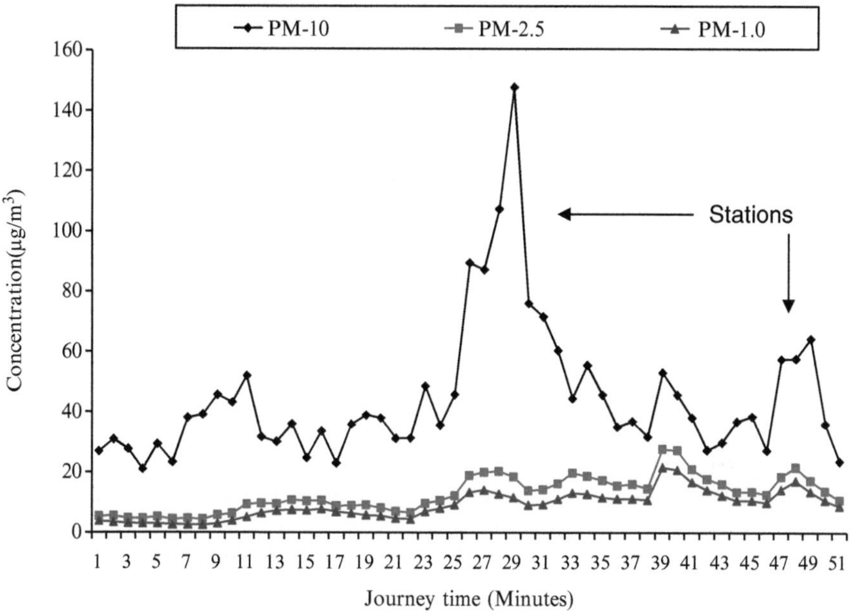

Fig. 2.9 Typical concentration profile of PM_{10}, $PM_{2.5}$ and PM_1 in air-conditioned train coaches

sociological factors are at play when considering exposure to indoor air pollution. Often the concentrations of air pollutants can be higher inside the dwelling of the poor. In developing countries women and their young children are likely to spend more time in the kitchen and so are at greatest risk from exposure of high levels of pollution from biofuels.

The indoor environment can be subdivided into different micro-environments (e.g. school, transport, restaurant and residential) and each may have a different source of indoor pollution.

There is a growing awareness of the importance of the indoor environment on health and exposure to indoor air pollution has been given higher attention in policy making (Harrison 2002). Several countries including Germany, Norway and Poland, have already established target concentrations for various indoor pollutants, and the UK has issued guidance on indoor air pollutants that includes numerical standards for nitrogen dioxide, carbon monoxide, formaldehyde, benzene and benzo(a)pyrene. Australia has adopted the approach of identifying indicators of good air quality rather than defining quantitative limits. The WHO Air Quality Guidelines 2005 recommended the development of guidelines specific for indoor air quality. These guidelines will cover three groups of issue: specific pollutants, biological agents and indoor combustion products.

On a global scale, the bulk of exposure to air pollution is experienced indoors, as most people spend most of their time there. Indoor concentrations are a complex interaction of various factors such as outdoor concentrations, indoor sources and sinks, pollutants, depletion, filtration and ventilation. The types, concentration, characteristics and sources of different air pollutants differ both in outdoor and indoor air. The scientific literature offers a broad database and case studies, which can be consulted to perform the necessary estimation of a real risk given in the particular environment.

References

Abt, E., Suh, H. H., Allen, G., & Koutrakis, P. (2000a) Characterization of indoor particle sources: A study conducted in the metropolitan Boston area. *Environmental Health Perspectives, 108*, 35–44.

Abt, E., Suh, H. H., Catalano, P., & Koutrakis, P. (2000b) Relative contribution of outdoor and indoor particle sources to indoor concentrations. *Environmental Science & Technology, 34*, 3579–3587.

ACGIH. (1999). *Bioaerosols: Assessment and control.* American Conference of Governmental Industrial Hygienists, Cincinnati, OH.

Adams, H. S., Nieuwenhuijsen, M. J., Colvile, R. N., McMullen, M. A. S., & Khandelwal, P. (2001). Fine particle (PM2.5) personal exposure levels in transport microenvironments, London, UK. *Science of the Total Environment, 279*, 29–44.

Adgate, J. L., Ramachandran, G., Pratt, G. C., Waller, L. A., & Sexton, K. (2003). Longitudinal variability in outdoor, indoor, and personal PM2.5 exposure in healthy non-smoking adults. *Atmospheric Environment, 37*, 993–1002.

Afshari, A., Matson, U., & Ekberg, L. E. (2005). Characterization of indoor sources of fine and ultrafine particles: A study conducted in a full-scale chamber. *Indoor Air, 15*, 141–150.

Ahmad, K., Jafary, F., Jehan, I., Hatcher, J., Khan, A. Q., Chaturvedi, N., et al. (2005). Prevalence and predictors of smoking in Pakistan: Results of the National Health Survey of Pakistan. *European Journal of Cardiovascular Prevention & Rehabilitation, 12*, 203–208.

Albalak, R., Bruce, N., McCracken, J. P., Smith, K. R., & De Gallardo, T. (2001). Indoor respirable particulate matter concentrations from an open fire, improved cookstove, and LPG/open fire combination in a rural Guatemalan community. *Environmental Science & Technology, 35*, 2650–2655.

Albalak, R., Keeler, G. J., Frisancho, A. R., & Haber, M. (1999). Assessment of PM10 concentrations from domestic biomass fuel combustion in two rural Bolivian highland villages. *Environmental Science & Technology, 33*, 2505–2509.

Ambroise, D., Greff-Merguet, G., Görner, P., Fabriès, J., & Hartemann, P. (1999). Measurement of indoor viable airborne bacteria with different bioaerosol samplers. *Journal of Aerosol Science, 30*, S669–S700.

Ashmore, M. R., & Dimitroulopoulou, C. (2009). Personal exposure of children to air pollution. *Atmospheric Environment, 43*, 128–141.

Awad, A. H. A. (2007). Airborne dust, bacteria, actinomycetes and fungi at a flourmill. *Aerobiologia, 23*, 59–69.

Aydogdu, H., Asan, A., Otkun, M. T., & Ture, M. (2005). Monitoring of fungi and bacteria in the indoor air of primary schools in Edirne city, Turkey. *Indoor and Built Environment, 14*, 411–425.

Balakrishnan, K., Parikh, J., Sankar, S., Padmavathi, R., Srividya, K., Venugopal, V., et al. (2002). Daily average exposures to respirable particulate matter from combustion of biomass fuels in rural households of southern India. *Environmental Health Perspectives, 110*, 1069–1075.

Balakrishnan, K., Sambandam, S., Ramaswamy, P., Mehta, S., & Smith, K. R. (2004). Exposure assessment for respirable particulates associated with household fuel use in rural districts of Andhra Pradesh, India. *Journal of Exposure Analysis and Environmental Epidemiology, 14*, S14–S25.

Bevan, M. A. J., Proctor, C. J., Bakerrogers, J., & Warren, N. D. (1991). Exposure to carbon-monoxide, respirable suspended particulates, and volatile organic-compounds while commuting by bicycle. *Environmental Science & Technology, 25*, 788–791.

Boleij, J. S. M., Ruigewaard, P., Hoek, F., Thairu, H., Wafula, E., Onyango, F., et al. (1989). Domestic air-pollution from biomass burning in Kenya. *Atmospheric Environment, 23*, 1677–1681.

Bornehag, C. G., Blomquist, G., Gyntelberg, F., Jarholm, B., Malmberg, P., Nordvall, L., et al. (2001). Dampness in buildings and health: Nordic interdisciplinary review of the scientific evidence on associations between exposure to "sampness" in buildings and health effects (NORDDAMP). *Indoor Air, 11*, 72–86.

Bouillard, L., Michel, O., Dramaix, M., & Devleeschouwer, M. (2005). Bacterial contamination of indoor air, surfaces, and settled dust, and related dust endotoxin concentrations in healthy office buildings. *Annals of Agricultural and Environmental Medicine, 12*, 187–192.

Brandl, H., von Daniken, A., Hitz, C., & Krebs, W. (2008). Short-term dynamic patterns of bio-aerosol generation and displacement in an indoor environment. *Aerobiologia, 24*, 203–209.

Branis, M. (2006). The contribution of ambient sources to particulate pollution in spaces and trains of the Prague underground transport system. *Atmospheric Environment, 40*, 348–356.

Briggs, D. J., de Hoogh, K., Morris, C., & Gulliver, J. (2008). Effects of travel mode on exposures to particulate air pollution. *Environment International, 34*, 12–22.

Brimblecombe, P. (1987). *The big smoke*. London: Methuen.

Bruce, N., McCracken, J., Albalak, R., Schei, M., Smith, K. R., Lopez, V., et al. (2004). Impact of improved stoves, house construction and child location on levels of indoor air pollution exposure in young Guatemalan children. *Journal of Exposure Analysis and Environmental Epidemiology, 14*, S26–S33.

Burroughs, B., & Hansen, S. J. (2008). *Managing indoor air quality* (4th ed.). Lilburn: Fairmont Press.

Buttner, M. P., & Stetzenbach, L. D. (1993). Monitoring airborne fungal spores in an experimental indoor environment to evaluate sampling methods and the effects of human activity on air sampling. *Applied and Environmental Microbiology, 59*, 219–226.

Capasso, L. (2000). Indoor pollution and respiratory diseases in ancient Rome. *Lancet, 356*, 1774–1774.

CARB. (2005). *Indoor air pollution in California: Report to the California legislature*. California: California Air Resources Board.

Carnelly, D., Haldane, J. S., & Anderson, A. M. (1887). The carbonic acid, organic matter and micro-organisms in air, more especially in dwellings and schools. *Philosophical Transactions of the Royal Society of Tropical Medicine and Hygiene (Series B), 178*, 61–111.

Chan, A. T. (2002). Indoor-outdoor relationships of particulate matter and nitrogen oxides under different outdoor meteorological conditions. *Atmospheric Environment, 36*, 1543–1551.

Chan, L. Y., Lau, W. L., Lee, S. C., & Chan, C. Y. (2002a). Commuter exposure to particulate matter in public transportation modes in Hong Kong. *Atmospheric Environment, 36*, 3363–3373.

Chan, L. Y., Lau, W. L., Zou, S. C., Cao, Z. X., & Lai, S. C. (2002b) Exposure level of carbon monoxide and respirable suspended particulate in public transportation modes while commuting in urban, area of Guangzhou, China. *Atmospheric Environment, 36*, 5831–5840.

Chillrud, S. N., Epstein, D., Ross, J. M., Sax, S. N., Pederson, D., Spengler, J. D., et al. (2004). Elevated airborne exposures of teenagers to manganese, chromium, and iron from steel dust and New York City's subway system. *Environmental Science & Technology, 38*, 732–737.

Colbeck, I., Nasir, Z.A., & Ali, Z. (2010). Characteristics of indoor/outdoor particulate pollution in urban and rural residential environment of Pakistan. *Indoor Air, 20*, 40–51.

Conner, T. L., Norris, G. A., Landis, M. S., & Williams, R. W. (2001). Individual particle analysis of indoor, outdoor, and community samples from the 1998 Baltimore particulate matter study. *Atmospheric Environment, 35*, 3935–3946.

Costa, C. D. N. (1997). *Dialogues and letters: Seneca*. London: Penguin Books.

Dasgupta, S., Huq, M., Khaliquzzaman, M., Pandey, K., & Wheeler, D. (2004). *Indoor air quality for poor families: New evidence from Bangladesh* (Policy Research Working Paper 3393). Washington, DC: World Bank.

Dasgupta, S., Huq, M., Khaliquzzaman, M., Pandey, K., & Wheeler, D. (2006). Indoor air quality for poor families: New evidence from Bangladesh. *Indoor Air, 16*, 426–444.

Destaillats H., Maddalena R.L., Singer B.C., Hodgson A.T. & McKone T.E. (2008) Indoor pollutants emitted by office equipment: A review of reported data and information needs. *Atmospheric Environment, 42*, 1371–1388.

De Koning, H. W., Smith, K. R., & Last, J. M. (1985). Biomass fuel combustion and health. *Bulletin of the World Health Organization, 63*, 11–26.

DeBruin, A. (1967). Carboxyhemoglobin levels due to traffic exhaust. *Archives of Environmental Health, 15*, 384–389.

Dennekamp, M., Howarth, S., Dick, C. A. J., Cherrie, J. W., Donaldson, K., & Seaton, A. (2001). Ultrafine particles and nitrogen oxides generated by gas and electric cooking. *Occupational and Environmental Medicine, 58*, 511–516.

Desai, M. R., & Ghosh, S. K. (2003). Occupational exposure to airborne fungi among rice mill workers with special reference to aflatoxin producing A-flavus strains. *Annals of Agricultural and Environmental Medicine, 10*, 159–162.

Desai, M. A., Mehta, S., & Smith, K. R. (2004). *Indoor smoke from solid fuels: Assessing the environmental burden of disease at national and local levels.* Geneva, Switzerland: World Health Organization.

Douwes, J., Thorne, P., Pearce, N., & Heederik, D. (2003). Bioaerosol health effects and exposure assessment: Progress and prospects. *Annals of Occupational Hygiene, 47*, 187–200.

Edwards, R. D., Li, Y., He, G., Yin, Z., Sinton, J., Peabody, J., et al. (2007). Household CO and PM measured as part of a review of China's National Improved Stove Program. *Indoor Air, 17*, 189–203.

Ellegard, A. (1996). Cooking fuel smoke and respiratory symptoms among women in low-income areas in Maputo. *Environmental Health Perspectives, 104*, 980–985.

Ezzati, M., & Kammen, D. M. (2001). Quantifying the effects of exposure to indoor air pollution from biomass combustion on acute respiratory infections in developing countries. *Environmental Health Perspectives, 109*, 481–488.

Ezzati, M., & Kammen, D. M. (2002). The health impacts of exposure to indoor air pollution from solid fuels in developing countries: Knowledge, gaps, and data needs. *Environmental Health Perspectives, 110*, 1057–1068.

Ezzati, M., Mbinda, B. M., & Kammen, D. M. (2000a). Comparison of emissions and residential exposure from traditional and improved cookstoves in Kenya. *Environmental Science & Technology, 34*, 578–583.

Ezzati, M., Saleh, H., & Kammen, D. M. (2000b). The contributions of emissions and spatial microenvironments to exposure to indoor air pollution from biomass combustion in Kenya. *Environmental Health Perspectives, 108*, 833–839.

Finkelman, R. B., Belkin, H. E., & Zheng, B. (1999). Health impacts of domestic coal use in China. *Proceedings of the National Academy of Sciences, 96*, 3427–3431.

Fischer, S. L., & Koshland, C. P. (2007). Daily and peak 1 h indoor air pollution and driving factors in a rural Chinese village. *Environmental Science & Technology, 41*, 3121–3126.

Foarde, K., Dulaney, P., Cole, E., VanOsdel, D., Ensor, D., & Chang, J. (1993). *Assessment of fungal growth on ceiling tiles under environmentally characterized conditions.* Paper presented at the Indoor Air 93, Jyväskylä, Finland.

Fondelli, M. C., Chellini, E., Yli-Tuomi, T., Cenni, I., Gasparrini, A., Nava, S., et al. (2008). Fine particle concentrations in buses and taxis in Florence, Italy. *Atmospheric Environment, 42*, 8185–8193.

Fortmann, R., Kariher, P., & Clayton, R. (2001). *Indoor air quality: Residential cooking exposure.* Report of "State of California Air Resources Board".

Fuentes-Leonarte, V., Tenias, J. M., & Ballester, F. (2009). Levels of pollutants in indoor air and respiratory health in preschool children: A systematic review. *Pediatric Pulmonology, 44*, 231–243.

Fullerton, D. G., Bruce, N., & Gordon, S. B. (2008). Indoor air pollution from biomass fuel smoke is a major health concern in the developing world. *Transactions of the Royal Society of Tropical Medicine and Hygiene, 102*, 843–851.

Gee, I. L., & Raper, D. W. (1999). Commuter exposure to respirable particles inside buses and by bicycle. *Science of the Total Environment, 235*, 403–405.

Gehin, E., Ramalho, O., & Kirchner, S. (2008). Size distribution and emission rate measurement of fine and ultrafine particle from indoor human activities. *Atmospheric Environment, 42*, 8341–8352.

Godwin, C., & Batterman, S. (2007). Indoor air quality in Michigan schools. *Indoor Air, 17*, 109–121.

Goh, I., Obbard, J. P., Viswanathan, S., & Huang, Y. (2000). Airborne bacteria and fungal spores in the indoor environment – A case study in Singapore. *Acta Biotechnologica, 20*, 67–73.

Gomez-Perales, J. E., Colvile, R. N., Nieuwenhuijsen, M. J., Fernandez-Bremauntz, A., Gutierrez-Avedoy, V. J., Paramo-Figueroa, V. H., et al. (2004). Commuters' exposure to PM2.5, CO, and benzene in public transport in the metropolitan area of Mexico city. *Atmospheric Environment, 38*, 1219–1229.

Gorny, R. L. (2004). Filamentous microorganisms and their fragments in indoor air – a review. *Annals of Agricultural and Environmental Medicine, 11*, 185–197.

Gorny, R. L., Dutkiewicz, J., & Krysinska-Traczyk, E. (1999). Size distribution of bacterial and fungal bioaerosols in indoor air. *Annals of Agricultural and Environmental Medicine, 6*, 105–113.

Granderson, J., Sandhu, J. S., Vasquez, D., Ramirez, E., & Smith, K. R. (2009). Fuel use and design analysis of improved woodburning cookstoves in the Guatemalan Highlands. *Biomass & Bioenergy, 33*, 306–315.

Griscom, J. H. (1848). *The uses and abuses of air: Showing its influence in sustaining life, and producing disease, with remarks on the ventilation of houses*. New York: Kessinger Publishing.

Gulliver, J., & Briggs, D. J. (2004). Personal exposure to particulate air pollution in transport microenvironments. *Atmospheric Environment, 38*, 1–8.

Gulliver, J., & Briggs, D. J. (2007). Journey-time exposure to particulate air pollution. *Atmospheric Environment, 41*, 7195–7207.

Haas, D., Habib, J., Galler, H., Buzina, W., Schlacher, R., Marth, E., et al. (2007). Assessment of indoor air in Austrian apartments with and without visible mold growth. *Atmospheric Environment, 41*, 5192–5201.

Hameed, A. A. A., Shakour, A. A., & Yasser, H. I. (2003). Evaluation of bio-aerosols at an animal feed manufacturing industry: A case study. *Aerobiologia, 19*, 1573–3025.

Han, X. L., Aguilar-Villalobos, M., Allen, J., Carlton, C. S., Robinson, R., Bayer, C., et al. (2005). Traffic-related occupational exposures to PM2.5, CO, and VOCs in Trujillo, Peru. *International Journal of Occupational and Environmental Health, 11*, 276–288.

Harrison, P. T. C. (2002). Indoor air quality guidelines. *Occupational and Environmental Medicine, 59*, 73–74.

He, L. Y., Hu, M., Huang, X. F., Yu, B. D., Zhang, Y. H., & Liu, D. Q. (2004b). Measurement of emissions of fine particulate organic matter from Chinese cooking. *Atmospheric Environment, 38*, 6557–6564.

He, C. R., Morawska, L. D., Hitchins, J., & Gilbert, D. (2004a). Contribution from indoor sources to particle number and mass concentrations in residential houses. *Atmospheric Environment, 38*, 3405–3415.

Hussein T., Hameri K., Heikkinen M.S.A. & Kulmala M. (2005). Indoor and outdoor particle size characterization at a family house in Espoo-Finland. *Atmospheric Environment, 39*, 3697–709.

He C., Morawska L. & Taplin L. (2007). Particle emission characteristics of office printers. *Environmental Science & Technology, 41*, 6039–6045.

Hughes, J. D. (1993). *Pan's travail: Environmental problems of the ancient Greeks and Romans*. Baltimore, MD: John Hopkins University Press.

Invernizzi, G., Ruprecht, A., Mazza, R., De Marco, C., & Boffi, R. (2004). Transfer of particulate matter pollution from smoking to non-smoking coaches: The explanation for the smoking ban on Italian trains. *Tobacco Control, 13*, 319–319.

IOM. (2004). *Institute of Medicine, board on health promotion and disease prevention committee on damp indoor spaces and health. Damp indoor spaces and health.* Washington, DC: The National Academies Press.
Jain, A. K. (2000). Survey of bioaerosol in different indoor working environments in central India. *Aerobiologia, 16*, 221–225.
Jiang, R., & Bell, M. (2008a). The characterization and health effects of indoor particulate matter pollution from biomass burning in Northeastern China. *Epidemiology, 19*, S89–S90.
Jiang, R., & Bell, M. L. (2008b). A comparison of particulate matter from biomass-burning rural and non-biomass-burning urban households in Northeastern China. *Environmental Health Perspectives, 116*, 907–914.
Jin, Y. L., Zhou, Z., He, G. L., Wei, H. Z., Liu, J., Liu, F., et al. (2005). Geographical, spatial, and temporal distributions of multiple indoor air pollutants in four Chinese provinces. *Environmental Science & Technology, 39*, 9431–9439.
Jothish, P. S., & Nayar, T. S. (2004). Airborne fungal spores in a sawmill environment in Palakkad District, Kerala, India. *Aerobiologia, 20*, 75–81.
Kalogerakis, N., Paschali, D., Lekaditis, V., Pantidou, A., Eleftheriadis, K., & Lazaridis, M. (2005). Indoor air quality – bioaerosol measurements in domestic and office premises. *Journal of Aerosol Science, 36*, 751–761.
Kaur, S., Nieuwenhuijsen, M., & Colvile, R. (2005). Personal exposure of street canyon intersection users to PM2.5, uhrafine particle counts and carbon monoxide in Central. London, UK. *Atmospheric Environment, 39*, 3629–3641.
Kim, K. Y., & Kim, C. N. (2007). Airborne microbiological characteristics in public buildings of Korea. *Building and Environment, 42*, 2188–2196.
Kulmala, M., Asmi, A., & Pirjola, L. (1999). Indoor air aerosol model: The effect of outdoor air, filtration and ventilation on indoor concentrations. *Atmospheric Environment, 33*, 2133–2144.
Kwon, S.-B., Cho, Y., Park, D.-S., & Park, E.-Y. (2008). *Study on the indoor air quality of Seoul Metropolitan Subway (SMS) during the rush hour*. Paper presented at the Indoor Air 2008, Copenhagen, Denmark.
Lazaridis, M., Aleksandropoulou, V., Hanssen, J. E., Dye, C., Eleftheriadis, K., & Katsivela, E. (2008). Inorganic and carbonaceous components in indoor/outdoor particulate matter in two residential houses in Oslo, Norway. *Journal of the Air & Waste Management Association, 58*, 346–356.
Lazaridis, M., Aleksandropoulou, V., Smolik, J., Hansen, J. E., Glytsos, T., Kalogerakis, N., et al. (2006). Physico-chemical characterization of indoor/outdoor particulate matter in two residential houses in Oslo, Norway: Measurements overview and physical properties – URBAN-AEROSOL Project. *Indoor Air, 16*, 282–295.
Lee, J. H., & Jo, W. K. (2006). Characteristics of indoor and outdoor bioaerosols at Korean high-rise apartment buildings. *Environmental Research, 101*, 11–17.
Lee, S. C., Li, W. M., & Chan, L. Y. (2001). Indoor air quality at restaurants with different styles of cooking in metropolitan Hong Kong. *Science of the Total Environment, 279*, 181–193.
Leutwyler, M., Siegmann, K., & Monn, C. (2002). Suspended particulate matter in railway coaches. *Atmospheric Environment, 36*, 1–7.
Lawrence A.J., Mask A. & Taneja A. (2005) Indoor/outdoor relationships of carbon monoxide and oxides of nitrogen in domestic homes with roadside, urban and rural locations in a central Indian region. *Indoor Air, 15*, 76–82.
Lewne, M., Nise, G., Lind, M. L., & Gustavsson, P. (2006). Exposure to particles and nitrogen dioxide among taxi, bus and lorry drivers. *International Archives of Occupational and Environmental Health, 79*, 220–226.
Liao, C. M., Luo, W. C., Chen, S. C., Chen, J. W., & Liang, H. M. (2004). Temporal/seasonal variations of size-dependent airborne fungi indoor/outdoor relationships for a wind-induced naturally ventilated airspace. *Atmospheric Environment, 38*, 4415–4419.
Lin, W. H., & Li, C. S. (1996). Size characteristics of fungus allergens in the subtropical climate. *Aerosol Science and Technology, 25*, 93–100.

Long, C. M., Suh, H. H., & Koutrakis, P. (2000). Characterization of indoor particle sources using continuous mass and size monitors. *Journal of the Air & Waste Management Association, 50*, 1236–1250.

Lung, S. C. C., & Hu, S. C. (2003). Generation rates and emission factors of particulate matter and particle-bound polycyclic aromatic hydrocarbons of incense sticks. *Chemosphere, 50*, 673–679.

Makra, L., & Brimblecombe, P. (2004). Selections from the history of environmental pollution, with special attention to air pollution. Part 1. *International Journal of Environment and Pollution, 22*, 641–656.

McBride, S. J., Ferro, A. R., Ott, W. R., Switzer, P., & Hildemann, L. M. (1999). Investigations of the proximity effect for pollutants in the indoor environment. *Journal of Exposure Analysis and Environmental Epidemiology, 9*, 602–621.

McCracken, J. P., Smith, K. R., Diaz, A., Mittleman, M. A., & Schwartz, J. (2007). Chimney stove intervention to reduce long-term wood smoke exposure lowers blood pressure among Guatemalan women. *Environmental Health Perspectives, 115*, 996–1001.

McNabola, A., Broderick, B. M., & Gill, L. W. (2008). Relative exposure to fine particulate matter and VOCs between transport microenvironments in Dublin: Personal exposure and uptake. *Atmospheric Environment, 42*, 6496–6512.

McNeill, J. R. (2001). *Something new under the sun. An environmental history of the twentieth century world*. New York: W.W. Norton & Company.

Mestl, H. E. S., Aunan, K., Seip, H. M., Wang, S., Zhao, Y., & Zhang, D. (2007). Urban and rural exposure to indoor air pollution from domestic biomass and coal burning across China. *Science of the Total Environment, 377*, 12–26.

Miller, J. (1992). Fungi as contaminants in indoor air. *Atmospheric Environment, 26a*, 2163–2172.

Mitchell, C. S., Zhang, J. F. J., Sigsgaard, T., Jantunen, M., Lioy, P. J., Samson, R., et al. (2007). Current state of the science: Health effects and indoor environmental quality. *Environmental Health Perspectives, 115*, 958–964.

Mitra, A., Morawska, L., Sharma, C., & Zhang, J. (2002). Chapter two: Methodologies for characterisation of combustion sources and for quantification of their emissions. *Chemosphere, 49*, 903–922.

Monkkonen, P., Pai, P., Maynard, A., Lehtinen, K. E. J., Hameri, K., Rechkemmer, P., et al. (2005). Fine particle number and mass concentration measurements in urban Indian households. *Science of the Total Environment, 347*, 131–147.

Morawska, L. (2008). Identifying and managing indoor-air problems. *Scandinavian Journal of Work Environment & Health, 4*, 5–9.

Morawska, L., He, C. R., Hitchins, J., Gilbert, D., & Parappukkaran, S. (2001). The relationship between indoor and outdoor airborne particles in the residential environment. *Atmospheric Environment, 35*, 3463–3473.

Mudarri, D., & Fisk, W. J. (2007). Public health and economic impact of dampness and mold. *Indoor Air, 17*, 226–235.

Mudway, I., Duggan, S., Venkataraman, C., Habib, G., Kelly, F., & Grigg, J. (2005). Combustion of dried animal dung as biofuel results in the generation of highly redox active fine particulates. *Particle and Fibre Toxicology, 2*. DOI: 10.1186/1743-8977-1182-1186.

Musk, A. W., Venables, K. M., Crook, B., Nunn, A. J., Hawkins, R., Crook, G. D. W., et al. (1989). Respiratory symptoms, lung-function, and sensitization to flour in a British bakery. *British Journal of Industrial Medicine, 46*, 636–642.

Naeher, L. P., Brauer, M., Lipsett, M., Zelikoff, J. T., Simpson, C. D., Koenig, J. Q., et al. (2007). Woodsmoke health effects: A review. *Inhalation Toxicology, 19*, 67–106.

Narayan, M. C. J., Ravichandran, V., & Sullia, S. B. (1982). Aeromycology of the atmosphere of Malleeswaram market, Bangalore. *Acta Botanica Indica, 10*, 196–200.

Nasir, Z. A., & Colbeck, I. (2009). Particulate air pollution in transport micro-environments. *Journal of Environmental Monitoring, 11*. DOI: 10.1039/b821824b.

Ott W.R., Steinemann A.C. & Wallace L.A. (2003). *Exposure Analysis*. CRC Press, Boca Raton.

Oppliger, A., Rusca, S., Charriere, N., Duc, T. V., & Droz, P. O. (2005). Assessment of bioaerosols and inhalable dust exposure in Swiss sawmills. *Annals of Occupational Hygiene, 49*, 385–391.

Pandey, M. R., Neupane, R. P., Gautam, A., & Shrestha, I. B. (1990). The effectiveness of smokeless stoves in reducing indoor air pollution in a rural hill region of Nepal. *Mountain Research and Development, 10*, 313–320.

Parat, S., Perdrix, A., FrickerHidalgo, H., Saude, I., Grillot, R., & Baconnier, P. (1997). Multivariate analysis comparing microbial air content of an air-conditioned building and a naturally ventilated building over one year. *Atmospheric Environment, 31*, 441–449.

Parikh, J., Balakrishnan, K., Laxmi, V., & Biswas, H. (2001). Exposure from cooking with biofuels: Pollution monitoring and analysis for rural Tamil Nadu, India. *Energy, 26*, 949–962.

Park, D.-S., Cho, Y.-M., Kwon, S.-B., Park, E.-Y., & Kim, D.-S. (2008). *Analysis of air quality and PM10 source in the railway vehicles*. Paper presented at the Indoor Air 2008, Copenhagen, Denmark.

Park, E., & Lee, K. (2003). Particulate exposure and size distribution from wood burning stoves in Costa Rica. *Indoor Air, 13*, 253–259.

Pasanen, A. L., Kasanen, J. P., Rautiala, S., Ikaheimo, M., Rantamaki, J., Kaariainen, H., et al. (2000). Fungal growth and survival in building materials under fluctuating moisture and temperature conditions. *International Biodeterioration & Biodegradation, 46*, 117–127.

Pasanen, A. L., Pasanen, P., Jantunen, M. J., & Kalliokoski, P. (1991). Significance of air humidity and air velocity for fungal spore release into the air. *Atmospheric Environment Part a-General Topics, 25*, 459–462.

Pastuszka, J. S., Paw, U. K. T., Lis, D. O., Wlazlo, A., & Ulfig, K. (2000). Bacterial and fungal aerosol in indoor environment in Upper Silesia, Poland. *Atmospheric Environment, 34*, 3833–3842.

Perera, F. P., Illman, S. M., Kinney, P. L., Whyatt, R. M., Kelvin, E. A., Shepard, P., et al. (2002). The challenge of preventing environmentally related disease in young children: Community-based research in New York city. *Environmental Health Perspectives, 110*, 197–204.

Pfeifer, G. D., Harrison, R. M., & Lynam, D. R. (1999). Personal exposures to airborne metals in London taxi drivers and office workers in 1995 and 1996. *Science of the Total Environment, 235*, 253–260.

Pluschke, P. (2004). *Indoor air pollution* (vol. 4.F). Berlin: Springer.

Practical-Action. (2004). *Smoke – The killer in the kitchen*. London: ITDG Publishing.

Praml, G., & Schierl, R. (2000). Dust exposure in Munich public transportation: A comprehensive 4-year survey in buses and trams. *International Archives of Occupational and Environmental Health, 73*, 209–214.

Raunemaa, T., Kulmala, M., Saari, H., Olin, M., & Kulmala, M. H. (1989). Indoor air aerosol model - transport indoors and deposition of fine and coarse particles. *Aerosol Science and Technology, 11*, 11–25.

Ren, P., Jankun, T. M., Belanger, K., Bracken, M. B., & Leaderer, B. P. (2001). The relation between fungal propagules in indoor air and home characteristics. *Allergy, 56*, 419–424.

Ren, P., Jankun, T. M., & Leaderer, B. P. (1999). Comparisons of seasonal fungal prevalence in indoor and outdoor air and in house dusts of dwellings in one Northeast American county. *Journal of Exposure Analysis and Environmental Epidemiology, 9*, 560–568.

Reponen, T., Hyvarinen, A., Ruuskanen, J., Raunemaa, T., & Nevalainen, A. (1994). Comparison of concentrations and size distributions of fungal spores in buildings with and without mold problems. *Journal of Aerosol Science, 25*, 1595–1603.

Reponen, T., Willeke, K., Ulevicius, V., Reponen, A., & Grinshpun, S. A. (1996). Effect of relative humidity on the aerodynamic diameter and respiratory deposition of fungal spores. *Atmospheric Environment, 30*, 3967–3974.

Riediker, M., Williams, R., Devlin, R., Griggs, T., & Bromberg, P. (2003). Exposure to particulate matter, volatile organic compounds, and other air pollutants inside patrol cars. *Environmental Science & Technology, 37*, 2084–2093.

Ritschkoff, A. C., Viitanen, H., & Koskela, K. (2000). *The response of building materials to the mould exposure at different humidity and temperature conditions*. Paper presented at the Healthy Buildings, Espoo.

Rodes C.E., Kamens R.M. & Wiener R.W. (1991). The significance and characteristics of the personal activity cloud on exposure assessment measurements for indoor contaminants. *Indoor Air, 1,* 123–145.

Robinson, A. L., Subramanian, R., Donahue, N. M., Bernardo-Bricker, A., & Rogge, W. F. (2006). Source apportionment of molecular markers and organic aerosol. 3. Food cooking emissions. *Environmental Science & Technology, 40*, 7820–7827.

Rolka, H., Krajewska-Kulak, E., Lukaszuk, C., Oksiejczuk, E., Jakoniuk, P., Leszczynska, K., et al. (2005). Indoor air studies of fungi contamination of social welfare home in Czerewki in north-east part of Poland. *Roczniki Akademii Medycznej w Bialymstoku, 50*(Suppl 1), 26–30.

Rosas, I., Calderon, C., Salinas, E., Martınez, L., Alfaro-Moreno, E., Milton, D. K., et al. (2001). Animal and worker exposure to dust and biological particles in animal care houses. *Aerobiologia, 17*, 49–59.

Rumchev, K., Spickett, J. T., Brown, H. L., & Mkhweli, B. (2007). Indoor air pollution from biomass combustion and respiratory symptoms of women and children in a Zimbabwean village. *Indoor Air, 17*, 468–474.

Saksena, S., Prasad, R., Pal, R. C., & Joshi, V. (1992). Patterns of daily exposure to Tsp and Co in the Garhwal Himalaya. *Atmospheric Environment Part a-General Topics, 26*, 2125–2134.

Saksena, S., Subida, R., Buttner, L., & Ahmed, L. (2007). Indoor air pollution in coastal houses of southern Philippines. *Indoor and Built Environment, 16*, 159–168.

Salthammer, T. (2009). *Organic indoor air pollutants: Occurrence, measurement, evaluation* (2nd ed.). Weinheim: Wiley VCH.

Savino, E., & Caretta, G. (1992). Airborne fungi in an Italian rice mill. *Aerobiologia, 8*, 267–275.

Seaton, A., Cherrie, J., Dennekamp, M., Donaldson, K., Hurley, J. F., & Tran, C. L. (2005). The London underground: Dust and hazards to health. *Occupational and Environmental Medicine, 62*, 355–362.

See, S. W., & Balasubramanian, R. (2008). Chemical characteristics of fine particles emitted from different gas cooking methods. *Atmospheric Environment, 42*, 8852–8862.

Shelton, B. G., Kirkland, K. H., Flanders, W. D., & Morris, G. K. (2002). Profiles of airborne fungi in buildings and outdoor environments in the United States. *Applied and Environmental Microbiology, 68*, 1743–1753.

Siegmann, K., & Sattler, K. (1996). Aerosol from hot cooking oil, a possible health hazard. *Journal of Aerosol Science, 27*, S493–S494.

Singh, A., & Singh, A. B. (1994). Airborne fungi in a bakery and the prevalence of respiratory dysfunction among workers. *Grana, 33*, 349–358.

Sjaastad, A. K., & Svendsen, K. (2008). Exposure to mutagenic aldehydes and particulate matter during panfrying of beefsteak with margarine, rapeseed oil, olive oil or soybean oil. *Annals of Occupational Hygiene, 52*, 739–745.

Smith, K. R. (1987). *Biofuels, air pollution and health: A global review*. New York: Plenum Press.

Smith, K. R., Apte, M. G., Ma, Y. Q., Wongsekiarttirat, W., & Kulkarni, A. (1994). Air-pollution and the energy ladder in Asian cities. *Energy, 19*, 587–600.

Smith, K. R., Uma, R., Kishore, V. V. N., Zhang, J., Joshi, V., & Khalil M. A. K. (2000). Greenhouse implications of household stoves: An analysis for India. *Annual Review of Energy and the Environment, 25*.

Statistics, N. (2006). *Transport Statistics Bulletin. National Travel Survey, 2005*. London: Department for Transport.

Sundell, J. (2004). On the history of indoor air quality and health. *Indoor Air, 14*, 51–58.

Tsai, F. C., & Macher, J. M. (2005). Concentrations of airborne culturable bacteria in 100 large US office buildings from the BASE study. *Indoor Air, 15*, 71–81.

Tsai, F. C., Macher, J. M., & Hung, Y. Y. (2007). Biodiversity and concentrations of airborne fungi in large US office buildings from the BASE study. *Atmospheric Environment, 41*, 5181–5191.

Viitanen, H., Hanhijärvi, A., Hukka, A., & Koskela, K. (2000). *Modelling mould growth and decay damages*. Paper presented at the Healthy Buildings, Espoo.

Wallace, L. (2005). Ultrafine particles from a vented gas clothes dryer. *Atmospheric Environment, 39*, 5777–5786.

Wallace, L. A., Emmerich, S. J., & Howard-Reed, C. (2004). Source strengths of ultrafine and fine particles due to cooking with a gas stove. *Environmental Science & Technology, 38*, 2304–2311.

Waring M.S., Siegel J.A. & Corsi R.L. (2008). Ultrafine particle removal and generation by portable air cleaners. *Atmospheric Environment, 42*, 5003–5014.

Wallace, L., Wang, F., Howard-Reed, C., & Persily, A. (2008). Contribution of gas and electric stoves to residential ultrafine particle concentrations between 2 and 64 nm: Size distributions and emission and coagulation rates. *Environmental Science & Technology, 42*, 8641–8647.

Wasson, S. J., Guo, Z. S., McBrian, J. A., & Beach, L. O. (2002). Lead in candle emissions. *Science of the Total Environment, 296*, 159–174.

Weschler C.J. & Shields H.C. (1999). Indoor ozone/terpene reactions as a source of indoor particles. *Atmospheric Environment, 33*, 2301–2312.

Weschler, C. J. (2009). Changes in indoor pollutants since the 1950s. *Atmospheric Environment, 43*, 153–169.

WHO (2007). Indoor air pollution: national burden of disease estimates WHO, Geneva.

WHO. (2006). *Fuel for life: Household energy and health*. Geneva: WHO.

WHO. (2008). *Evaluating household energy and health interventions: A catalogue of methods*. Geneva: WHO.

Wohrnschimmel, H., Zuk, M., Martinez-Villa, G., Ceron, J., Cardenas, B., Rojas-Bracho, L., et al. (2008). The impact of a bus rapid transit system on commuters' exposure to benzene, CO, PM2.5 and PM10 in Mexico City. *Atmospheric Environment, 42*, 8194–8203.

Wu, P. C., Su, H. J., & Lin, C. Y. (2000). Characteristics of indoor and outdoor airborne fungi at suburban and urban homes in two seasons. *Science of the Total Environment, 253*, 111–118.

Yu, C. H., Morandi, M. T., & Weisel, C. P. (2008). Passive dosimeters for nitrogen dioxide in personal/indoor air sampling: A review. *Journal of Exposure Science and Environmental Epidemiology, 18*, 441–451.

Zhang, J. F., & Morawska, L. (2002). Combustion sources of particles: 2. Emission factors and measurement methods. *Chemosphere, 49*, 1059–1074.

Zorman, T., & Jersek, B. (2008). Assessment of bioaerosol concentrations in different indoor environments. *Indoor and Built Environment, 17*, 155–163.

Zuk, M., Rojas, L., Blanco, S., Serrano, P., Cruz, J., Angeles, F., et al. (2007). The impact of improved wood-burning stoves on fine particulate matter concentrations in rural Mexican homes. *Journal of Exposure Science and Environmental Epidemiology, 17*, 224–232.

Zuraimi, M. S., & Tham, K. W. (2008). Indoor air quality and its determinants in tropical child care centers. *Atmospheric Environment, 42*, 2225–2239.

Chapter 3
Chemical Reactions Among Indoor Pollutants

Glenn Morrison

Abstract Chemistry takes place all around us, regulating the intensity and nature of our exposure to pollutants in water, air and soil. In indoor environments, chemistry can significantly alter the composition of the air we breathe. Transformations reduce our exposure to reactants and increase our exposure to products. If this reaction takes place on or in a surface, the relative exposure depends on the nature of the species; exposure to reactants and products may depend not only on rates and mechanisms but also on volatility.

3.1 Introduction

The nature of indoor microenvironments complicates the task of unraveling indoor chemistry. The composition of air in one room, or even one part of a room may be significantly different from an adjacent room or space. Air has many different pathways into buildings that influence its composition. There are many different types of surfaces in a single room, each coated with varying degrees of dirt, oils, cleaners and other materials. There is much we have to learn about the composition of indoor spaces and the reactions that alter these compositions. However, several decades of research have begun to reveal the important chemistry, unique microenvironmental characteristics, and reasonable simplifications that allow us to better estimate the influence of chemistry on human exposure. We may even begin to apply chemistry to improve indoor environments.

This chapter begins with a description of how chemistry is incorporated into indoor air quality models, and is then organized by type of reaction. Sections 3.3

G. Morrison (✉)
Missouri University of Science and Technology, 221 Butler-Carlton Hall, Rolla MO 65409, USA
e-mail: gcm@mst.edu

M. Lazaridis and I. Colbeck (eds.), *Human Exposure to Pollutants via Dermal Absorption and Inhalation*, Environmental Pollution 17, DOI 10.1007/978-90-481-8663-1_3,

and 3.4 focus on reactions involving ozone since much indoor chemistry is driven by ozone from smog. Auto-oxidation, hydrolysis, acid-base and chlorine chemistry are reviewed in Section 3.5. Throughout, the reactions are further categorized by the reaction phase. Homogeneous reactions include purely gas-phase reactions and reactions that take place within condensed phases such as resins. Heterogeneous reactions, here, refer to interfacial chemistry such as might take place at the surface of a painted wall. This categorization is helpful because there are useful simplifications, in either phase, for understanding reaction rates and the resulting relevance to exposure. These reactions can be further categorized as inorganic or organic. The final sections consider how our present knowledge of chemistry can be applied to improve indoor air quality.

3.2 Indoor Air Quality Models That Incorporate Chemistry

Indoor air quality models that include chemistry and all other important source and loss mechanisms have been in development for decades (Nazaroff and Cass 1986; Weschler et al. 1989). These models, some described fully in Chapter 8, balance pollutant source rates with removal rates and can be used to determine if chemistry results in a significant alteration in pollutant concentrations. Key components are shown here for a steady-state, "well-mixed" mass-conservation model of a building:

$$C = \frac{E_p/V + E_s/V + \lambda C_0}{\lambda + k} \tag{3.1}$$

where C is the concentration of the compound of interest, E_p is the primary emission rate of the compound into the building from all sources, E_s is the secondary emission (or generation) rate of the compound into the building as a result of transformative chemistry, V is the volume of the building, λ is the building air exchange rate (ACH), C_o is the outdoor concentration of the compound and k is the sum of all first-order (or pseudo-first-order) removal rate constants for that compound.

Although somewhat simplified, this model captures most of the relevant processes, allows for comparison of rates and allows for an estimation of the influence a particular chemical transformation has on indoor concentrations. If chemistry results in a first-order removal rate constant (k) that is a substantial fraction of the air exchange rate (λ), then this chemistry is an important removal mechanism. With typical air exchange rates ranging from 0.1 to 2 h^{-1}, the sum of all first-order rate constants (k) needs to be greater than ~0.03 h^{-1} to significantly decrease indoor *reactant* concentrations. As will be demonstrated in the main body of the chapter, homogeneous reactions between ozone and many fragrances are fast enough to result in a substantial decrease in ozone and the fragrance. Likewise, heterogeneous reactions of ozone with oils on the surface of new carpet and other furnishings result in large first order removal rates of ozone. Similarly, if the generation rate of the compound due to chemistry, E_s, is a significant fraction of primary emission rates or the rate of entry by ventilation, then chemistry must be considered when estimating indoor exposure to that compound. Increases in reaction *products* may be relevant even if rate constants are small. For example, the pseudo first-order rate constant for the reaction of ·OH with gas-phase organics is predicted to be small (Nazaroff and Weschler 2004), but the reaction has been shown to be responsible for much aerosol generation (Fan et al. 2003).

3.3 Ozone and Related Chemistry

3.3.1 Homogeneous Chemistry

3.3.1.1 Indoor Smog Chemistry

Any gas-phase reactions that occur outdoors will also take place indoors if the reactants are present. The outdoor air pollution literature is rich and can be mined to understand what inorganic or organic reactions may occur indoors. A key distinction between indoor and outdoor smog chemistry is that photolysis is minimal indoors. Outdoor daytime chemistry is driven by the energy of sunlight; nighttime chemistry results in a release of chemical energy generated during daytime photolysis. Without intense sunlight, these "dark reactions" are more important indoors. Ozone, O_3, is the primary carrier of photolysis energy into buildings and is directly responsible for most of the indoor transformations studied in the last 3 decades.

Several inorganic dark reactions have been studied in indoor environments; chief among them are the reactions among ozone, nitrogen oxides and organics (Weschler et al. 1994)

$$O_3 + NO \rightarrow NO_2 + O_2 \tag{3.2}$$

$$O_3 + NO_2 \rightarrow NO_3 + O_2 \tag{3.3}$$

$$NO_3 + ORG \rightarrow HNO3 + ORG \tag{3.4}$$

$$O_3 + ORG \rightarrow OH, HO_2, \quad \text{oxidized organics, aerosols} \tag{3.5}$$

$$OH + ORG \rightarrow H_2O + ORG\cdot \tag{3.6}$$

In polluted urban air, reaction 2 titrates O_3 out of air in regions near NO sources, such as roadways. In indoor environments, NO can remove ozone at rates faster than air exchange, especially if NO sources such as open flame are present. Weschler estimates that reaction 3 does not substantially reduce either ozone or NO_2 indoors; instead it generates the nitrate radical, NO_3, which rapidly abstracts a hydrogen atom from an organic compound, ORG. Because of high levels of organic compounds found indoors, reaction 4 contributes substantially to nitric acid formation, HNO_3. Nitric acid has also been shown to be generated indoors by the evaporation of ammonium nitrate aerosols (Lunden et al. 2003),

$$NH_4NO_3 \rightarrow NH_3 + HNO_3 \tag{3.7}$$

3.3.1.2 Ozone and Organic Compounds

Ozone can react directly with some unsaturated organic compounds (Eq. 3.5) at substantial rates. These reactions can significantly reduce indoor ozone levels and generate a suite of oxidized products. Early on, Weschler and Shields suggested ozone reactions with terpenes could lead to a substantial increase in aldehydes, radicals and even aerosols. This work initiated many investigations into the outcomes of this chemistry including laboratory, field, modeling and health studies.

Terpenes are emitted naturally from plants and are responsible for the fresh scent of lemons, pine, and camphor, to name just a few. Because of their desirable scents, they are commonly added to consumer products as fragrance, or to mask off-odors from other constituents of the product. Terpenes are emitted into buildings from many sources including wood products (Hodgson et al. 2002), cleaners (Nazaroff and Weschler 2004) and air fresheners (Sarwar et al. 2004). Limonene, α and β-pinene, 3-carene, isoprene and p-cymene are among the more commonly detected terpenes in surveys of VOCs in indoor air (Brown et al. 1994; Hodgson and Levin 2003).

These terpenes can react rapidly with ozone, substantially reducing levels of the reactants (Nazaroff and Weschler 2004). For example, α-pinene reacts with ozone with a second-order rate constant equal to 2.1×10^{-6} ppb^{-1} s^{-1} (Nazaroff and Weschler 2004). For a typical indoor concentration of ozone of ~20 ppb (Weschler and Shields 2000), a pseudo-first order rate constant for this system is 0.15 h^{-1}. This rate constant is comparable to the low end of air exchange rates and therefore can moderately reduce indoor concentrations of the reactants. While quantification of kinetic rate constants has primarily been performed to understand outdoor pollution (Atkinson and Arey 2003), indoor pollution concerns prompted the recent experimental determination of the kinetics of ozone reactions with α-terpineol (Wells 2005) and citronellol (Ham et al. 2006).

Ozone reactions can generate irritants, carcinogens and even aerosols. Some of the reaction products of the α-pinene-ozone reaction are shown in Fig. 3.1. Formaldehyde, a carcinogen and irritant, is formed with a "yield" of ~0.15 (Atkinson and Arey 2003) (and references therein). This means that for every 100 molecules of ozone that react with α-pinene, about 15 molecules of formaldehyde

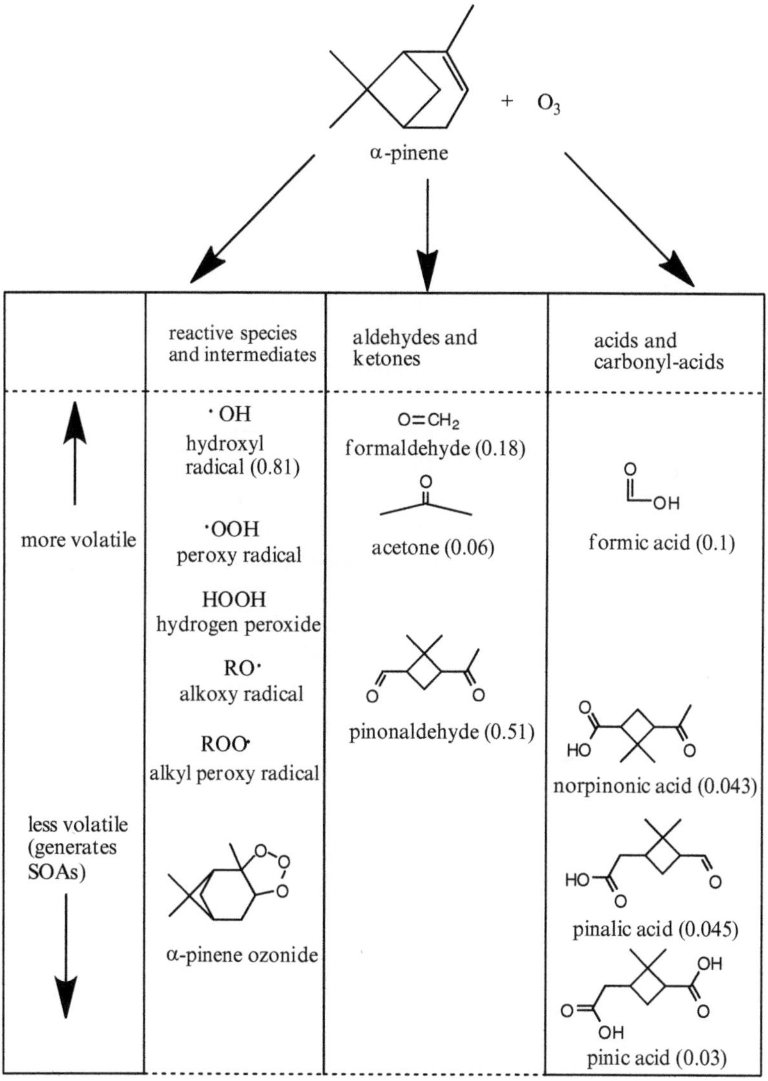

Fig. 3.1 Selected products and yields ( ) of the reaction of α-pinene with ozone

will form. Yields depend the terpene and other conditions; isoprene and β-pinene have formaldehyde yields of ~0.7 an 0.9, respectively. Destaillats et al. (2005) and Singer et al. (2006) show that, when relatively high levels of ozone react with terpenes emitted from cleaners, formaldehyde rises to levels (10–35 ppb) much greater than the California reference exposure level of 2 ppb.

The reaction products include other carbonyl compounds and carboxylic acids which have low odor and irritancy thresholds (Cometto-Muñiz et al. 1998). Pinonaldehyde, a stable di-aldehyde that is formed at high yield in ozone-α-pinene reaction, is a suspected irritant. Organic acids are about ten-times more irritating than their analogous aldehydes; a number of acids (formic acid), di-acids (pinic acid) and acid/aldehyde compounds (norpinonic acid) are also formed.

Several animal and human studies have evaluated the subjective and objective health effects of these ozone-terpene reaction mixtures. In animal studies of reactions of ozone with isoprene (Wolkoff et al. 2000), α-pinene (Wolkoff et al. 1999) and d-limonene (Clausen et al. 2001), identified products that acted as airway irritants. Neither the individual reactants nor the aged products were as irritating as the mixture during reaction. Further, the more aged reaction mixtures were less irritating than younger mixtures (Wilkins et al. 2003), suggesting that unidentified products, such as short-lived radicals, were responsible for the irritation. Similarly, Tamás et al. (2006) found that the sensory load, for 20 human subjects exposed to an ozone-limonene reaction mixture, was much greater than for either compound alone. The reaction mixture also caused eye-blink frequency to increase in human subjects (Kleno and Wolkoff 2004). Rohr et al. (2002) observed limited respiratory flow and possible long-term sensitization in mice exposed to reaction products. In contrast to these studies, a recent study incorporating 130 female human subjects found no significant subjective or objective health effects of a mixture of ozone and VOCs (including α-pinene and d-limonene) (Fiedler et al. 2005; Laumbach et al. 2005). Even though aldehydes, peroxides, acids and aerosols were formed by these reactions, only stress was shown to be correlated with adverse health effects.

Low-vapor pressure oxidation products can self-nucleate to form small aerosols, or condense on and increase the mass of existing aerosols. Collectively, these are known as secondary organic aerosols (SOAs). Weschler and Shields (1999) observed increases in submicron SOA concentrations in an office with elevated ozone and either limonene, α-terpinene or a terpene mixture from a cleaner. For experiments using outdoor ozone (instead of injected ozone), particle concentrations followed the rise and fall of indoor ozone concentrations. Similarly, Long et al. (2000) showed that a pine-scented cleaner increased submicron particle concentrations in the presence of infiltrated ozone. In a laboratory chamber, the limonene-ozone reaction resulted in a 7–100 fold increase in the number concentration of particles with diameters less than 0.5 μm (Rohr et al. 2003; Wainman et al. 2000). Sarwar et al. (2004) showed, in laboratory chamber experiments, that SOAs increase substantially when realistic levels of ozone combine with terpenes from cleaners, air fresheners and perfumes. Shown in Fig. 3.2 is an experiment (Sarwar et al. 2004) in which a solid air freshener is introduced into a chamber with an air exchange rate of 0.62 h^{-1}, and with ozone initially at ~110 ppb. A sharp increase in aerosols occurs due to the reaction of the terpene mixture and ozone, and is sustained for several hours. Note that the reported particle mass concentration rises above outdoor standards. Similarly, Singer et al. (2006) observed an increase of as much as 100 $\mu g\ m^{-3}$ in aerosol mass concentration due to the use of terpenoid containing cleaners in the presence of ozone. Hubbard et al. (2005a) recently demonstrated that ozone generating particle filters, advertised as "air cleaners" designed to reduce indoor particle concentrations, actually increase particle mass substantially in the presence of terpene fragrances.

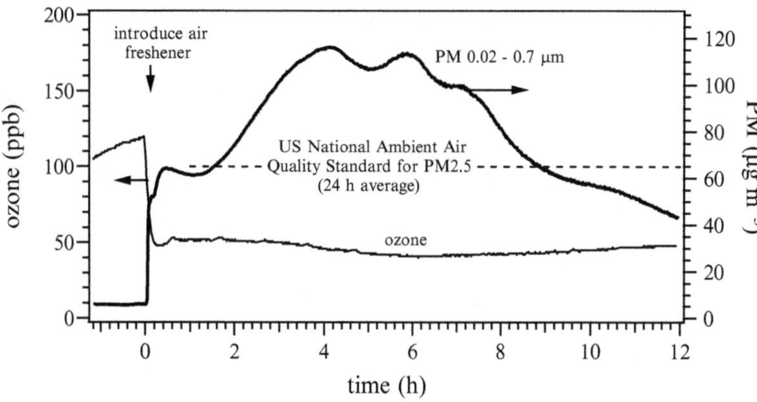

Fig. 3.2 Aerosol formation due to ozone reactions with terpene emitted from a solid air-freshener in a large laboratory chamber (Adapted from Sarwar et al. 2004)

Aerosol generation by the ozone-terpene reaction can significantly increase indoor levels of small aerosols. A large body of epidemiological literature indicates that increases in ambient aerosol concentrations are associated with increased mortality. In particular, an increase of 25 µg m^{-3} in particle mass concentration for PM2.5 (total particle mass for particles with an aerodynamic diameter <2.5 µm) is correlated with a 1–3.5% increase over baseline mortality and a 1–12% increase in hospitalization for respiratory and cardiovascular disease (USEPA 2004). For typical indoor ozone concentrations, in the presence of a pine cleaner, Weschler and Shields (1998) observed a 15–20 µg m^{-3} increase in particles <1.0 mm in diameter. Similarly, Wainman et al. (2000) found that a 20 µg m^{-3} increase due to these reaction is probable under typical indoor conditions. Where ozone is introduced by an ozone generating air filter in field homes, Hubbard et al. (2005a) observed a 10–20 µg m^{-3} increase in particles <1.0 µm in diameter due to the ozone reaction with a variety of terpene sources. The more extreme, but not unrealistic, conditions simulated by Sarwar et al. (2004) in a chamber show that these reactions can result in particle mass concentrations (for 0.02–0.7 µm diameter particles) greater than ambient PM2.5 standards (Fig. 3.2). A recent screening study indicates that perfume wearers will be subjected to a "personal reactive cloud" of reaction products, such as fine aerosols (Karamalegos et al. 2005). Increasing ventilation rates tends to decrease SOA concentrations by dilution, but also tends to shift the respirable particle size distribution to smaller sizes (Weschler and Shields 2003) because the particles have less time to grow or coagulate. Although these reactions result in increased respirable aerosol mass, little is known about human health effects associated with these particular condensed reaction products. Tamás et al. (2006) recently observed a correlation between the number concentration of SOAs, from the limonene-ozone reaction in an office, and sensory load reported by 20 human subjects.

3.3.1.3 Hydroxyl Radical Chemistry

The hydroxyl radical, OH, is formed in the general reaction (3) as demonstrated in the α-pinene-ozone chemistry (Fig. 3.1). Hydroxyl radical yields for the ozone-terpene reaction can be quite high, nearly unity for α-pinene (Atkinson and Arey 2003). Based on the prior observations of ·OH formation (Atkinson et al. 1992), Weschler and Shields (1996) predicted that indoor ·OH could reach ~10^{-5} ppb, a level greater than outdoor nighttime mixing ratios. They confirmed this prediction in an office environment by measuring ·OH formed from the reaction of ozone and limonene, using cyclohexane as an ·OH scavenger (Weschler and Shields 1997a). By modifying an outdoor atmospheric chemistry model, Sarwar et al. (2002) was able to simulate indoor ·OH for typical indoor VOC mixtures. They predicted ·OH mixing ratios similar to those measured by Weschler and Shields. Destaillats et al. (2005) recently showed that the mixing ratio of ·OH ranged from $1–9 \times 10^{-5}$ for the reaction of ozone with terpenes emitted from a variety of cleaning products.

The hydroxyl radical reacts rapidly with almost any hydrocarbon. Reactions of ozone with unsaturated organic compounds are thought to be the primary source of ·OH in indoor environments; thus, its further reactions with terpenes are of great interest. That is, if ·OH is present, terpenes and similar compounds are probably also present at high concentrations. Pseudo-first order rate constants for the reaction of ·OH (at 10^{-5} ppb) with unsaturated organic compounds are much lower than that for ozone. This is because the hydroxyl radical concentration is much lower than ozone (Sarwar et al. 2002; Weschler and Shields 1996; Weschler and Shields 1997b). Yet, ·OH reactions can significantly increase levels of specific reaction products. In the presence of nitric oxide, the hydroxyl radical can generate organic nitrates with yields as high as 0.18 for α-pinene (Noziere et al. 1999). Organic nitrates, specifically peroxyacetylnitrate, are potent irritants. Fan et al. (2003) showed that ·OH, generated from ozone-terpene reactions in the presence of 21 saturated VOCs, was responsible for most of the formaldehyde and p-tolualdehyde production and 19–29% of the organic aerosol formation. The hydroxyl radical can also enhance concentrations of products initially formed in the ozone reaction. In the reaction of ozone with d-limonene, Sarwar et al. (2002) predicted that ·OH is responsible for about a four-fold increase in 3-isopropenyl-6-oxoheptanal, over ozone alone. They also predicted roughly a two-fold increase in pinaldehyde due the reaction of ·OH with α-pinene, over ozone alone.

3.3.1.4 Nitrate Chemistry

As noted in Reaction (4) the nitrate radical, NO_3, is also potent agent of change in indoor environments. Like OH, the nitrate radical abstracts a hydrogen from organics, initiating free-radical oxidation. It also adds to double-bonds, making terpenes easy targets for nitration (Kirchner and Stockwell 1996). As Weschler and Shields (1997b) noted, the combination of low lighting and high concentrations of unsaturated VOCs makes this chemistry potentially important in indoor environments.

Indeed, organic nitrates have been detected at substantial yield in the reactions with isoprene and α-pinene (Atkinson and Arey 2003). Many of the same radical intermediates associated with ozone and ·OH chemistry have been observed as the result of NO_3· terpene chemistry, including the peroxy radical, HOO·, and secondary aerosols. Beyond terpenes, Nazaroff and Weschler (2004) specifically identify the reaction of NO_3 with 2-butoxyethanol as troubling due to the formation of propylnitrate, and other unique products.

Referring to Fig. 3.1, we see that other radical products are formed in the reaction of α-pinene (and other terpenes) with ozone. Secondary or tertiary reactions also occur as the result of reactions between VOCs and these radical products. Complexity increases many-fold as we consider the tens to thousands of organic species present to react with OH, NO_3 and other secondary oxidants. Further, product yields depend strongly on the environmental factors. Capouet et al. (2004) predict that pinonaldehyde yields from the α-pinene-·OH reaction are six time higher in a polluted atmosphere than in a more pristine setting. Pinonaldehyde yield is also sensitive to relative humidity (Warscheid and Hoffmann 2001).

3.4 Heterogeneous Ozone Chemistry

Chemistry that occurs at interfaces is termed heterogeneous chemistry and is remarkably important in indoor environments. This is because (1) the available surface-area is large relative to the building volume, (2) surfaces tend to enhance reaction rates relative to that in the gas-phase, and (3) unique compositions and morphologies at surfaces can promote some reactions (that is, catalytically) or promote selectivity in reaction pathways.

Condensed phases are roughly 1,000 times more concentrated (by mass) than the overlying gas, meaning that surface reactions can be fast. Surface reactions proceed so rapidly that the effective conversion rates are often limited by the availability of a gas-phase reactant such as ozone. The primary loss mechanism for ozone in indoor volumes is surface reaction, causing the I/O ratio of ozone to be typically 0.2 or lower (Weschler et al. 1989). Using a mass-conservation approach, and assuming that the outdoor mixing ratio is at the US 24-h standard (80 ppb) and that every reacted ozone molecule creates a volatile product, then the total indoor mixing ratio of products could be of the order of ~0.8 × 80 ppb or 64 ppb. This is a substantial fraction of typical VOC levels, especially for the types of products that tend to be generated: aldehydes, carboxylic acids, nicotine oxidation products, and, indirectly, aerosols.

On indoor surfaces, carbonyls (aldehydes and ketone) are the prominent products observed, probably because they are more volatile and easier to detect than other possible products. Fats, left over from manufacturing or deposited during day-to-day activities, are probable targets of ozone on surfaces. As an example, in Fig. 3.3 ozone splits the linoleic ester into C_6 and C_9 aldehydes and acids. In ozone reactions with carpet, Morrison and Nazaroff (2002) found the aldehyde yield to

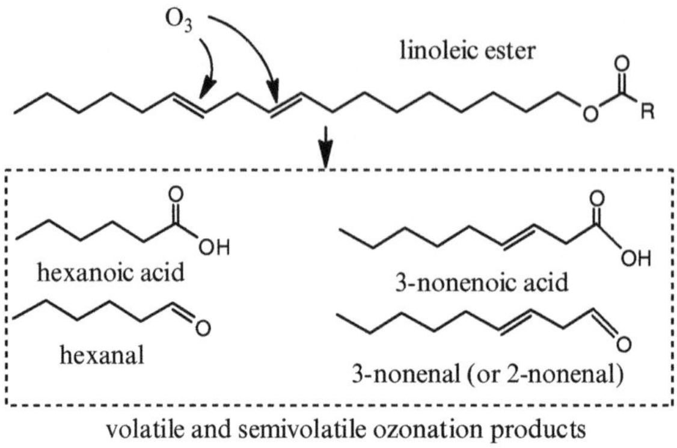

Fig. 3.3 Heterogeneous ozone oxidation of linoleic ester to form volatile aldehydes and less volatile acids

range from 0.1 to 0.74 for summed aldehydes. A yield of 0.5 would be expected if aldehydes and acids were formed equally in Fig. 3.3. Low yields may indicate that surface conditions favor the formation of acids or other low-volatility products, or that other reactions consume ozone (such as homogenous decomposition). High yields suggest that surface conditions favor aldehydes. Interestingly, lower-volatility products may also contribute to aerosol growth by partitioning from the indoor surface to existing aerosols (Aoki et al. 2005; Bekö et al. 2005).

Heterogeneous ozone chemistry has been studied on carpets, paint, HVAC surfaces, and a variety of soiled indoor materials.

3.4.1 Ozone and Carpet

In the first study of its kind, Weschler et al. (1992) discovered that ozone reacted with carpet surfaces generating formaldehyde, acetaldehyde and C4–C10 aldehydes. Ozone converted some of the condensed phase material on the carpet to volatile species, demonstrating for the first time that surface chemistry can increase the gas-phase VOC concentration. Morrison and Nazaroff (2002) observed similar reaction products, but also observed unsaturated aldehydes generated on new carpet. Based on product yield studies, 2-nonenal was predicted to significantly exceed odor thresholds under typical conditions and that odorous conditions could persist for years. Field measurements of secondary emissions from carpet (Wang and Morrison 2006) showed that 1–2-year-old carpet is more reactive and aldehyde yields are higher than for carpet that had been installed for over 10 years. Thus, secondary emissions can persist for at least 2 years, but surface reactants may eventually become depleted. Knudsen et al. (2003) performed sensory evaluations on carpet and other materials that had been exposed to ozone. They found that carpet in particular exhibited significantly higher odor intensity as a result of secondary emissions.

3.4.2 *Latex Paint*

Reiss et al. (1995a) showed that ozone reactions with latex paint generates formaldehyde, acetone and acetaldehyde. Older painted surfaces were not as reactive. They estimated that ozone reactions on newly applied paint would constitute 10–15% of the total emission rate of formaldehyde measured in field homes (Reiss et al. 1995b). By isolating wall sections in field homes and exposing the sections to ozone, Wang and Morrison (2006) confirmed a small generation rate of aldehydes from latex painted walls.

3.4.3 *HVAC Materials*

Ventilation duct liners also react with ozone forming formaldehyde, acetone and $C_5 - C_{10}$ aldehydes. Morrison et al. (1998) subjected new and used duct liners, air filters, sealants, sheet metal and other HVAC materials to ozone in small chambers. They observed secondary emissions of C5–C10 aldehydes from a new duct liner, a neoprene gasket and duct sealants. They predicted that secondary emissions from these materials could increase indoor aldehyde concentrations to levels comparable with odor thresholds. As will be discussed later, soiled HVAC materials also generate secondary products.

3.4.4 *Surface Aging and Regeneration by Soiling*

Over time, original surface reactants will eventually become depleted. Therefore, ozone reactivity and secondary emissions of byproducts should decrease as the cumulative ozone exposure increases. Many studies have shown that ozone uptake on indoor surfaces tends to decrease, a phenomenon known as "aging" (Morrison and Nazaroff 2000; Morrison et al. 1998; Reiss et al. 1995a; Sabersky et al. 1973; Simmons and Colbeck 1990). Further, there is evidence that secondary emission rates also decrease with time. Morrison and Nazaroff (2002) showed that secondary aldehydes on new carpet fibers, in a fixed-bed reactor, could be depleted in a day; however, the reactivity of whole carpet was not substantially decreased over the relatively short time periods studied. Wang and Morrison (2006) showed that carpet in older homes exhibited significantly lower secondary emission rates of aldehydes, than carpet in newer homes.

Although original surface reactants become depleted, these surfaces may become regenerated. Consumer activities repeatedly coat surfaces with cooking oils, soaps, tobacco products and even human skin oils, all of which have been shown to react with ozone to form volatile products.

The surfaces in HVAC systems may be regenerated by continuous deposition of reactive aerosols. Morrison et al. (1998) observed that used duct liners from commercial HVAC systems consumed more significantly ozone than new duct liners. In related work, several studies have recently shown that soiled HVAC filters are more ozone reactive than clean filters (Hyttinen et al. 2006; Hyttinen et al. 2003; Zhao et al. 2007). Hyttinen et al. (2003) observed formaldehyde as the only reaction product from a subset of these filters. Bekö et al. (2006) inferred that reaction byproducts were formed on used filters that were treated with ozone because they perceived as “less acceptable” than filters treated with plain air or nitrogen. They suggest that this chemistry may account for previous observations that sick building syndrome symptoms and poor occupant performance correlate with the presence of loaded HVAC filters (Clausen et al. 2002; Wargocki et al. 2003; Wyon et al. 2000). Fine aerosols have also been observed as indicators that semi-volatile reaction products form and partition from the soiled filter surface to aerosols (Bekö et al. 2005).

Wang et al. (2005) showed that countertops that become coated with cooking oils or soaps will generate volatile aldehydes in the presence of ozone, in chemistry akin to Fig. 3.3. They verified in field studies (Wang and Morrison 2006) that kitchen counters are a major emitter of secondary aldehydes (on a unit area basis).

Wisthaler et al. (2005) studied ozone-surface reactions in a simulated aircraft environment that included used carpet, seats and other inner airplane surfaces. They also evaluated the presence of t-shirts worn by volunteers for a day. A key finding was that ozone consumption increases due to reactions with human skin oils, specifically squalene, on t-shirts and other airplane surfaces. Using a very sensitive proton-transfer mass spectrometry system, they observed volatile oxidation products of the ozone-squalene reaction including acetone and 6-methyl-5-heptenone (Fig. 3.4). They identified aldehydes, similar to those observed from earlier carpet studies, which may be generated from reactions with airline carpet and upholstery. They also tentatively identified organic acids and organic nitrates as volatile reaction products.

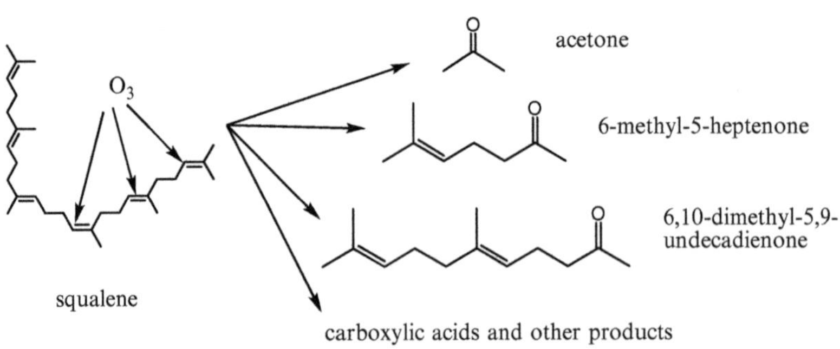

Fig. 3.4 Ozone oxidation of squalene in human skin oils (Wisthaler et al. 2005)

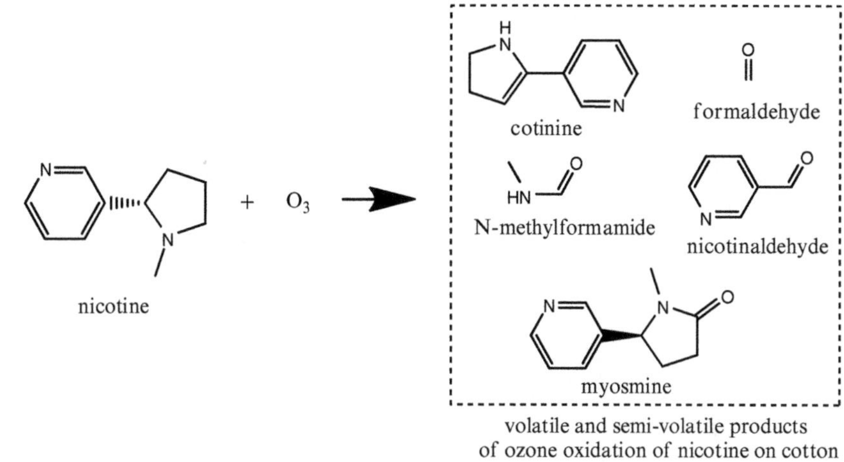

Fig. 3.5 Ozone oxidation of nicotine on cotton and Teflon (Destaillats et al. 2006b)

Ozone-nicotine surface chemistry has recently been identified by Destaillats et al. (2006b). Nicotine and other products of tobacco smoking adsorb strongly to indoor surfaces. In the presence of ozone, surface nicotine can form a variety of products (Fig. 3.5). On cotton, high humidity inhibits this reaction. Ozone attacks the pyrrolidinic N, suggesting that other indoor surface bound species with this functionality may be oxidized in this manner. Not only does this chemistry generate volatile byproducts, it may explain why indoor nicotine concentrations correlate poorly with exposure to tobacco smoke. This raises the question, what other, similarly functionalized, surface amines may also react with ozone in indoor environments?

Homogeneous reactions with terpenes have already been discussed, but surface chemistry of these species may also be important. Pommer (2003) and Fick (2003) inferred heterogeneous terpene reactions with NO_2 and ozone and radicals on teflon tubing because observed rate constants were much larger than expected. Similarly, oxidation of terpenes on surface of HVAC heat exchangers was inferred from a larger-than-expected reduction in the concentration between the upstream and downstream locations (Fick et al. 2005). They also observed that the fraction of terpenes oxidized correlated with surface area. Increased ozone deposition rates to surfaces are observed long after terpene primary emissions have ceased (Destaillats et al. 2006a). Together, these studies indicate that adsorbed terpenes promote ozone uptake on indoor surfaces and that overall terpene decomposition is enhanced over homogeneous processes.

Taken as a whole, it is probable that some fraction of ozone uptake and secondary emission in commercial and residential buildings is due to reactions with soaps, cooking oils, human skin oils, terpenes and other products of human inhabitation. This may partially answer the question posed by Nazaroff et al. (1993) in their evaluation of ozone deposition velocities on indoor surfaces: why does the area-averaged ozone reactivity of indoor surfaces in most field sites tend to be roughly the same, even though laboratory studies show that individual surfaces are very different and that surfaces age and lose reactivity with time? They suggested that some sort of uniform soiling of surfaces with reactive species may explain this observation. Here we see that human skin oils, cooking oils, and even cleaning all contribute to enhancing the surface reactivity of indoor surfaces. It is probable that these materials also contribute to the enhanced reactivity of HVAC filters and surfaces.

3.5 Auto-oxidation, Hydrolysis, Acid-Base, and Chlorine Chemistry

3.5.1 Auto-oxidation

A great deal of chemical transformation takes place before consumers receive their furniture, flooring, and other indoor building materials. Polymerization of glues, coatings and plastics, or heat processing of manufactured wood products, generates volatile byproducts that can continue to be emitted from the material after it has been installed. Once installed, further transformations increase the load of odorous or toxic gas-phase species (Uhde and Salthammer 2006).

The autooxidation and photoxidation of fatty acids and triglycerides is a well studied phenomenon (rancidity), and has been recognized to occur indoors in linoleum (Jensen et al. 1996), furniture coatings (Salthammer et al. 1999) and in cooking oils. This oxidation results in the formation of volatile aldehydes, methyl-esters, alcohols and more. See Fig. 3.6 for oxidation chemistry of linolenic acid, a major constituent in linseed and other oils. The molecule is truncated to show only the last nine carbons of the 18 carbon chain. Oxidation is initiated when hydrogen is abstracted, generating a radical-stabilized resonance structure that may shift the radical position along the carbon backbone of the fatty acid. This is followed by addition of oxygen to the radical location and abstraction of hydrogen from a neighboring ester, propagating the oxidation chemistry from one ester to another. The resulting hydroperoxy compound is semi-stable, but may eventually decompose forming a variety of short-chain, volatile, compounds such as odorous aldehydes and carboxylic acids.

The relevance of this chemistry is demonstrated by the sensory impressions of linseed-oil based linoleum. Jensen et al. (1995) studied the auto-oxidation products emitted from linoleum and identified saturated and unsaturated aldehydes and fatty acids. An odor evaluation of the identified 2-alkenals and the fatty acids as contributing most to odor intensity (Jensen et al. 1995), although many odorous products may not be quantified by traditional analytical methods (Clausen et al. 2005).

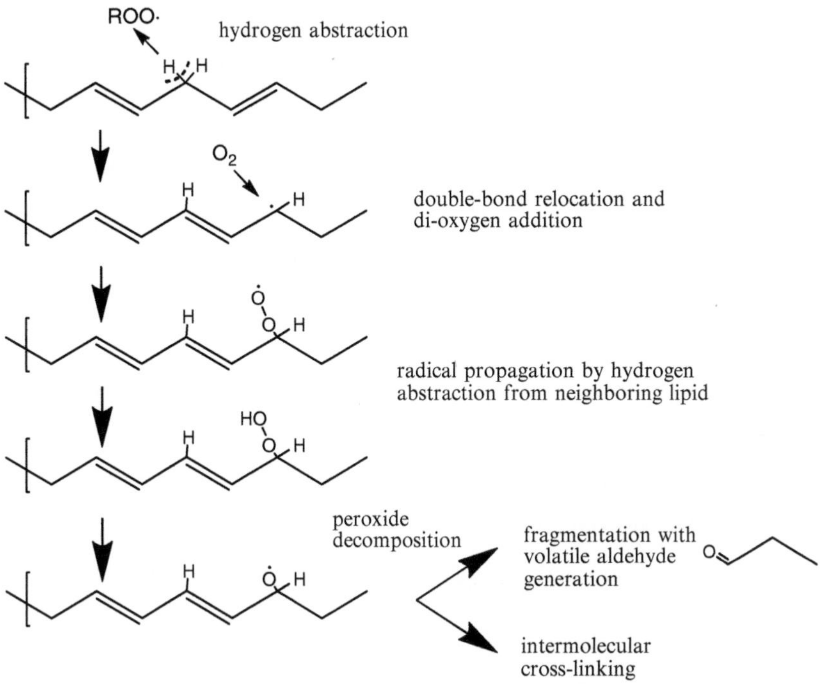

Fig. 3.6 Auto-oxidation of linolenic acid (truncated) leading to radical propagation, intermolecular cross-linking, and volatile aldehyde emissions

Knudsen et al. showed that linseed oil based products exhibited a more negative sensory perception than similar product not containing linseed oil, and that the negative perception persisted for at least a year (Knudsen et al. 2005).

3.5.2 *Hydrolysis*

Excessive dampness in buildings can lead to a host of problems, mold growth being one of the more obvious ones. However, water also promotes hydrolysis of glues, plasticizers and flame retardants. A classic example is the hydrolysis of urea-formaldehyde resins with the carcinogen formaldehyde as a product (Fig. 3.7). These resins are widely used in manufactured wood products (plywood, particle board and so forth). In western countries, manufacturers have reduced formaldehyde emission rates by reducing the amount of formaldehyde in the resin, adding formaldehyde scavengers and using coatings as barriers to emissions (Conner 1996). However, formaldehyde continues to be a problem in residences in China (Zheng et al. 2005), perhaps due to emissions from furnishings (Wang and Lee 2005).

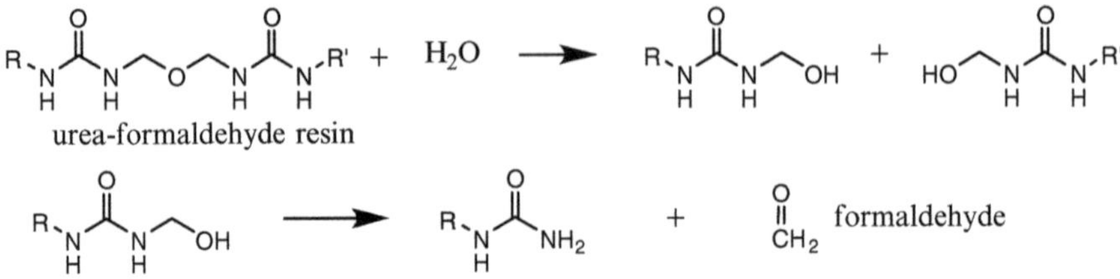

Fig. 3.7 Hydrolysis of urea-formaldehyde resin, generating formaldehyde

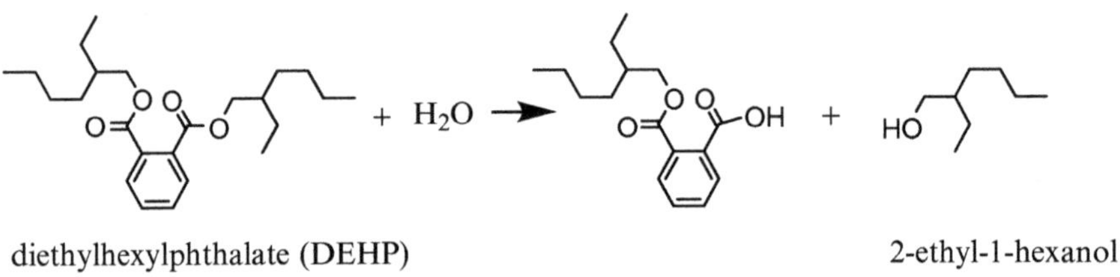

Fig. 3.8 Hydrolysis of diethylhexylphthalate to form 2-ethyl-1-hexanol

As shown in Fig. 3.7, hydrolysis has the effect of breaking down condensed materials to more volatile molecules. Uhde and Salthammer (2006) explain that esters are cracked in this way and converted to alcohols and carboxylic acids. Esters known to hydrolyze indoors include the plasticizers diethylhexylphthalate (DEHP) and di/iso-butylphthalate (DBP/DIBP). DEHP forms the odorous and volatile 2-ethyl-1-hexanol by hydrolysis (Fig. 3.8).

Although hydrolysis is generally slow, Uhde and Salthammer (2006) explain that by installing PVC flooring on poorly cured concrete, the secondary emission rate of 2-ethyl-1-hexanol is enhanced. The flooring traps moisture and the alkaline concrete catalyzes the hydrolysis. Weschler (2004) has suggested that observed associations between damp buildings and health complaints (Bornehag et al. 2001) may be due to products of hydrolysis chemistry. In support of this hypothesis, Bornehag et al. (2005) recently observed that PVC flooring, in combination with water leakage in residences, was associated with a higher prevalence of childhood respiratory complaints such as wheezing and asthma.

Hydrolysis of flame retardants can generate a variety of odorous compounds, including halogenated alcohols. Halogenated and non-halogenated organophosphates have been added to building materials such as polyurethane construction foam. Products of organophosphate hydrolysis were detected by Salthammer et al. (2003) from recycled flooring materials, including 1-chloro-2-propanol and 2-chloro-1-proponol. Hydrolysis of tris (1,3-dichloro-2-propyl)phosphate (TDCPP) is shown in Fig. 3.9 (Uhde and Salthammer 2006). The product, 1,3–dichloro-2-propanol is considered to be a human carcinogen by the Deutsche Forschungsgemeinschaft (German Research Foundation) based on animal studies. Newer products contain primarily non-halogenated organophosphates,

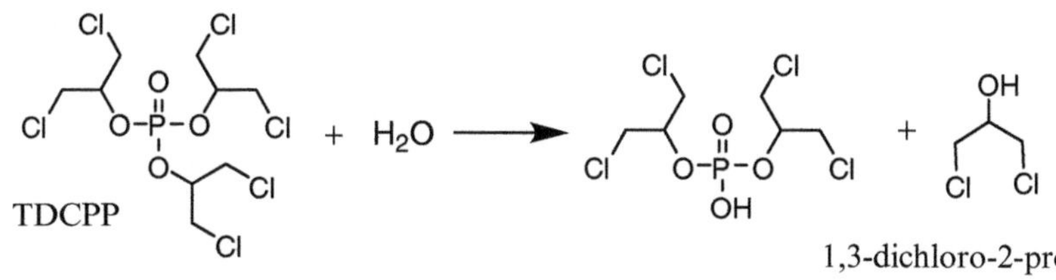

Fig. 3.9 Flame retardant TDCPP hydrolysis forming 1,3-dichloro-2-propanol

due to concerns over the toxicity of halogenated flame retardants. However, recycled materials retain the older, chlorinated compounds (Salthammer et al. 2003).

3.5.3 Acid-Base Chemistry

Acid-base chemistry on a surface has the effect of changing the form of the surface bound species, instead of directly generating new volatile compounds. This can influence the overall sorptive capacity of indoor surfaces or even catalyze transformative reactions. Compound sorption influences the timing and intensity of exposure by temporarily storing these species on indoor surfaces (Tichenor et al. 1991; Won et al. 2001). For example, if a compound adsorbs strongly during an emission event, the peak concentration during the event will be lower than anticipated. However, desorption of those compounds will cause occupants to be exposed over an extended time period.

Acid base chemistry taking place on surfaces may influence adsorption of basic or acidic compounds. It is well established in aqueous systems that pH influences the sorption of organic compounds to solids. However, indoor surfaces are not uniformly covered with bulk water, nor are they uniformly dry. Instead surfaces are complex and exhibit a range of molecular water coverage depending on humidity, surface composition and morphology. Therefore, the influence of acid-base chemistry on sorption is non-obvious. A volatile or semi-volatile organic compound that has basic functionality, such as nicotine (Van Loy et al. 1998), may bind strongly to an acidic surface site, or perhaps convert to its conjugate acid salt in acidic surface water. If sufficient acidity exists, the total surface capacity for that compound can be substantial. Organic acids accumulate in an analogous fashion on basic (alkaline) surfaces. The acidity/basicity of a surface depends on the composition of that surface, including the substrate (for example, a carpet fiber), surface amendments (salts, oils and so forth), surface moisture. Therefore, sorption can also be influenced by adsorption of overlying gases such as acidic carbon dioxide and basic ammonia.

Evidence of this behavior was reported by Webb et al. (2002). They opened a bottle of an ammonia based cleaning solution in a room that had been preconditioned with 3,200 mg of nicotine. They observed an immediate increase in the gas-phase nicotine concentration that they ascribed to conversion of non-volatile nicotine salts to semi-volatile free-base nicotine.

Ongwandee et al. studied the adsorption behavior of trimethylamine (TMA) on three surfaces: $ZrSiO_2$ (representative of mineral surfaces) (Ongwandee et al. 2005), painted gypsum board and carpet (Ongwandee, 2006). They found that ammonia, a base, decreases the surface capacity of all materials. Carbon dioxide, an acid, tends to increase the surface capacity, but only moderately so. This is to be expected if aqueous-phase chemistry is taking place, converting TMA to its less volatile, protonated form. Instead, competitive adsorption, rather than aqueous-phase acid-base chemistry appears to drive most of these dynamics. Changes in surface site acidity may also contribute to this observed behavior.

Combined, these studies demonstrate the importance of evaluating sorption of VOCs under the appropriate conditions, that is, with typical indoor levels of CO_2, NH_3, humidity, temperature, and so forth. Since most polar organic compounds are Lewis bases, or are amphoteric, they may also be influenced by changes in surface acid sites or the presence of competitive gas-phase species. Destaillets et al. (2006) note that only the free-base form of nicotine is susceptible to oxidation. This may mean that interactions with acid sites on the surface could also suppress the oxidation reaction of this and other amines.

3.5.4 Chlorine Chemistry

The formation of chloroform ($CHCl_3$) during disinfection of municipal drinking water led Olson and Corsi (2004) to hypothesize that this chemistry could also take place in residential appliances. Hypochlorite containing detergent was used in a dishwasher and the aqueous and gas-phase chloroform concentration was assessed during a normal dishwashing cycle. They found that a substantial amount of chloroform is generated by chlorine-organic chemistry. Further, they estimated that emissions from appliances could account for the chloroform concentrations that have been observed in residences. While chloroform concentrations are tightly controlled in the water delivered from water treatment plants, they estimate that exposure by volatilization and inhalation in residences is more important than ingestion.

3.6 Control of Indoor Chemistry and Control in Indoor Air Quality Using Chemistry

3.6.1 Control of Ozone Chemistry

Controlling ozone chemistry may be a simple as controlling ozone itself. Efforts to reduce smog have been moderately successful, but outdoor ozone levels remain problematic across the globe. Therefore, the key will be to control ozone in indoor microenvironments. Activated carbon filtration has been shown to reduce ozone levels in buildings with ventilation or recirculation systems (Weschler et al. 1994), yet reactions of ozone with particles on filters themselves (Bekö et al. 2005, 2006; Hyttinen et al. 2006; Zhao et al. 2007) will need to be considered. For buildings without filtered recirculation air, other approaches could be taken. Perhaps paint or flooring surfaces could be modified to catalytically consume ozone. Alternatively, ozone could be scrubbed with activated carbon materials at the most accessible outdoor infiltration points such as around doors and windows.

Controlling the other reactant, be it terpenes or surface materials, could also be worth considering, especially if a specific reaction product is identified as being particularly dangerous. Some terpenes are not nearly as reactive as α-pinene, or limonene and may be useful as alternative fragrances in consumer products. It may also be possible to use different spinning oils in the manufacture of textiles and carpet that do not generate aldehydes, etc. However, given the ubiquity of many of these products and the inevitability of surface soiling with cooking oils, cleaners and skin oils, a focus on reducing ozone will probably be more productive.

3.6.2 Misguided Chemistry for Controlling Indoor Air Quality

As our understanding of indoor chemistry expands, we may discover chemical mechanisms that we can engineer to clean the air. One possibility, already discussed, is the removal of ozone using filtration. However, most commercial efforts to chemically improve indoor air have been misguided at best, dangerous at their worst.

The fact that ozone can react rapidly to remove some VOCs have led manufacturers of ozone generating "air cleaners" to claim that their products effectively destroy organic pollutants (Boeniger 1995). The reaction rates with most indoor VOCs are far too small to effect any significant change in exposure. For those compounds that are "destroyed", such as the terpenes already discussed, the products of these reactions appear to be worse than the original compound.

The need to kill-off biological contaminants in commercial and residential buildings has engendered what Hubbard et al. (2005b) have coined as "extreme heterogeneous oxidation chemistry". Anthrax mail attacks led to contamination of several government buildings in 2002. These buildings were effectively decontaminated by generating and maintaining very high ozone levels. After re-entering the building, occupants complained of irritated eyes and other problems, indicating that the treatment may generate unanticipated byproducts. This application of extreme chemistry is not limited to isolated instances of bioterrorism: mold infested residences in the United States are now being treated with ozone in a similar fashion. Hubbard et al. (2005b) have shown that subjecting typical indoor materials to these high levels of ozone generated substantial levels of oxidized organic compounds. They also showed that using chlorine dioxide, an alternative to ozone, generates similar compounds, but also volatile chlorinated organic species.

3.7 Summary

Buildings are dynamic environments, long on heterogeneity, short on simplicity. Thousands of compounds interact in the air or on a myriad of surfaces while temperature, humidity and other conditions simultaneously vary. The "chemical soup" that is our primary living space needs the same attention that we have given to our outdoor air for the past 50 years. We have studied indoor ozone chemistry a great deal because much of the fundamental leg-work has been done for us by the pioneering ambient air scientists. Yet, our discipline is making strides towards a distinctive understanding of chemical transformations in this unique environment.

References

Aoki, T., Tanabe, S., Funaki, R., Tanaka, H., Nakagawa, T., Kihara, I., et al. (2005). *Generation of sub-micron particles and secondary emissions from building materials by ozone reaction.* The 10th International Conference on Indoor Air Quality and Climate, INDOOR AIR '05, Beijing, China.

Atkinson, R., & Arey, J. (2003). Gas-phase tropospheric chemistry of biogenic volatile organic compounds: A review. *Atmospheric Environment, 37*, S197–S219.

Atkinson, R., Aschmann, S. M., Arey, J., & Shorees, B. (1992). Formation of OH radicals in the gas phase reactions of O_3 with a series of terpenes. *Journal of Geophysical Research, 97*, 6065–6073.

Bekö, G., Halás, O., Clausen, G., & Weschler, C. J. (2006). Initial studies of oxidation processes on filter surfaces and their impact on perceived air quality. *Indoor Air, 16*, 56–64.

Bekö, G., Tamás, G., Halás, O., Clausen, G., & Weschler, C. J. (2005). *Ultra-fine particles as indicators of the generation of oxidized products on the surface of used air filters.* The 10th International Conference on Indoor Air Quality and Climater, INDOOR AIR '05, Beijing, China.

Boeniger, M. F. (1995). Use of ozone generating devices to improve indoor air quality. *American Industrial Hygiene Association, 56*, 590–598.

Bornehag, C. G., Blomquist, G., Gyntelberg, F., Jarvholm, B., Malmberg, P., Nordvall, L., et al. (2001). Dampness in buildings and health – Nordic interdisciplinary review of the scientific evidence on associations between exposure to "dampness" in buildings and health effects (NORDDAMP). *Indoor Air, 11*, 72–86.

Bornehag, C. G., Sundell, J., Hagerhed-Engman, L., Sigsggard, T., Janson, S., & Aberg, N. (2005). Dampness' at home and its association with airway, nose, and skin symptoms among 10,851 preschool children in Sweden: A cross-sectional study. *Indoor Air, 15*, 48–55.

Brown, S. K., Sim, M. R., Abramson, M. J., & Gray, C. N. (1994). Concentrations of volatile organic compounds in indoor air – A review. *Indoor Air, 4*, 123–134.

Capouet, M., Peeters, J., Noziere, B., & Muller, J. F. (2004). Alpha-pinene oxidation by OH: Simulations of laboratory experiments. *Atmospheric Chemistry and Physics, 4*, 2285–2311.

Clausen, G., Alm, O., & Fanger, P. O. (2002). *Sensory strength of used ventilation filters.* The 9th International Conference on Indoor Air Quality and Climate. INDOOR AIR '02, Monterey, CA.

Clausen, P. A., Knudsen, H. N., Larsen, K., Kofoed-Sørensen, V., Wolkoff, P., & Wilkins, C. K. (2005). *Use of gas chromatography olfactometry (GC-O) to detect unknown emissions from building products containing linseed oil.* The 10th International Conference on Indoor Air Quality and Climate, INDOOR AIR '05, Tsingua University Press, Beijing, China.

Clausen, P. A., Wilkins, C. K., Wolkoff, P., & Nielsen, G. D. (2001). Chemical and biological evaluation of a reaction mixture of R-(+)-limonene/ozone – Formation of strong airway irritants. *Environment International, 26*, 511–522.

Cometto-Muñiz, J. E., Cain, W. S., & Abraham, M. H. (1998). Nasal pungency and odor of homologous aldehydes and carboxylic acids. *Experimental Brain Research, 118*, 180–188.

Conner, A. H. (1996). Urea formaldehyde adhesive resins. In J. C. Salamone (Ed.), *Polymeric materials encyclopedia* (Vol. 11, pp. 8497–8501). Boca Raton, FL: CRC Press.

Destaillats, H., Lunden, M. M., Singer, B. C., Coleman, B. K., Hodgson, A. T., Weschler, C. J., et al. (2006a). Indoor secondary pollutants from household product emissions in the presence of ozone. A bench-scale chamber study. *Environmental Science & Technology, 40*, 4421–4428.

Destaillats, H., Singer, B. C., Coleman, B. K., Lunden, M. M., Hodgson, A. T., Weschler, C. J., et al. (2005). *Secondary pollutants from cleaning products and air fresheners in the presence of ozone*. 10th International Conference on Indoor Air Quality and Climate (Indoor Air 2005), Tsinghua University Press, Beijing, China.

Destaillats, H., Singer, B. C., Lee, S. K., & Gundel, L. A. (2006b). The effect of ozone on nicotine desorption from model surfaces: Evidence for heterogeneous chemistry. *Environmental Science & Technology, 40(6)*,1799–1805.

Fan, Z., Lioy, P., Weschler, C., Fiedler, N., Kipen, H., & Zhang, J. (2003). Ozone-initiated reactions with mixtures of volatile organic compounds under simulated indoor conditions. *Environmental Science & Technology, 37*, 1811–1821.

Fick, J. (2003). *Chemical reactions in ventilation systems: Ozonolysis of monoterpenes*. Ph.D. thesis, Department of Chemistry, Umeå University, Umeå, Sweden.

Fick, J., Pommer, L., Åstrand, A., Östin, R., Nilsson, C., & Andersson, B. (2005). Ozonolysis of monoterpenes in mechanical ventilation systems. *Atmospheric Environment, 39*, 6315–6325.

Fiedler, N., Laumbach, R., Kelly-McNeil, K., Lioy, P., Fan, Z. H., Zhang, J. F., et al. (2005). Health effects of a mixture of indoor air volatile organics, their ozone oxidation products, and stress. *Environmental Health Perspectives, 113*, 1542–1548.

Ham, J. E., Proper, S. P., & Wells, J. R. (2006). Gas-phase chemistry of citronellol with ozone and OH radical: Rate constants and products. *Atmospheric Environment, 40*, 726–735.

Hodgson, A. T., Beal, D., & McIlvaine, J. E. R. (2002). Sources of formaldehyde, other aldehydes and terpenes in a new manufactured house. *Indoor Air, 12*, 235–242.

Hodgson, A. T., & Levin, H. (2003). *Volatile organic compounds in indoor air: A review of concentrations measured in North America since 1990*. Berkeley, CA: Lawrence Berkeley National Laboratory.

Hubbard, H. F., Coleman, B. K., Sarwar, G., Corsi, R. L. (2005a). Effects of an ozone-generating air purifier on indoor secondary particles in three residential dwellings. *Indoor Air, 15*, 432.

Hubbard, H. F., Poppendieck, D., Ward, M., Weschler, C., & Corsi, R. L. (2005b). Extreme heterogeneous oxidation chemistry (EHOC): reaction kinetics and product formation. *Indoor Air 2005*. Beijing, China: Tsinghua University Press.

Hyttinen, M., Pasanen, P., & Kalliokoski, P. (2006). Removal of ozone on clean, dusty, and sooty supply air filters. *Atmospheric Environment, 40*, 315–325.

Hyttinen, M., Pasanen, P., Salo, J., Björkroth, M., Vartiainen, M., & Kalliokoski, P. (2003). Reactions of ozone on ventilation filters. *Indoor and Built Environment, 12*, 151–158.

Jensen, B., Wolkoff, P., & Wilkins, C. K. (1995). Characterization of Linoleum: Part 2: preliminary odor evaluation. *Indoor Air, 5*, 38–43.

Jensen, B., Wolkoff, P., & Wilkins, C. K. (1996). *Characterization of linoleum: Identification of oxidative emission processes* (pp. 145–152). Philadelphia, PA: ASTM, American Society for Testing and Materials.

Karamalegos, A., Simon, H., Zhao, P., Morrison, G. C., Siegel, J. A., & Corsi, R. L. (2005). *Personal reactive clouds: Introducing the concept of near head chemistry.* The 10th International Conference on Indoor Air Quality and Climate, INDOOR AIR '05, Tsinghua University Press, Beijing, China.

Kirchner, F., & Stockwell, W. R. (1996). Effect of peroxy radical reactions on the predicted concentrations of ozone, nitrogenous compounds, and radicals. *Journal of Geophysical Research-Atmospheres, 101*, 21007–21022.

Kleno, J., & Wolkoff, P. (2004). Changes in eye blink frequency as a measure of trigeminal stimulation by exposure to limonene oxidation products, isoprene oxidation products and nitrate radicals. *International Archives of Occupational and Environmental Health, 77*, 235–243.

Knudsen, H. N., Clausen, P. A., Wilkins, C. K., & Wolkoff, P. (2005). *Sensory and chemical evaluation of odorous emissions from building products containing linseed oil*. The 10th International Conference on Indoor Air Quality and Climate, INDOOR AIR '05, Tsingua University Press, Beijing, China.

Knudsen, H. N., Nielsen, P. A., Clausen, P. A., Wilkins, C. K., & Wolkoff, P. (2003). Sensory evaluation of emissions from selected building products exposed to ozone. *Indoor Air, 13*, 223–231.

Laumbach, R. J., Fiedler, N., Gardner, C. R., Laskin, D. L., Fan, Z., Zhang, J., et al. (2005). Nasal effects of a mixture of volatile organic compounds and their ozone oxidation products. *Journal of Occupational and Environmental Medicine, 47*, 1182–1189.

Long, C. M., Suh, H. H., & Koutrakis, P. (2000). Characterization of indoor particle sources using continuous mass and size monitors. *Journal of the Air and Waste Management Association, 50*, 1236–1250.

Lunden, M. M., Revzan, K. L., Fischer, M. L., Thatcher, T. L., Littlejohn, D., Hering, S. V., et al. (2003). The transformation of outdoor ammonium nitrate aerosols in the indoor environment. *Atmospheric Environment, 37*, 5633–5644.

Morrison, G. C., & Nazaroff, W. W. (2000). The rate of ozone uptake on carpets: Experimental studies. *Environmental Science & Technology, 34*, 4963–4968.

Morrison, G. C., & Nazaroff, W. W. (2002). Ozone interactions with carpet: secondary emissions of aldehydes. *Environmental Science & Technology, 36*, 2185–2192.

Morrison, G. C., Nazaroff, W. W., Cano-Ruiz, J. A., Hodgson, A. T., & Modera, M. P. (1998). Indoor air quality impacts of ventilation ducts: Ozone removal and emissions of volatile organic compounds. *Journal of the Air and Waste Management Association, 48*, 941–952.

Nazaroff, W. W., Cass, G., (1986). Mathematical modeling of chemically reactive pollutants in indoor air. *Environmental Science & Technology, 20(9)*, 924–934.

Nazaroff, W. W., Gadgil, A. J., & Weschler, C. J. (1993). Critique of the use of deposition velocity in modeling indoor air quality. In N. L. Nagda (Ed.), *Modeling of indoor air quality and exposure* (pp. 81–104). Philadelphia, PA: American Society for Testing and Materials. *ASTM STP* 1205.

Nazaroff, W. W., & Weschler, C. J. (2004). Cleaning products and air fresheners: Exposure to primary and secondary air pollutants. *Atmospheric Environment, 38*, 2841–2865.

Noziere, B., Barnes, I., & Becker, K. H. (1999). Product study and mechanisms of the reactions of alpha-pinene and of pinonaldehyde with OH radicals. *Journal of Geophysical Research-Atmospheres, 104*, 23645–23656.

Olson, D. A., & Corsi, R. L. (2004). In-home formation and emissions of trihalomethanes: The role of residential dishwashers. *The Journal of Exposure Analysis and Environmental Epidemiology, 14*, 109–119.

Ongwandee, M. (2006). *Adsorption of basic compounds on indoor surfaces: influence of pH, CO2 and NH3*. Ph.D. thesis, Department of Civil, Architectural and Environmental Engineering, University of Missouri-Rolla, Rolla, MO.

Ongwandee, M., Bettinger, S. S., & Morrison, G. C. (2005). The influence of ammonia and carbon dioxide on the sorption of a basic organic pollutant to a mineral surface. *Indoor Air, 15*, 408–419.

Pommer, L. (2003). Oxidation of terpenes in indoor environments: A study of influencing factors. Ph.D. thesis, Department of Chemistry, Umeå University, Umeå, Sweden.

Reiss, R., Ryan, P., Koutrakis, P., & Tibbetts, S. (1995a). Ozone reactive chemistry on interior latex paint. *Environmental Science & Technology, 29*, 1906–1912.

Reiss, R., Ryan, P. B., Tibbetts, S., & Koutrakis, P. (1995b). Measurement of organic acids, aldehydes, and ketones in residential environments and their relation to ozone. *Journal of Air & Waste Management Association, 45*, 811–822.

Rohr, A. C., Weschler, C. J., Koutrakis, P., & Spengler, J. D. (2003). Generation and quantification of ultrafine particles through terpene/ozone reaction in a chamber setting. *Aerosol Science and Technology, 37*, 65–78.

Rohr, A. C., Wilkins, C. K., Clausen, P. A., Hammer, M., Nielsen, G. D., Wolkoff, P., et al. (2002). Upper airway and pulmonary effects of oxidation products of (+)-alpha-pinene, d-limonene, and isoprene in BALB/c mice. *Inhalation Toxicology, 14*, 663–684.

Sabersky, R. H., Sinema, D. A., & Shair, F. A. (1973). Concentrations, decay rates and removal of ozone and their relation to establishing clean indoor air. *Environmental Science & Technology, 7*, 347–353.

Salthammer, T., Fuhrmann, F., & Uhde, E. (2003). Flame retardants in the indoor environment – Part II: Release of VOCs (triethylphosphate and halogenated degradation products) from polyurethane. *Indoor Air, 13*, 49–52.

Salthammer, T., Schwarz, A., & Fuhrmann, F. (1999). Emission of reactive compounds and secondary products from wood-based furniture coatings. *Atmospheric Environment, 33*, 75–84.

Sarwar, G., Corsi, R. L., Kimura, Y., Allen, D. E., & Weschler, C. J. (2002). Hydroxyl radicals in indoor environments. *Atmospheric Environment, 36*, 3973–3988.

Sarwar, G., Olson, D. A., Corsi, R. L., & Weschler, C. J. (2004). Indoor fine particles: The role of terpene emissions from consumer products. *Journal of the Air & Waste Management Association, 54*, 367–377.

Simmons, A., & Colbeck, I. (1990). Resistance of various building materials to ozone deposition. *Environmental Technology, 11*, 973–978.

Singer, B. C., Coleman, B. K., Destaillats, H., Hodgson, A. T., Lunden, M. M., Weschler, C. J., et al. (2006). Indoor secondary pollutants from cleaning product and air freshener use in the presence of ozone. *Atmospheric Environment, 40*, 6696–6710.

Tamás, G., Weschler, C. J., Toftum, J., & Fanger, P. O. (2006). Influence of ozone-limonene reactions on perceived air quality. *Indoor Air, 16*, 168–179.

Tichenor, B. A., Guo, Z., Dunn, J. E., Sparks, L. E., & Mason, M. A. (1991). The interaction of vapour phase organic compounds with indoor sinks. *Indoor Air, 1*, 23–35.

Uhde, E., & Salthammer, T. (2006). Impact of reaction products from building materials and furnishings on indoor air quality-a review of recent advances in indoor chemistry. *Atmospheric Environment, 41(15)*, 3111–3128.

USEPA. (2004). Integrative synthesis. in Air quality criteria for particulate matter. *US Environmental Protection Agency, II*, 9.1–9.121.

Van Loy, M., Nazaroff, W. W., & Daisey, J. M. (1998). Nicotine as a marker for environmental tobacco smoke: Implications of sorption on indoor surface materials. *Journal of Air & Waste Management Association, 48*, 959–968.

Wainman, T., Zhang, J., Weschler, C. J., & Lioy, P. J. (2000). Ozone and limonene in indoor air: a source of submicron particle exposure. *Environmental Health Perspectives, 108*, 1139–1145.

Wang, B., & Lee, S. C. (2005). *Carbonyl levels in indoor and outdoor air at four cities, China.* The 10th International Conference on Indoor Air Quality and Climate, INDOOR AIR '05, Tsingua University Press, Beijing, China.

Wang, H., & Morrison, G. C. (2006). Ozone initiated secondary emission rates of aldehydes from indoor surfaces in four homes. *Environmental Science & Technology, 40*, 5263–5268.

Wang, H., Springs, M., & Morrison, G. C. (2005). *Ozone initiated secondary emissions of aldehydes from indoor surfaces.* Air and Waste Management Association 2005 Meeting, Minneapolis, MN.

Wargocki, P., Wyon, D. P., & Fanger, P. O. (2003). Call-centre operator performance with new and used filters at two different outdoor air supply rates. *Healthy Buildings 2003*, Singapore.

Warscheid, B., & Hoffmann, T. (2001). On-line measurements of alpha-pinene ozonolysis products using an atmospheric pressure chemical ionisation ion-trap mass spectrometer. *Atmospheric Environment, 35*, 2927–2940.

Webb, A. M., Singer, B. C., & Nazaroff, W. W. (2002). *Effects of gaseous ammonia on nicotine sorption.* The 9th International Conference on Indoor Air Quality and Climate, INDOOR AIR '02, Monterey, CA.

Wells, J. R. (2005). Gas-phase chemistry of alpha-terpineol with ozone and OH radical: rate constants and products. *Environmental Science & Technology, 39*, 6937–6943.

Weschler, C. J. (2004). Chemical reactions among indoor pollutants: what we've learned in the new millennium. *Indoor Air, 14*, 184–194.

Weschler, C. J., Hodgson, A. T., & Wooley, J. D. (1992). Indoor chemistry: ozone volatile organic compounds, and carpets. *Environmental Science & Technology, 26*, 2371–2377.

Weschler, C. J., Sheilds, H. C., & Naik, D. V. (1994). Ozone-removal efficiencies of activated carbon filters after more than three years of continuous service. *ASHRAE Transactions, 100*, 1121–1129.

Weschler, C. J., Shields, H., & Naik, D. V. (1989). Indoor ozone exposures. *JAPCA: Journal of the Air and Waste Management Association, 39*, 1562–1568.

Weschler, C. J., & Shields, H. C. (1996). Production of the hydroxyl radical in indoor air. *Environmental Science & Technology, 30*, 3250–3258.

Weschler, C. J., & Shields, H. C. (1997a). Measurements of the hydroxyl radical in a manipulated but realistic indoor environment. *Environmental Science & Technology, 31*, 3719–3722.

Weschler, C. J., & Shields, H. C. (1997b). Potential reactions among indoor pollutants. *Atmospheric Environment, 31*, 3487–3495.

Weschler, C. J., & Shields, H. C. (1999). Indoor ozone/terpene reactions as a source of indoor particles. *Atmospheric Environment, 33*, 2301–2312.

Weschler, C. J., & Shields, H. C. (2000). The influence of ventilation on reactions among indoor pollutants: Modeling and experimental observations. *Indoor Air, 10*, 92–100.

Weschler, C. J., & Shields, H. C. (2003). Experiments probing the influence of air exchange rates on secondary organic aerosols derived from indoor air chemistry. *Atmospheric Environment, 37*, 5621–5631.

Weschler, C. J., Shields, H. C., & Naik, D. V. (1994). Indoor chemistry involving O_3, NO, and NO_2 as evidenced by 14 months of measurements at a site in Southern California. *Environmental Science & Technology, 28*, 2120–2132.

Wilkins, C. K., Wolkoff, P., Clausen, P. A., Hammer, M., & Nielsen, G. D. (2003). Upper airway irritation of terpene/ozone oxidation products (TOPS). Dependence on reaction time, relative humidity and initial ozone concentration. *Toxicology Letters, 143*, 109–114.

Wisthaler, A., Tamas, G., Wyon, D. P., Strom-Tejsen, P., Space, D., Beauchamp, J., et al. (2005). Products of ozone-initiated chemistry in a simulated aircraft environment. *Environmental Science & Technology, 39*, 4823–4832.

Wolkoff, P., Clausen, P. A., Wilkins, C. K., Hougaard, K. S., & Nielsen, G. D. (1999). Formation of strong airway irritants in a model mixture of (+)-alpha-pinene/ozone. *Atmospheric Environment, 33*, 693–698.

Wolkoff, P., Clausen, P. A., Wilkins, C. K., & Nielsen, G. (2000). Formation of strong airway irritants in terpene/ozone mixtures. *Indoor Air, 10*, 82–91.

Won, D., Sander, D. M., Shaw, C. Y., & Corsi, R. L. (2001). Validation of the surface sink model for sorptive interactions between VOCs and indoor materials. *Atmospheric Environment, 35*, 4479–4488.

Wyon, D. P., Tham, K. W., Croxford, B., Young, A., & Oreszczyn, T. (2000). The effects on health and self-estimated productivity of two experimental interventions which reduced airborne dust levels in office premises. *Healthy Buildings 2000*, Espoo, Finland.

Zhao, P., Siegel, J. A., & Corsi, R. L. (2006). Ozone removal by HVAC filters. *Atmospheric Environment, 41(15)*, 3151–3160.

Zheng, C., Zhang, G. Q., Hao, J. H., Lin, J. P., & Liu, Y. H. (2005). *Risk assessment of exposure to benzene, toluene, and formaldehyde in residential environment in China*. The 10th International Conference on Indoor Air Quality and Climate, INDOOR AIR '05, Tsingua University Press, Beijing, China.

Chapter 4
Personal Exposure Measurements

Martin Braniš

Abstract Human exposure to air pollutants is ubiquitous. Once a pollutant has been discharged into or has been formed in the air, exposure to this pollutant can hardly be avoided as people have to breathe continuously. Because people move, commute, and frequently change their positions, they can be exposed daily to various kinds and mixtures of gases and airborne particles. In addition to the diversity of the environments where exposure may occur, the many different activities and the potential number of chemicals present all pose a challenge in investigating the health risks posed by air pollutants. Not only do the daily activities and social behaviour of modern humans vary substantially. Air chemistry (species, ions, elements, mixtures), physics (temperature, pressure, radiation) and biology (fungal spores, viruses, bacteria, mites) all change in space and time as well. The air also changes dynamically in connection with differences in the meteorological, microclimatic, and other environmental characteristics.

4.1 Introduction

Assessment of human exposure to air pollutants is not an easy matter. Even though most chemical species, basic chemical reactions and the behaviour of pollutants in the air have already been described in detail and numerous studies about the effect of poor air quality on human health have been published, a number of questions remain unanswered and some relationships between air pollution and human health are unclear and/or uncertain.

M. Braniš (✉)
Faculty of Science, Institute for Environmental Studies, Charles University in Prague, Albertov 6, 128 43 Prague 2, Czech Republic
e-mail: branis@natur.cuni.cz

M. Lazaridis and I. Colbeck (eds.), *Human Exposure to Pollutants via Dermal Absorption and Inhalation*, Environmental Pollution 17, DOI 10.1007/978-90-481-8663-1_4,

Even though understanding of the association between poor air quality and adverse health effects dates back to the times of severe London smogs (Logan 1953; Bell and Davis 2001), until the 1980s epidemiologists possessed few accurate data on the actual exposure of the population to air pollutants. Almost nothing or very little was known about the variations in the exposure to a given pollutant from one individual to another or about the causes of these variations.

Pioneering observations performed at the end of the 1970s and at the beginning of the 1980s showed that personal exposures to pollutants during various types of activities or in various types of environments (at home, at work, in means of transport, during shopping, etc.) were frequently not closely correlated with measurements provided by fixed-site ambient monitors (FSM) of official networks (Sexton et al. 1984; Wilson et al. 2000; Monn 2001; Jantunen et al. 2002). Most of the studies focussing on comparison between personal and ambient concentrations (e.g., of particulate matter) showed that the ambient (outdoor) concentrations obtained from fixed-site monitors correlate poorly with total personal exposure to particulate matter, but are highly correlated with personal exposure to particles that have their origin outdoors regardless of whether they were inhaled indoors or outdoors (Sexton et al. 1984; Wallace 2000; Boudet et al. 2001; Adgate et al. 2002; Meng et al. 2005b). The ability to correlate ambient gases with personal exposure to gases, such as O_3, NO_2, SO_2 is, however, quite small (Sarnat et al. 2006).

Paradoxically, to date, the vast majority of exposure assessment and epidemiological studies have relied only on ambient air quality data obtained at a single or a few stationary air monitoring site(s). Typically, a population residing in a defined area (city, region, state) was treated as a set of individuals – receptors – homogenously distributed in the vicinity of a fixed monitoring station or network of stations. A growing number of studies experimenting with small personal monitors in the last 15–20 years have shown that this approach is inadequate.

Because people in industrialized countries spend a major part of their time indoors, personal exposure is generally predominated by indoor levels, which contain a mixture of indoor and outdoor pollutants. In many cases, exposure to air pollution in an indoor environment may differ considerably from ambient pollution, because of various types of indoor sources of pollution and the different atmospheric chemistry of the two microenvironments, various types of ventilation, etc.

The ongoing debate about the suitability of ambient air quality data as a surrogate for personal exposure data cannot detract from a number of studies documenting that there is a association between ambient (particulate) air quality and human health (the authors can recommend a recent review by Pope and Dockery (2006), where other references are given).

The subject of the similarity between ambient and individual levels of pollutants remains important for scientific as well as for regulatory purposes. Emission controls placed on major stationary combustion sources in developed industrial countries during the last 20–30 years have significantly reduced the concentrations of "principal pollutants", namely sulfur dioxide (SO_2), nitrogen dioxide (NO_2), carbon monoxide (CO), and suspended particulate matter. Regulation of the emission of excess or toxic pollutants in working environments began even earlier. Because of these developments, more efforts are being exerted towards a detailed description of individual or subpopulation exposure characteristics. Attention has recently been focussed on populations of sensitive individuals, for example, on people with asthma, chronic obstructive pulmonary disease, heart disease, the elderly, children etc. (Janssen et al. 2000; Wheeler et al. 2000; Williams et al. 2000; Rojas-Bracho et al. 2000; Sarnat et al. 2000; Gauvin et al. 2002; Harrison et al. 2002; Liu et al. 2003; Adgate et al. 2004; Wu et al. 2005b and many others).

4.2 Concepts and Definitions of Personal Exposure

4.2.1 Basic Risk Model

Human risk or exposure models (e.g. Ott 1985) are usually described as a sequence of events comprising five distinct components:

1. *Emission* – description of a source and specification of a pollutant
2. *Transport and transformation* – physical and chemical processes in the atmosphere between the source and the target
3. *Exposure* – the contact of the pollutant with a target
4. *Dose* – the amount of the pollutant received by a particular individual
5. *Effect* – health outcome – resulting from the dose (or if the dose is not known, it is inferred from the exposure)

Each component of this series constitutes the input of the following one, and thus the absence of information on any component in the sequence (1) compromises our ability to make an accurate assessment of human health risk and (2) has serious adverse implications for regulatory policies designed to protect public health. Of the five components, the three for which the least information is available for most pollutants are "transport and transformation", "exposure" and "dose". Technically and theoretically, each component can be studied independently of the others. This chapter will deal only with the third item – exposure – though this does not mean that it is the most important aspect. However, none of the above-mentioned components exists independently of the others nor are they all together sufficient for resolving the problem of the impacts on human health. The whole sequence should be extended to include another item:

6. *Policies* – regulations designed and measures taken to minimize emissions, prevent exposure, or mitigate/treat health impacts (if the effect is considered unacceptable)

Incorporation of the "policy" component transforms the linearity of the 'original' conceptual risk model into a cycle Fig. 4.1.

The content of the policy paragraph may not be important from a scientific point of view, but it is crucial to understand that, in reality, only this component can

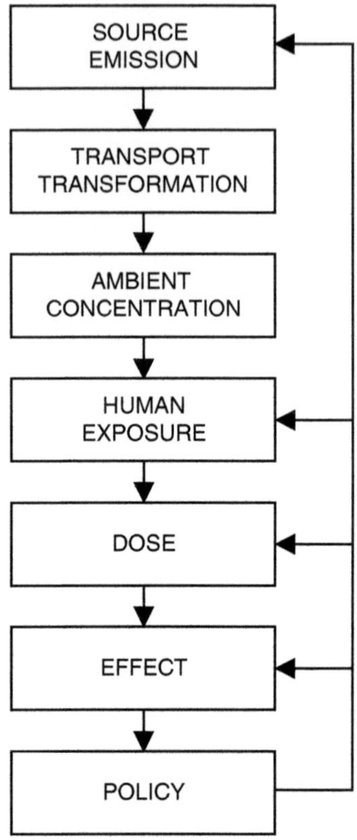

Fig. 4.1 Basic model of source-to-effect-policy sequence of events in the human exposure concept

quantitatively or qualitatively change the behaviour of sources (by means of financial or legal constraints), the characteristics of exposure (e.g. by means of prescribed use of protective instruments), the dose level or the severity of the effect (by application of medical treatment). It should be taken into account that, without limiting the source of pollution, policy can reliably influence neither the transport and transformation processes nor the ambient concentration of a pollutant (pollutants) of concern.

4.2.2 *Exposure*

Personal exposure can be defined as measurement of a pollutant of concern performed by a monitor (or sampler) worn by a person while the sample is taken from a point near the breathing zone of the person (but not affected by her/his exhaled breath). Personal exposure naturally includes pollutants of concern from ambient air while the person is outdoors, those from indoor air while the person is indoors and pollutants generated by personal activities.

In reality, the situation is not as simple as suggested above. If, for example, we intend to measure and subsequently assess exposure to particulate matter (PM), we must consider that total exposure (E_t) consists of exposure to ambient (outdoor) airborne pollutants (E_{out}), indoor airborne pollutants (E_{in}) and pollutants that are generated by the person himself (E_{pers}). The indoor exposure (E_{in}) can be further divided into exposure to outdoor-generated PM (E_{og}) and to indoor-generated PM (E_{ig}) and to PM generated by the person him/herself (E_{pers}). While the exposure to outdoor PM can be readily separated and studied (virtually all outdoor PM is generated or formed outdoors), this is not true for exposure to indoor particles, mainly because of the various sources that contribute to it's levels ($E_{in} = E_{ig} + E_{og} + E_{pers}$). It is also true that E_{pers} can be a complicated variable. In its narrow sense, it comprises aerosol generated by friction of clothing or the body surface (skin flakes and hair fragments) and coughing. In a broader sense, it can be caused by resuspension of floor dust or can be created by various types of activities, such as hobbies, smoking, etc. The E_{pers} "sensu stricto" is also termed a *personal cloud.* It is given by the difference between the readings of a personal monitor (placed in the breathing zone of the person) and the stationary area (microenvironmental) monitor. (For a detailed theoretical and semantic discussion of this issue, the reader is referred to Wilson et al. 2000).

Before the personal measurement of exposure to air pollutants will be discussed, it is necessary to describe some basic concepts related to the term "exposure". In the beginning of the 1980s, Ott (1982) and Duan (1982) elaborated a system of definitions for human exposure and stated that "exposure is an event that occurs when a person comes into contact with a pollutant". In other words, for personal exposure to take place, two events have to occur simultaneously: the pollution concentration at a particular place and time must be above zero and the person must be present in that same place and at that time. The meaning of the definition does not necessarily state that exposure to an air pollutant results in a dose. Theoretically, when a pollutant does not cross the organismal boundary (in inhalation exposure, the cell membranes in the lungs) we can speak about non-effective exposure. There can be exposure without a dose but there cannot be a dose without exposure (Ott 1985).

Under normal conditions, when a pollutant is dispersed or suspended in the air in the presence of freely breathing individuals, exposure to this pollutant is inevitable. The above mentioned elementary definition describes an instantaneous contact between a pollutant and a single person (or group of persons). It simply means that, at a certain instant, a person is exposed to for example 100 $\mu g\ m^{-3}$ (micrograms per cubic meter) of SO_2. It should be mentioned, however, that no studied pollutant exists in the air as an independent compound, but is always mixed with other (gaseous or particulate) compounds. A breathing person therefore does not inhale the single pollutant of concern (measured) but a complicated mixture, of which many components are not or cannot be registered.

Instantaneous, that is very short, contacts are very improbable (or are most often not important for risk assessment). Usually people spend a certain – measurable – portion of time in a polluted environment. The exposure "E" of a person "i" can therefore be mathematically defined as (Eq. 4.1):

$$E_i = \int_0^{t_1} c(t)dt \tag{4.1}$$

Where *E* is the exposure, *c(t)* is the concentration of the pollutant of concern, which varies as a function of time, and *dt* is the increment of time from 0 to t_1. Including time in the calculation results in the "integral exposure". The integral exposure is expressed in μg m^{-3}.h (or ppm.h). For practical reasons, the average or "time-weighted average exposure" is calculated by dividing the integral by the time of the contact between the person and the pollutant concentration (Eq. 4.2):

$$\overline{E}_i = \frac{\int_0^{t_1} c(t)dt}{t_1} \left(\text{in } \mu \text{ g.m}^{-3} \text{ or ppm or ppb}\right) \tag{4.2}$$

The time-weighted average exposure is almost never calculated by the complicated formula given above but is frequently the output of a much simpler and practical averaging of the pollutant concentration obtained from the measuring systems over a certain period of time (hour, day, week, etc.), for which a number of data-points are available. Instruments used for monitoring the air quality and analytical methods used to ascertain pollutant concentrations and compositions have inherent limitations which, in many instances, make it impossible to obtain instantaneous results or continuous readings. Thus, we usually use the term "average concentration" or "average exposure" even for a single value obtained from one measurement of longer duration. For example, a filter through which air is drawn for 24 h offers a "daily average concentration" of particulate matter, a passive sampler exposed for a week in a kitchen offers a "7-day average" for NO_2, etc.

According to Sexton and Ryan (1988) there are three important aspects connected with estimation of exposure of humans to air pollutants:

1. Magnitude (pollutant concentration)
2. Duration (time during which an individual is in contact with the pollutant)
3. Frequency (number of exposures – how often the exposure to a certain pollutant occurs during the period of interest)

For some compounds, health effects may be observed only when certain threshold values are exceeded. For others, health effects may be related to long time exposure due to accumulation and sometimes it is important that a person or group of persons be exposed repeatedly. Measurement of the total exposure may therefore not always be the ideal way of predicting adverse health outcomes and uncertainties exist as to the importance of the factors described above.

Exposure assessment based on averaged or integrated measurements does not fully reflect individual short peaks to which a person or community of interest can be exposed in specific types of environment during the day. For example, when indoor personal measurements are averaged over a time period longer than 1 h, the output data cannot register important short events, such as flushing a toilet, frying in the kitchen, vacuuming and other cleaning activities, etc. In integrated measurements, the personal exposure profile is smoothed out and manifestation of extremes is suppressed (Quintana et al. 2001).

There are basically three general approaches/concepts for determining the exposure: Direct, indirect and biological monitoring. (Lioy 1990; Duan 1982; Tardiff and Goldstein 1991; Ott 1985).

4.3 Direct Approach

This method, which should be considered to be "true personal exposure measurement", seeks to measure a person's exposure by measuring the concentration of a given pollutant (or several pollutants simultaneously) in the air breathed for a defined time using a device that can be worn by the selected person. Put simply, the idea of this concept is to "directly measure" the concentration of the pollutants of concern in the closest possible vicinity of the person of concern.

In many instances, personal measurement does not mean that a single person is measured and the results of the measurement are generalized. As almost all of the measurements have a clear purpose to assess health risks from exposure to pollutants in a community, it is necessary to monitor a relatively large number of individuals. It is, however, almost impossible to supply measuring devices to all the members of the studied population. Therefore, it is necessary to select a representative subpopulation or "sample" population (or panel) representing the whole community for which the exposure is to be determined.

4.3.1 Selection of a Sample Population

The easiest (and originally a frequent) way of selecting a sample population is to ask volunteers to take part in the experiment (Dockery and Spengler 1981; Sexton et al. 1984; Lioy et al. 1990). This empirical way usually involves students, colleagues, family members, friends or other people who are somehow interested in the experiment or motivated to take part in it. The advantage of this method is that training of these individuals to handle measuring devices is relatively easy and the number of rejected samples or days when samples could not be taken is low. This method of selection is suitable for pilot sampling campaigns but is far from being a correct statistical method of selecting a sample population.

The most suitable method of selection is to find individuals according to the well-established procedure of probability selection/sampling. This selection yields a random sample, which is highly representative of the entire population. It is also possible to intentionally select more individuals from the group that usually has higher contact with the pollutant. This method of selection is also called a "stratified random sample" (Tardiff and Goldstein 1991). A purposefully selected population can also be recruited from the subpopulation of vulnerable or highly exposed individuals alone. In this case, the researchers seek out, for example, older people, children with asthma, specific professional workers (policemen, postmen, bus, taxi or lorry drivers, etc.) with anticipated high exposure levels (Watts et al. 1994; Zagury et al. 2000; Perico et al. 2001; Hansen et al. 2004; Jones et al. 2006).

Personal measurement based on a correctly selected population of individuals who represent the community to be assessed consists in a combination of statistical survey techniques used for many years to determine public opinion and state-of-the-art technologies for air-quality measurement and analysis. In many panel or community studies, the main problem does not lie in selection of a suitable sampler/monitor but in obtaining the necessary number of individuals who randomly represent the community of concern. As was recently shown, a selection bias is likely to occur when recruiting participants, not only for studies including demanding personal exposure assessment, but also for detailed time-activity-pattern surveys (Oglesby et al. 2000b).

However, even though respondents may finally agree to participate in a field study, they might not be able to do so effectively for any number of reasons, such as unclear or inadequate instructions, inability to follow instructions, lack of incentive or desire to participate beyond a certain point, lack of clarity or understanding of specific questions, lack of information required to respond to specific questions. Some of these reasons relate to poor design of the study, while other reasons may relate to the survey designer's failure to understand the social and cultural habits of community members or human behaviour. How the participants are approached and asked to participate is very important. What is said and who instructs the individuals to be monitored both significantly influence the outcome (Tardiff and Goldstein 1991).

4.3.2 Personal Samplers

As there are a number of chemical compounds and mixtures recognized as 'pollutants', correspondingly, there are a number of methods and devices to measure them.

Several technical terms referring to a device that can quantify the presence of a selected compound in the air that can be used in exposure calculations can be found in literature. Often the terms are used as synonyms. These devices are most frequently described as *monitors*. In a narrow sense, a monitor is a device that provides (mainly numerical) information about the pollutant concentration but does not necessarily provide, as an output, a material sample that can be subsequently analyzed, weighed, or measured. Similarly, the terms *sensor* and *detector* are used for devices yielding only a non-material numerical or qualitative (colour) or semi-quantitative (scale) output. The term sensor is usually used for equipment measuring the temperature, humidity or gases. A *sampler* is generally understood to be a device providing the researcher with a sample that can be processed using various techniques, such as chemical analysis, gravimetrical analysis (weighing), crystallography, reflectance, etc.

It should be born in mind that personal monitors and samplers must be small, easily transportable devices. Their dimensions and independence of power networks compromise their ability to monitor low concentrations and/or collect large amounts of pollutants for subsequent analysis. A single personal monitor is very rarely able to record several pollutants simultaneously. The diversity of environments where exposure may occur and the number of pollutants that may be important for human health pose a challenge in selecting an appropriate method for air pollution measurement. In most indoor and outdoor environments, human activities result in exposure to complex mixtures of gases, vapours and particles (e.g. ETS – environmental tobacco smoke, or oxidative smog compounds). Lack of information on the characteristics of these mixtures complicates investigation of the health effects.

Because of the limitations inherent in the dimensions and simplicity, personal samplers are not suitable for simultaneous monitoring of several or tens of pollutants that may appear in the same place at the same time in one type of environment. As an alternative to full characterization of mixtures, *indicators* (tracers, markers, proxies or surrogates) have been used to represent exposures to complex airborne pollutant mixtures. Such indicators may be elements, isotopes, chemical compounds, size-fractionated airborne particles, etc. Optimally, an indicator of exposure to a complex mixture should be unique to the mixture source, present in a more or less stable ratio to other (non-detected) constituents and easily detectable itself.

In some cases, instead of specific concentrations that can be expressed as amount per volume of air breathed, relative, semi-quantitative or integrated values enter into exposure assessment calculations (e.g. light absorbing carbon – black carbon detected by reflectance or a mixture of volatile organic compounds detected as a group of chemicals by gas chromatography). Unfortunately, markers, indicators, surrogates and proxies reflect neither toxicity of the whole mixture, nor synergistic interactions among the mixture components. Aspects related to assessment of exposures to inhaled complex mixtures have been in detail covered namely by Leaderer et al. (1993) and Scherer (2005).

An ideal personal monitor should fulfil a number of requirements. It should be:

- *Portable*: is small and light, to be worn without (substantially) modifying the normal behaviour of the individual monitored, namely his/her working and moving habits, should be independent – light and long-life battery operated – or even without a power supply (passive gas or aerosol samplers).
- *Selective*: has no (or negligible) response to similar compounds which might be present in the sampled air simultaneously with the pollutant of interest.

- *Sensitive*: detects pollutants at concentrations below those causing adverse health effects (the detection limit is approximately 0.1 times the level of interest and the range is between 0.1–10 times the level of interest, the precision and accuracy are maximum ±5%).
- *Fast responding*: has short sampling times and response time (order of tens of seconds or minutes) compared to the biological response of the target organism.
- *Inexpensive*: the cost of the sampling should not be prohibitive, consumables (sampling media, filters, fillings, tubing, accessories) commercially available, where possible not instrument-specific (i.e. versatile).
- *Flexible, resistant and user friendly*: can operate in a wide range of temperatures and relative humidities enabling samples to be taken in extreme indoor and/or outdoor conditions, is resistant to frequent handling, shaking or turning. An ideal active personal sampler is silent, is easy to operate, allows easy and accurate calibration and has sufficient datalogging capacity (Leaderer et al. 1993; Jantunen et al. 2002).

Small portable personal exposure monitors sensitive enough for measuring ambient concentrations are now available for a number of pollutants and even better ones will undoubtedly be available in the near future. Personal as well as microenvironmental (see below) monitoring and sampling devices can be classified according to a variety of aspects, for example according to the substance measured (gas, vapours, particles, bioaerosol), according to the manner the sampled air reaches the sensor or sampling medium – forced air flow versus diffusion (active, passive) and according to the length of the integration time (continuous, time-integrated). Most of the personal portable monitors available today are integrated samplers.

4.3.2.1 Personal Monitoring of Particulate Matter

There are many types of instruments used for sampling and monitoring aerosol, depending on the purpose of the study and the availability of post-sampling analysis. They are able to offer a variety of particulate matter characteristics, such as mass, number, size, surface, reflectance, etc. (For review se also Chow et al. 2002 or Jantunen et al. 2002). Most generally, the particulate sampling devices are divided into two types – *time integrated* and *continuous* monitors. Another division can be made according to the ability to operate with or without a pump – *active* and *passive* monitors.

Time-integrated gravimetrical samplers. Generally, time-integrated samplers collect pollutants over a specific period of time on a collecting medium (usually filter), which is returned after exposure to the laboratory for analysis. As they use a pump and power source to move the air through a collector, integrated samplers are active devices. Determination of the concentration of particulate matter in the air by small personal samplers is almost always based on a relatively small amount of particles collected on a sampling medium (most frequently a filter). Due to the small dimensions of samplers, the filters are also small (usually 2.5 or 3.7 mm in diameter) and the air flow is low (maximum of up to 10 l min^{-1} or less). Higher air flow rates result in greater noise of the pump which compromises not only the comfort of the person wearing the sampler but disturbs others in the close vicinity. Because of these constraints, the time of exposure should be long enough to collect the needed sample mass. The standard requirement in personal sampling is that a sufficient amount of the aerosol for subsequent weighing is collected during 12 h or a shorter sampling period. The low sample mass forms the primary boundary condition, not only for gravimetric, but also for elemental and chemical analyses of the aerosol collected.

Over the last 2 decades, measurements of total suspended particulates (TSP) have been replaced by PM_{10} and $PM_{2.5}$ (particles smaller than 10 and 2.5 μm in aerodynamic diameter respectively) measurement, not only in ambient air monitors but also for personal exposure monitoring. Currently, a choice of commercially available personal monitors is supplied with relatively quiet portable pumps with sufficient air flow. Most of the devices use sharp cut impactors or cyclones to separate particles that are larger than the required diameter (SKC PEM impactors; BGI cyclone; KTL GK 2.05 cyclones and others). Impactor samplers are small, of light weight and easy to operate, but may suffer overloading in highly polluted spaces. Cyclones are heavier and larger than impactors but tolerate a wide range of aerosol loading.

Recently, personal cascade impactors have been designed and field tested. These devices allow separation of particulate matter in several fractions - size ranges. The time-weighted mass-size concentration of aerosol is provided as output from following gravimetric analysis. Due to the low air flow and low amount of aerosol deposited on the impaction filters, cascade impactors must be exposed for a relatively long time, at least 24 h, under normal conditions (urban indoor/outdoor/transport pattern). Nevertheless, the aerosol load on individual filters is sufficient to enable chemical analysis of the size-fractionated particles. The original invention The Marple Personal Cascade Impactor (Sierra Andersen) can be used with four, six, or eight stages with cut points ranging from 0.5 to 21 μm aerodynamic diameter (Rubow et al. 1987). Operating air flow of 2 l min^{-1} and the related low material deposition on the individual stages may be the main limitation for detailed chemical analysis using standard analytical procedures. The more recent Personal Cascade Impactor Sampler (PCIS) consists of four stages followed by an after-filter. Particles are separated in five diameter ranges (<0.25, 0.25–0.5, 0.5–1.0, 1.0–2.5 and 2.5–10 μm) (Singh et al. 2003; Misra et al. 2002). In both types of cascade impactors, laboratory and/or field tests showed that the instruments have sharp cut-off characteristics and good agreement with reference methods.

Continuous aerosol monitors. In contrast to time-integrated samplers suffering from the inability to register the spatial and temporal characteristics of personal exposure, continuous monitors have very fast response times (seconds to tens of minutes). Their self-contained analytical system enables measurement directly at the location. Information about the aerosol concentration is available as a direct reading output; however, in modern commercial products, the concentration is usually also recorded in a built-in datalogger. Continuous monitors are mostly photometers (more exactly nephelometers).

The principle of operation of nephelometers is that air passes through a measurement chamber where the particles are illuminated by a light source and the resulting scattered light is detected under a certain detection angle (usually 90°). The resulting signal of the scattered light is transformed by a built in processor to mass-per-volume values (μg m^{-3}). Nephelometers are usually combined with a preclassifier (impactor or cyclone), which separates particles that are larger than the required diameter, yielding the output readings as PM_{10}, $PM_{2.5}$, or some other fraction, such as the "respirable" suspended particulate (RSP – corresponding to the fraction with a cut point of 4 μm in aerodynamic diameter).

An aerosol photometer measures a single parameter that is dependent on many variables, for example particle size distribution, particle agglomeration, or refractive index; these values can and do change under various conditions and in different environments. Field studies have shown that photometers systematically over-report the real (gravimetric) concentrations. As has been repeatedly shown, the bias of photometers is linearly proportional to the selected reference methods and the readings of the instrument can be recalculated according to a linear regression equation (Ramachandran et al. 2003; Wu et al. 2005a) or even by a simple ratio between the reference and the photometer readings (Jenkins et al. 2004). Nephelometers are calibrated in the factory against certain aerosol standards (e.g. ISO 12103-1, A1 test dust – Arizona road dust). Factory calibration is primarily useful to ensure that the instrument is operating properly and responds in the same as other instruments of this type. It does not ensure that the photometer will respond accurately to a different aerosol.

The difference between the reference and photometer readings may be source-, microenvironment- or season-specific according to temperature, humidity and especially particular aerosol characteristics. Because of theses problems, nephelometers should be calibrated against reference methods during collocated campaigns (Yanoski et al. 2002; Ramachandran et al. 2003; Jenkins et al. 2004; Wu et al. 2005b; Branis 2006). Due to the influence of the relative humidity on the light scattering response of photometers, statistical corrections or heated inlets are recommended in cases when nephelometers are used in microenvironments with high relative humidities (Sioutas et al. 2000; Lanki et al. 2002; Ramachandran et al. 2003; Wu et al. 2005a). Sometimes it is even recommended that readings from light scattering devices should be calibrated separately for different types of indoor and outdoor microenvironments, as well as for different seasons (Brauer 1995). As stated by Görner et al. (1995), nephelometers can be fairly simple and compact instruments with excellent sensitivity and time resolution; however, scattering per unit mass is a strong function of particle size and refractive index. According to our recent unpublished study (Branis and Kolomaznikova in m.s.), the average nephelometer/β-gauge ratio was strongly influenced by extreme levels of both temperature and relative humidity. As can be seen from Fig. 4.2, a ratio of between 2 and 3 (as found in most of the above mentioned studies) is obtained at temperatures above approximately 15°C and relative humidities below 80%. When the

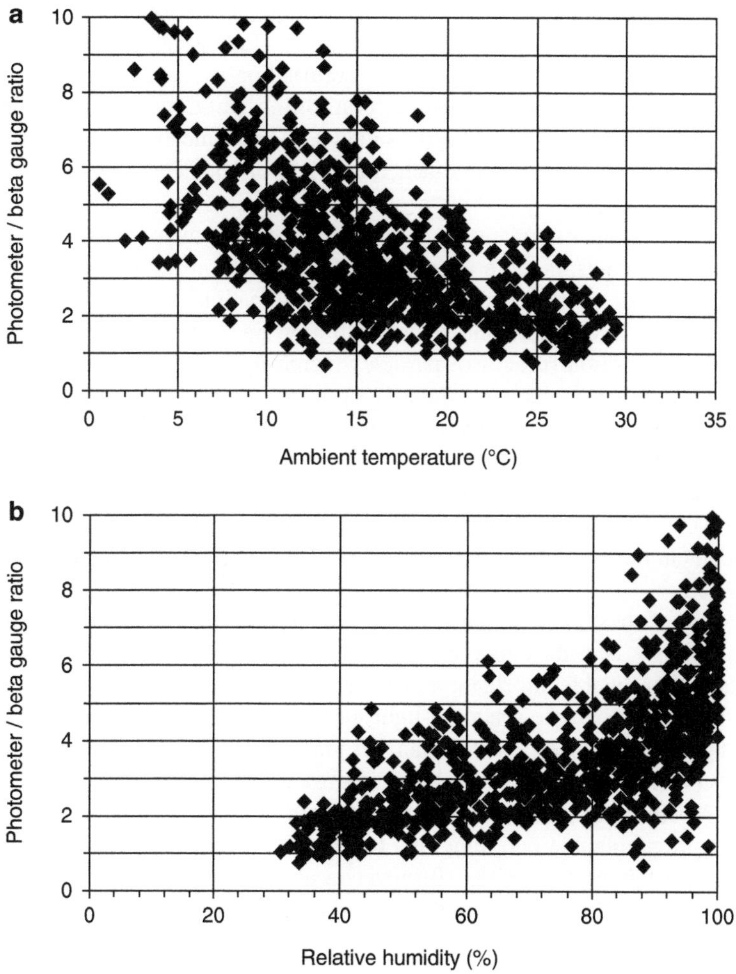

Fig. 4.2 Ratio between 1-h DustTrak[TSI] nephelometer and fixed site monitor (beta attenuation method) $PM_{2.5}$ concentrations plotted against the ambient temperature (**a**) and relative humidity (**b**) each data point represents a 1-h average. The measurements were performed in September–October period in Prague, Czech Republic (temperate zone)

temperature drops and/or the relative humidity increases, the ratio steeply increases to values over five or more.

Despite these shortcomings, the greatest advantage is that photometers can provide very good relative measurements of rapid changes in aerosol concentrations in relation to the source or changes in human behaviour and that they are battery-operated instruments allowing continuous personal monitoring over more than 24 h.

A number of photometers suitable for personal monitoring are now commercially available. Others are being developed or are in the prototype stage. Of the many light-scattering devices, the personal DataRAM (Thermo Electron), DustTrak and Sidepack (TSI), and the Radiance nephelometer (Radiance Research) are the most widely used (Howard-Reed et al. 2000; Liu et al. 2002; Morawska et al. 2003; Chakrabarti et al. 2004; Branis 2006). For the working environment, Haz-Dust, SPLIT2 (SKC) monitors (combining photometric principle with backup filter) and Microdust Pro (Casella CEL) instruments are frequently used. These instruments can be worn attached to a belt or shoulder strap or placed in a custom-made backpack.

For both dust measurement in the workplace environment and for community epidemiology studies, the importance of the total dust impact has been decreasing since approximately 1980. At the same time, occupational medicine is taking the number-, size-, fractionized mass- concentration and surface characteristics and chemical composition of aerosol particles increasingly into account, as limits in mass per volume (µg m^{-3}) are not the only significant values for the health hazards of the aerosol. The potential impact on human health depends on the composition of the dust, the size distribution and on the surface bound substances. Small particles contribute very little to the total mass but, due to their large number, they increase the inhaled particulate matter surface area dramatically. In addition, they augment the health hazard because they penetrate deep into the human respiratory system.

Because of these requirements, more sophisticated measurements – using real spectrometers – became important in personal exposure assessment. These spectrometers offer reliable information on the size distribution and number of particles using the "single-particle scattered-light method". For total particle counts in the size range of 0.01 µm to greater than 1.0 µm (concentration range of 0–100,000 particles per 1 cm^3), the TSI portable Condensation Particle Counter (CPC) Model 3007 can be used in personal monitoring. It can operate on alkaline batteries and includes programmable data logging. The lack of size discrimination is a major limitation of this portable CPC (Brouwer et al. 2004). For size-resolved particle counts, portable, battery-powered (or main power supply 110/220 operating) GRIMM 1.10x monitors can be used (e.g. Grimm 1.105 Workcheck 5 and higher). These instruments measure the individual particle concentration and size using of light scattering. The measured sample is also collected on a removable ø 47 mm Teflon filter for gravimetric or chemical analysis (dual technology). Brauer et al. (1999) also successfully used the APC-1000 hand-held continuous particle counter (Biotest Diagnostics) for assessment of personal exposure to size-resolved particle-number concentrations.

Most photometers have a sampling (battery operated) pump that draws the sampled air into an inlet and through a duct to the detector. These instruments are called active because an energy supply is required to take the samples. Some instruments rely on aerosol diffusion, on air convection or motion of the photometer to bring the sampled air with aerosol to the detector. This principle is called passive. Active personal monitors need strong batteries to power the pump drawing the air into the detection chamber or through the filter. Such devices need frequent recharging or battery replacement, usually once every 24 h. Passive devices, which can function with smaller batteries, are quieter, lighter and smaller than their active counterparts and therefore more suitable for personal monitoring. Of the commercially available instruments, DataRAM and SPLIT2 can operate in the active as well as in the passive mode. So far, only the DataRAM units were repeatedly and successfully used in the passive and active modes for research in epidemiological studies (Liu et al. 2002; Wu et al. 2005a, b; Wallace et al. 2003; Howard-Reed et al. 2000; Lanki et al. 2002). Limited experience has been gained with passive dust monitors working without any power supply. The principle of operation of these samplers is based on free deposition (by gravitation and diffusion) of particles on sticky transparent foils and subsequent light extinction and microscopic analysis (Vinzents 1996).

Few years ago, a new type of personal continuous aerosol mass detecting monitor has been developed by Rupprecht and Patashnick Co. (the Personal Dust Monitor – PDM model 3600). This instrument is a portable battery-operated miniaturized TEOM (tapered element oscillating microbalance system). Because the instrument measures the true particulate mass collected on its filter, its results do not exhibit systematic errors like photometers when monitoring various types of aerosol. So far, this instrument has not been widely used in research but rather has been tested in work places with high concentrations of aerosol (coal mines).

Bioaerosol samplers. Bioaerosol is a complex mixture that may include viable or dead microorganisms (bacteria, viruses), mould spores, pollen and fragments of their bodies or products of their metabolism (mycotoxins, endotoxins). An inflammatory, allergenic or toxic response can be caused by both viable and non-viable components of bioaerosol. Much of our understanding of the nature of human exposure to airborne microorganisms is derived from measurements performed regularly or randomly at a single site within a community and the majority of conventional bioaerosol monitors designed to collect particles of biological origin are not adaptable to personal monitoring. Efforts have been repeatedly exerted to overcome this problem.

Agranovski et al. (2002) invented a personal sampler for culturable microorganisms (bacteria and fungal spores). In the sampler, the air is bubbled through a porous medium submerged in a liquid layer. Particles including the viable microorganisms are collected in the wet medium. The unit can be attached to the monitored person's lapel by a pen-type clamp. The standard spread-plate technique is used for subsequent determination of counts of the collection fluid. A different type of sampler (microcentrifuge tube sampler) for collecting fungal spores was designed by Chen et al. (2004). The collecting method is based on the cyclone principle. The air is tangentially drawn in at the top of the microcentrifuge and extracted from the tube at its axis, creating a double vortex flow within the tube. Particles having enough inertia impact on the walls and gradually migrate to the bottom of the tube.

Recently personal sampling of allergenic substrates, such as pollen, mould spores, or mites and cat and dog allergens, has been tested and performed by a simple nasal bioaerosol sampler. This is a small moulded device that is sealed into the nostrils of the monitored person and collects particles by impacting them onto an internal collection plate covered by a sticky film of silicon grease.

The air is drawn onto the collecting substrate by the monitored person's breath alone. The collected spores are then counted and determined using light microscopy or by immunological assays (Aizenberg et al. 2000; Gore et al. 2002, 2006; Graham et al. 2000; Mitakakis et al. 2000). The Button Personal Inhalable Sampler was also repeatedly and successfully tested under field conditions for measurement of outdoor aeroallergens (pollen, fungal spores) in the range of 1–100 μm aerodynamic diameter (Adhikari et al. 2003).

It is clear that the above-mentioned samplers, techniques and models are only examples of a wide variety of personal samplers of particulate matter concentrations. Other commercially available or custom-made instruments exist and are being used in research and in practical applications.

The next generation of classical aerosol samplers will have to meet a number of requirements:

- The sampler should continuously record the aerosol mass/number and/or size-resolved particle mass or particle count instead of only an integrated time-weighted average mass.
- The sampler should be able to record other supporting variables, such as the temperature and relative humidity to allow direct linkage of exposures to general microenvironmental characteristics.
- Current developments in GPS (global positioning system) will require inclusion of miniaturized telemetry sensors in personal monitors as well.
- Pumps should be silent and programmable for scheduled operation, should control their flow rate at the set level and should record irregularities in the sampling process. An ideal pump should be light, small and able to run at least for 1 week.
- Reducing size and weight will necessarily result in reduction of sample size, which would impose other requirements on subsequent weighing and/or analytical techniques.
- The whole sampler should be designed with additional free datalogging channels for connection from other accessorial sensors, such as heart-rate monitors (surrogate for respiratory rate), light intensity, noise, etc.

4.3.2.2 Personal Monitors for Gases and Vapours

In contrast to aerosol sampling, with some exceptions (e.g. Geyh et al. 1999; Staimer et al. 2005), most of the methods employed for gas and vapour monitoring are based on the passive mode. Passive or diffusive sampling of hazardous substances in the air has been recognized as an efficient method for personal exposure assessment, particularly in occupational hygiene. In recent years, diffusive sampling has also been increasingly utilized in regional-scale ambient air-quality studies, mainly in the form of stationary microenvironmental sampling. Diffusive samplers offer an array of advantages: they are small, simple and convenient to use (and often re-use), inexpensive, well characterized and appropriate for a wide range of analytes. They are user-friendly, can be operated by inexperienced personnel and do not require electricity for operation. The simplicity of use and small dimensions make passive samplers suitable for personal exposure measurements (for review see, e.g. Namiesnik et al. 2005). It should be noted that, in the United States, the adjective "passive" is preferred in describing these dosimeters and should be regarded as synonymous with "diffusive". Similarly, the terms sampler, dosimeter and monitor should be also regarded as synonyms, because of confusion of the terms in the scientific and technical literature.

Passive gas samplers (or passive dosimeters) can be defined as devices that are capable of collecting samples of the pollutant of concern from the atmosphere without forced air flow. Passive sampling is therefore based only on molecular diffusion – free flow of the pollutant molecules to the collecting medium as a result of differences in chemical potential. Put simply, a passive dosimeter is a hollow tube of specified length and cross section constructed with one end sealed, containing a contaminant collector (sorbent), where the other end is fitted with a removable cap. Immediately before the use of the device, the cap is removed for a specified time period, while the sampler is exposed to the test environment. After the exposure, the cap is closed, the sampler taken in a lab and the sorbent is analyzed quantitatively for the contaminant. The time-weighted average exposure concentration can then be calculated.

Passive dosimeters can be used for determination of various airborne pollutants, including gases, vapours and their mixtures, or aerosols. One of the main advantages is that they also integrate sampling, pollutant isolation and preconcentration into a single step. Currently, passive samplers are being used to assess air pollution in various ambient microenvironments including urban and rural localities or remote vegetation canopy, workplaces and the indoor living environment. Personal workplace air monitoring is typically performed with sampling times from several tens of minutes to the whole shift (usually between six and 12 h) (Herber et al. 2001). Since personal and/or non-occupational microenvironmental concentrations of pollutants may be several rimes (or even orders of magnitude) lower than occupational ones, prolonged sampling times, at least 24 h but typically several days to 1–4 weeks, are often required.

The passive sampling method is based on one principle which almost all designed devices share. This is the presence of a barrier between the sampled and the collecting medium. The barrier defines the sampling rate, which is crucial for final quantitative analysis. The barrier should also eliminate or minimize the effect of external factors, such as air turbulence at the inlet (opening) of the sampler, temperature and humidity (Gorecki and Namiesnik 2002).

There are several conditions to be met or aspects to be considered for this approach to work:

- The receiving medium must act as a "zero sink", that is that it should not release the trapped molecules even if the concentration of the pollutant in the vicinity of the sampler drops to zero
- The amount of the pollutant collected by the sampler per unit time at constant concentration in the surrounding air must remain constant throughout the sampling period

- Linearity in the reaction response of the sampler to varying pollutant concentrations must be assured
- Turbulence of the air around the sampler must be prevented
- The absorbed or adsorbed pollutant should not degrade during sampling or during transport and/or storage of the sampler before analysis
- The sorbent medium must be specific in reaction to the pollutant of concern and should be resistant to interference from other compounds/pollutants in the atmosphere

Passive gas samplers can be divided according to many characteristics and the division given below does not claim to be the only and/or the best one.

- According to mass transfer – diffusion and permeation dosimeters
- According to the geometry – tube and badge type
- According to the principle of collection – absorption (chemical reaction) and adsorption (physical collection) dosimeters
- According to the medium sampled – gas, vapours, aerosol
- According to the information required – direct (alert) and indirect readings

In the following text, some of the above given types will be described in more detail.

Diffusion and permeation dosimeters. The difference between these two similar types of passive dosimeters lies in the presence or absence of a diffusional barrier in the form of a membrane.

In diffusional dosimeters the diffusional mass transfer of gas 1 through gas 2 is given by Fick's first law of diffusion (Eq. 4.3):

$$J' = -D_{12} dx / dz \tag{4.3}$$

where J is the flux of the gas (in mol cm^{-2} s^{-1}) D_{12} is the diffusion coefficient of gas 1 in gas 2 (in cm^{-2} s^{-1}), x is the concentration of the pollutant and z is the length of diffusion (in cm). When the concentration of the pollutant at the surface of the sorption medium approaches zero (the reaction between the absorbent and the pollutant is very fast) and the opening of the sampling tube is freely exposed to the surrounding air, then: when $z = L$, then $x = 0$, and when $z = 0$, then $x = c$ (L is the length of the tube – diffusion distance, c is the concentration of the pollutant in the atmosphere). Application of these conditions to Eq. (4.3) results in Eq. (4.4):

$$J = D.c / L \tag{4.4}$$

If the cross-sectional area of the diffusion tube is A (in cm^2), the quantity of the gas transferred and collected by the dosimeter can be expressed as (Eq. 4.5):

$$Q + J.A.t = D.c.t.A / L \tag{4.5}$$

The concentration of the pollutant *c*, can then be calculated as follows Eq. (4.6):

$$c = Q.L / D.A.t \quad (4.6)$$

In Eqs. (4.5) and (4.6), factors *A* (cross-sectional area) and *L* (diffusion length of the tube) are given by the geometry of the sampler, *D* (diffusion coefficient) is dependent on the diffusing gas of concern. Factors *Q* (quantity of the gas transferred) and *t* (time – duration of sampling) must be measured.

In permeation dosimeters, the diffusion barrier is represented by a non-porous membrane fixed at the face of the sampler. In other words, the inner diffusion space of the sampler is divided from the open atmosphere by a thin membrane that is semi-permeable for gases. In the permeation dosimeter, the flow of the analyte across the membrane can also be described by Fick's first law of diffusion as Eq. (4.7):

$$F = D.A.(P_2 - P_1) / S \quad (4.7)$$

where *F* is the permeation flow of the analyte across the membrane ($\mu g\ s^{-1}$), *D* is the diffusion coefficient of the analyte ($m^2\ s^{-1}$), *A* is the area of the membrane (m^2), *S* is the membrane thickness (m), and P_1 and P_2 are the partial pressures of the analyte on the opposite sides of the membrane (Pa). The amount of the analyte diffusing across the membrane and collected on the filter can be described by an empirical formula Eq. (4.8):

$$Q = c.A.t / k \quad (4.8)$$

where *Q* is the amount of the analyte collected on the filter (μg), *c* is the concentration of the analyte in the air ($\mu g\ m^{-3}$), *A* is the area of the membrane (m^2), *t* is the time – duration of sampling (s) and *k* is a constant termed the total resistance ($s\ m^{-1}$) or calibration constant. This constant is usually obtained experimentally by a span gas (gas with known concentration of the analyte) or during collocated exposure of the passive sampler and an active reference method.

Tubes and badges. Generally, a tube sampler is a dosimeter with long diffusion length and small sorption area, while a badge sampler is a dosimeter with short diffusion distance and large sorption area. In tube-type devices, the tubular part defines the diffusion barrier, with the length of the tube determining the diffusion path (*L*) and the diameter of the tube determining the *A* term (Eqs. (4.7) and (4.8)). The geometry of tube-type samplers is very important. If the diffusion barrier of the sampler is not efficient enough, the sampling rate can be so high that the analyte around the tube can be depleted, especially when the air in the vicinity of the sampler inlet is stagnant. In consequence, the sampling rate decreases and the accuracy of the measurement is reduced. This problem can be eliminated either by increasing the air movement around the passive sampler (condition well met in personal exposure measurements when the person moves) or by changing the geometry of the sampler. Generally the effect of the ambient air velocity disappears when the diameter/length ratio of the tube reaches 1:7 (Gorecki and Namiesnik 2002).

Tube-type samplers have relatively low sampling rates and it is necessary to expose the sampler for several days to collect a sufficient amount of analyte for accurate analysis. An alternative is to increase the surface area of the collection medium and to reduce the diffusive path length, both of which cause a corresponding increase in the sampling rate. An increase in the sensitivity of typically 100 is achieved by increasing the cross-sectional diameter by a factor of 3 and decreasing the path length by a factor of 10. The geometry of the sampler changes from a tube to a badge. Increasing the sampling rate still further is counter-productive, because boundary layer effects begin to predominate.

As mentioned above, "tube-type" devices are usually diffusive – low sensitivity samplers suitable for long-term monitoring, whereas "badge-type" samplers are usually permeation dosimeters with a thin semi-permeable membrane at the face of the sampler. The badge type has faster sampling rates more suitable for short-term monitoring.

Even though diffusive sampling in the occupational environment dates back at least to the 1930s when qualitative devices were described, one of the best-known tube-shaped passive samplers is the Palmes-tube (Palmes et al. 1976), designed to collect NO_2. The device consisted of a simple acrylic tube, about 8 cm long and 1 cm in diameter, with one closed end. The sampled gas enters the tube at the open end, migrates down the tube by molecular diffusion, and is collected at the closed end by triethanolamine coated on stainless steel grids. NO_2 is trapped as the nitrate ion and is measured spectrophotometrically after reduction to nitrite and reaction with sulphanilamide and naphthyl ethylene diamine. Numerous passive samplers based on the "Palmes-tube" model have been developed for ambient and microenvironmental air monitoring since its invention. In each case, the diffusion path length and/or cross-sectional area were modified to achieve the desired sampling rate. Of badge type monitors, the Willems badge (NO_2) or OGAWA (NO_2, SO_2, O_3) badge are well-known types for ambient air measurements, with tested modifications for personal exposure measurements (e.g. Hagenbjörk-Gustafsson et al. 2002; Liu et al. 1994).

Recently, a micro-tube-type diffusion sampler was designed and used commercially. The technique is based on the newly proposed SPME (Solid Phase Microextraction) analytical method. SPME is a solvent-free technology based on the use of a fused silica fibre coated with a liquid (polymer), a solid (sorbent), or a combination of both placed inside a thin hollow needle – microtube. The length of the diffusion barrier (and, consequently, the sampling rate) can be adjusted by retracting or protracting the fibre more or less inside the needle. The analyte diffuses accordingly along a well-defined diffusion barrier. The gas or vapours in the air are selectively collected by absorption (in the case of a liquid coating) or adsorption (in the case of a solid coating). After exposure, the SPME fibre is inserted directly into the Gas Chromatograph for thermal desorption and analysis. SPME has gained wide-spread acceptance as the technique of preference for many applications. For many volatile organic compounds, SPME can be an interesting alternative to conventional passive dosimeters because of the very small area of the microtube opening, which almost completely eliminates the effect of air turbulence on the sampling rate (Pawliszyn 1997; Martos and Pawlyszyn 1999; Ouyang and Pawliszyn 2006).

Absorption and adsorption samplers. Passive collection of a given analyte can be achieved by chemical absorption (chemical reaction between the analyte and the collecting medium) or by physical adsorption onto an inert medium.

The principle of operation of *absorption* dosimeters is based on a selective chemical reaction between the pollutant and the sorbent. After exposure, the reaction product collected on the substrate (usually a filter) coated by the sorbent must be extracted from the filter and properly analyzed by appropriate methods (colorimetry, spectrophotometry, gravimetry, conductometry, titrimetry, gas or liquid chromatography, ion chromatography, reflectance, etc.). Types requiring specific chemical analysis are also called dosimeters with *indirect reading*. Quantification of the analyte concentration in the air is relatively good, depending on the sampling conditions.

Absorption dosimeters can, however, provide useful information without requiring complicated chemical analysis. These types of dosimeters are called *direct reading samplers*. With increasing cumulative exposure, the collecting medium as a whole changes colour or an increasing amount of it becomes stained. Subjective or ordinal estimates of the cumulative exposure to a given pollutant over a given sampling time can be based on the colour change. The concentration of the pollutant in the air is expressed either by the length of the colour change in the column of the sorbent read on a relative scale marked on the tube surface (tube-type samplers) or by changes in the colour intensity of the sorbent, which is compared with a standard scale (badge-type samplers). Direct-reading personal samplers are frequently used in workplace personal monitoring as alert indicators for toxic or irritant gases and vapours. The majority of direct-reading dosimeters are disposable badges designed to be worn by individuals who are potentially at risk from acute exposure to toxic gases and vapours in their working environment. A new badge is issued to the employees at risk at the beginning of each shift, so each badge serves to record an individual's daily exposure. After the shift, a colour wheel or colour table with a viewing window is used to assess the exposure. The dosimeter is held in place behind the colour table so that a window overlays the stained area of the dosimeter. The colour wheel is rotated or table moved until a colour patch matching the dosimeter stain in density is found. Then the exposure is assessed.

Personal exposure measurement using direct-reading passive samplers is an invaluable method at the workplace. These monitors provide immediate visual warning, are easy to dispense and use and do not need maintenance and the costs are low compared to active sampling techniques. These monitors have been tested in the laboratory and in the field and are available for tens of chemical compounds. Unfortunately, they do not provide exact information about the real (quantitatively accuratc) cxposure.

Physical *adsorption* sampling is based on a reversible sorption process. In most cases, the adsorption dosimeters are designed to monitor volatile organic compounds and are of the tube type. After the exposure, the content of the tube is subjected to desorption either by means of extraction by a solvent or by thermal desorption and subsequent gas chromatography.

Many (tens of) absorbents and adsorbents are used as collecting media in passive samplers for coating filters of reaction dosimeters or as fillings for adsorbing dosimeters. Some are species-specific – can be used for only one gas species while others can be used for collecting more than one pollutant. For example, triethanolamine (2,2',2"-nitrilotriethanol) can be used to collect NO_2 as well as SO_2. Similarly, sodium carbonate can be used to collect SO_2 but also permits collection of NO_2. Many adsorption media (TENAX, or activated charcoal) can also be used for collecting a number of pollutants. Interested readers can consult some of the numerous reviews, specialized papers and commercial information published on this topic and the references therein (see e.g. Kruppa and Legge 2000; Raynie 2000; Gorecki and Namiesnik 2002; Namiesnik et al. 2005).

At the present time, a number of passive samplers are commercially available for collecting a variety of air pollutants and are manufactured or sold by numerous suppliers such as 3M, Dräger, DuPont, Perkin-Elmer, SKC, Supelco, etc. These are only a few examples of the most used and most recent personal and microenvironmental passive gas and vapour samplers. Other manufacturers and models exist and newer ones are being developed.

Comprehensive description of the theory of passive sampling and with many examples of absorbent/adsorbent systems and their corresponding analytical methods used for quantification of a number of pollutants are available in a number of papers (e.g. Palmes and Gunnison 1973; Kruppa 1998; Brown 1993; Kruppa and Legge 2000; Gorecki and Namiesnik 2002).

It should be mentioned that gaseous pollutants can also be monitored by using miniaturized active samplers. These devices use a small battery-operated pump connected by silicon rubber tubing to a tube coated with a sorbent, for example triethanolamine for NO_2 or potassium nitrite for ozone (Geyh et al. 1999; O'Neill et al. 2003; Staimer et al. 2005). Devices of this type can measure personal exposure during significantly shorter times than the corresponding passive samplers. However, active gas samplers have disadvantages in their dependence on a power supply, more complicated handling and maintenance and longer time required to instruct the people carrying the device.

4.3.2.3 Future Trends in Personal Exposure Measurements

Over the past 20 years, there has been significant progress in the development of small portable, selective, fast-response devices for monitoring and/or sampling a variety of chemical and biological agents which may be present in the air and potentially cause adverse health effects. New technologies to measure concentrations at the personal level have recently included automatic lab-on-a-chip measuring devices, allowing detection of nanoscale volumes by detection sensors based on electricity, fluorescence, chemical affinity or cell function. The monitors can be very small, inexpensive, easy to use and offer a potential for continuous monitoring (Hood et al. 2004; Kwok 2004; Weis et al. 2005). However, designers and scientists should bear in mind that all the new devices will also be worn by people and therefore they should not compromise the comfort of participants in exposure studies (Lioy 1993).

In addition to small, sophisticated and highly sensitive personal samplers, simpler, cheaper and easy to handle instruments will be needed for monitoring of personal exposure in specific miroenvironmnents, such as rural and urban locations in developing countries where high indoor and outdoor concentrations of combustion particles have already been recorded (Balakrishnan et al. 2002; Smith 2002; Andresen et al. 2005). These robust and simple monitors will also be needed for pilot experiments or for large-scale studies of personal exposure (e.g. school children). These devices, however, will necessarily have to produce data with sufficient precision and detection limits allowing comparison with other existing data and projects.

Efforts to obtain information on several pollutants at a time and to minimize errors associated with combining personal monitoring and area or microenvironmental monitoring have resulted in the development of multipollutant personal samplers (Demokritou et al. 2001; Chang et al. 2001). These samplers have the capability to simultaneously measure particulate mass (PM_{10} or $PM_{2.5}$), sulfate and nitrate contents, elemental and organic carbon, and principal gaseous pollutants such as nitrogen dioxide, sulfur dioxide, carbon monoxide, ozone and VOCs. However, so far, few exposure studies have been published using these types of personal sampler (e.g. Chang et al. 2000; Sarnat et al. 2001; Kim et al. 2006; Sarnat et al. 2006). Integrated multipollutant samplers provide fast, complex and valuable information on simultaneous personal exposure to a set of pollutants but they require good laboratory and personnel infrastructure for post-exposure analysis of the samples and subsequent data processing. One of the few in use and commercially available is the ChemPas, Rupprecht and Patashnick model 3400-T multipollutant personal sampler. Another system, the custom made RAS (Roll-Around System) type, permits measurement of 1-h personal concentrations of particles ($PM_{2.5}$) and gases (Chang et al. 2001). The sampling heads, weighing about 1 kg, are worn on the chest, while the rest of the equipment weighing several kilograms including the pumps and bodies of the samplers, connected to the sampling heads by tubing, have to be carried out in a separate luggage case. This arrangement, however, prevents the system from being used in population studies, namely because of the weight and discomfort for the target person while sampling.

4.4 Indirect Approach

Usually it is difficult and costly to provide personal monitors for many members of a community of concern or it may happen that a large proportion of the community members are not willing to participate in such a demanding survey. In this case, indirect methods must be employed. In the indirect approach, the exposure is determined by combining information about pollutant concentrations at all or certain locations with information about time spent in these locations. The indirect approach can be realized by three methods:

1. Relation of the individual's or community exposure to outdoor pollutant concentrations obtained from the nearest FSM (the front-door concentration method)
2. Collection of human activity information (surveys, questionnaires, etc.)
3. The microenvironmental method

4.4.1 Fixed-Site Monitoring Approach

Measuring an individual's exposure to a given pollutant may be a relatively straightforward procedure. It is sufficient to provide the person with a portable sampler and ask the person to fill in a time-activity diary. Obtaining information about exposure to a pollutant of interest for a community or a cohort of sensitive individuals is a much more complex task. Usually not all the individuals are keen to participate in personal measurements, and the number of personal samplers to be distributed among the panel participants is limited. For these reasons, when the assessor needs to determine exposure – effect relationships, the simplest method is to use the fixed site monitoring data from the area of residence of the subpopulation in question. Exposure assessment data are provided for the whole urban population from the FSM network and are averaged over the whole city or region. To date, this has been the most frequent way of assessing morbidity and mortality associated with air pollution (namely particulate pollution) on an urban or regional scale (Dockery and Pope 1994; Katsouyani et al. 1997; Beeson et al. 1998; for an extensive review, see Pope and Dockery 2006).

The problem of relating fixed-site/ambient monitoring data with population exposure levels has been extensively discussed in the scientific literature with no clear result (Sexton et al. 1984; Monn et al. 1997; Oglesby et al. 2000a; Wallace 2000; Wilson et al. 2000; Boudet et al. 2001; Adgate et al. 2002; Gauvin et al. 2002; Jantunen et al. 2002; Meng et al. 2005a; Sarnat et al. 2006; and many others). The validity of outdoor particulate concentrations as a measure of exposure has been examined in some studies, with most of these studies focusing on the relationship between personal exposures and ambient PM_{10} and $PM_{2.5}$ concentrations. In these studies, daytime personal PM_{10} and $PM_{2.5}$ exposures were consistently higher than the corresponding outdoor levels. For example, the Particle Total Exposure Assessment Methodology (PTEAM) study (Thomas et al., 1993) showed daytime personal PM_{10} exposures to be, on an average, 50% higher than the corresponding ambient levels, while the Harvard Six City study (Dockery et al. 1993) found mean personal PM_{10} exposures to be more than 100% greater than mean ambient levels. Differences between personal and FSM concentrations were also found in European studies (Kousa et al. 2002). On the other hand, some scientific papers suggest that ambient data can be used for estimation of personal exposure to particulate air pollutants, namely to fine particles (Buckley et al. 1991; Janssen et al. 1999). Sarnat et al. (2006) shoved that ambient particle concentrations were much better surrogates for personal exposure than gases.

Results from recent exposure assessment studies suggest that correlations between personal exposures and outdoor concentrations are significant only for approximately half of the individuals studied. In addition, significant intra- and interpersonal variability in exposures was found. These differences can be attributed to differences in activity patterns and home characteristics. As expected, the personal exposures of individuals who spend most of their time outdoors are more closely associated with the ambient concentrations. Activity patterns have an additional influence on particle exposures, as several commonly-performed activities such as smoking, cooking, cleaning, and even walking can generate (or resuspend) particles and other pollutants in the air.

Ambient monitoring (or FSM) cannot be fully considered to be an indirect personal approach as the target receptor is not an individual but a whole community or a selected subpopulation from that community.

In situations when ambient monitoring, usually based on FSM data, is not available, exposure assessment can be based on matching the demographic data (number, age, socio-economic status, etc.) of people from a certain region with those about the location and intensity of major sources of air pollution (such as industrial emission sources, power plants). Population information in the above-mentioned type of study is usually retrieved from census databases and emission source information can be obtained, for example, from emission or toxic release inventories, depending on what is available to the assessor (Sexton et al. 1992). Studies of this type have been performed in the past on minority communities that are difficult to access through surveys or questionnaires (poor, illiterate, marginal populations) (Perlin et al. 1999; Sexton et al. 1993).

An obvious disadvantage of using the indirect method based on FSM or generally outdoor monitors is the already-mentioned fact that people spend a significant portion of their time indoors. Some knowledge about indoor pollutant concentrations or at least the presence of important indoor sources of pollution is therefore vital in assessing exposure.

4.4.2 Microenvironmental Approach

A common term "microenvironment" was adopted and widely used for specific locations where an individual spends a certain (non-negligible) portion of time during the day. The microenvironment can be the indoor space at home, the interior of a bus, a working place, a shopping centre, an urban street or a rural space in the countryside. As an example, the microenvironment exposure profile of a person commuting by the underground from home to work can encompass various types of environments (indoor home – outdoor between home and the underground station, underground spaces before getting on the train – trains – again underground spaces where the person leaves the underground, walking between the station and his/her working place – working place and vice versa [Fig. 4.3]).

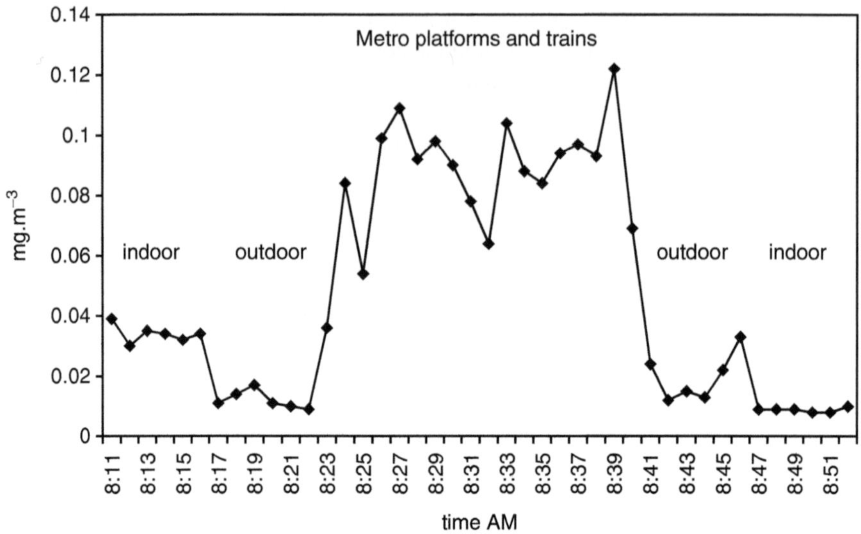

Fig. 4.3 Concentration of PM_{10} during one journey in the Prague Underground on 7 October 2003 in the morning (for details see Branis 2006)

The whole sequence may even be complicated by including other places (such as restaurants, leisure activity spaces, etc.) where the person spends a certain portion of time during the day. In theory, a microenvironment is defined as a three-dimensional space where the concentration of a pollutant is spatially and temporarily uniform or has constant statistical properties. The above-mentioned exposure scenario can be described by a microenvironmental model or integrated exposure as follows (Eq. 4.9):

$$E_i = \sum_{j}^{J} C_j t_{ij} \tag{4.9}$$

Where E_j is the total exposure for person i over the specific period of time, C_j is the pollutant concentration in microenvironment j, t_{ij} the residence time of the person i in microenvironment j and J is the total number of microenvironments visited during the analyzed period (usually day). The total exposure is therefore the sum of the exposures during a given time. The dimension of total exposure is the same that of integral exposure, μg m^{-3}.h or ppm.h. The formula is based on the assumption that the assessor has information about time spent by the relevant individual or group of individuals (also called time-activity pattern) and about the pollutant concentration in each of the particular microenvironments.

In practice, the total exposure is calculated as the sum of the time-weighted averages of pollutants from all of the microenvironments visited during the whole studied period T (Eq. 4.10):

$$\overline{E}_i = \frac{\sum_{j}^{J} C_j t_{ij}}{T} \tag{4.10}$$

According to the formulas given above, the concept of personal exposure to air pollutants assessed by means of the microenvironmental approach may seem simple. However, it is a rather complicated issue. The total exposure to a given pollutant is composed of multiple sub-exposures taking place in various spaces to pollutants of various types and origin. Theoretically, the research worker should have provided the same measuring device for each of the microenvironments where the person(s) of concern spent a non-negligible period of time during the day.

As mentioned by Monn (2001), it is demanding and expensive to obtain refined data for dozens of microenvironments visited during the day (week, month) by a selected population of individuals. The most feasible way is to aggregate the most similar microenvironments into microenvironmental types – MET. In this approach, personal activity profiles and household, outdoor and workplace characteristics are merged and 5–10 MET are defined. Several studies suggest different MET and MET groups. For example, in the EXPOLIS study (Jantunen et al. 1998), the following MET were selected to cover most of the daily activities of adult individuals:

- *Outdoor* (around the home, in the study region)
- *Indoor at home* (kitchen, bedroom, living room, workshop)
- *In transit* (car, train, bus, pedestrian from home area to work or to other places)
- *Other* (shopping mall, restaurant, theatre, indoor sports)
- *Workplace* (office, factory, … various places and professions)

As shown, a constraint for using the indirect microenvironmental method lies in the fact that the residence time of the persons of concern must be known, together with the pollutant concentration of in at least some of the microenvironments (the most important in terms of pollution concentration) at the time when the persons are present. For this purpose, information about human activities should be collected to describe the behavioural patterns of persons studied.

4.4.2.1 Human Activity Information

The fastest and technically simplest way of approximately assessing individual or community exposure to air pollutants is to acquire human activity/behaviour data through surveys, questionnaires and other similar techniques. They can be used to identify the presence of important sources of pollution and also the frequency of contacts with these sources in the indoor environment. Through questionnaires, perception of traffic near the home can be used to estimate the traffic intensity and hence the pollution levels in the ambient air. Collection of personal activity data can provide surrogates for exposure, can identify sources of exposure and can identify intervening variables and mitigating events between the source and the individual(s) of concern.

The most important advantage of surveys, interviews, questionnaires and other "diary type" data acquisition techniques lies in their low cost and the ability to involve a high number of individuals in a community in the study. On the other hand, these methods offer only qualitative data of a categorical character (polluted/unpolluted, low/medium/high pollution, etc.) and therefore cannot fully replace sampling or monitoring of airborne pollutants.

Critical questions, that have to be addressed (and answered) through personal activity/behaviour data collection, include:

- *Who* is exposed (individual or group of individuals represented by the subject – his/her/their gender, social status, age, handicap, etc.).
- *What* is the pollutant/source of concern (pollutant or source of pollution relevant to the research question and health outcome – chemical compound, mixture of compounds, specific work or household technology, specific activity, such as smoking, etc.).
- *Where* the exposure takes place (the place of contact of the pollutants(s) with the subject – e.g. home, street, work, transport).
- *When* the exposure takes place (time, period, season or any fraction of time specifying the contact between the subject and the pollutant – in the summer, indoors, at work, etc.).
- *How* the persons come into contact with the pollutant (the situation or activity that results in contact with the pollutant of concern – e.g. during cooking, smoking, while working).
- *How* often the exposure takes place (frequency of exposure during the reference period – e.g. daily, twice a week, seldom).
- *How* much of the pollutant in total the person(s) of concern receive(s) during the day/year/lifetime (quantity of the pollutant to which the person is exposed within the reference period or quantity of the pollutant inhaled per unit of time – e.g. 24-h exposure to 100 μg m^{-3} of PM_{10}).

All the above-mentioned questions about human activity patterns can be answered through several methods:

- Focus groups
- Surveys
- Interviews
- Questionnaires
- Time-activity diaries
- Observations
- Telemetry

The method used depends on the breadth of information to be collected, the degree of participant burden that is acceptable and the resources available for this aspect of the study (Freeman and de Tejada 2002). The approach depends upon the characteristics of the communities being sampled, as well as on budgetary constraints. However, in most studies on human exposure to air pollutants, a combination of methods is employed, such as in the NHEXAS study developed by US EPA, where almost all forms of instruments to collect information on human activities and exposure were employed (Whitmore et al. 1999; Pellizzari et al. 1999; Freeman et al. 1999).

Focus groups. Can provide community information that helps identify the sources or location of pollution, or identify community leaders and social issues that may facilitate or interfere with conducting the study. They are made up of typically 10–20 persons drawn from the community of interest. These groups can identify items in a questionnaire that are not appropriate for their community. They can also help the investigator identify social classes within the community and regional or linguistic idioms that will make questions understandable to members of the community. The facilitator (who communicates with the selected group) needs to be trained to keep the discussion focussed on the issues of concern. Often a tape recorder could be used to capture the discussion accurately.

Surveys. Are useful when large population-based studies are performed and the objective is to characterize general aspects of exposure within the community. Surveys can be performed by personal visits to participants' homes, by phone, by mail or recently by e-mail and various Internet applications. Surveys typically focus on aspects related to households rather than to individuals. The choice of method depends on many aspects, such as budgetary limitations or type of community being assessed.

Interviews. Interviews are usually conducted to address aspects related to the personal habits or work environment characteristics as efforts to visit an employer's private property to obtain information about the individual's occupational environment are not successful. However, interviews can also provide the assessor with relevant information about household characteristics. Interviews are essential in communities or countries where literacy may be low. In these cases, interviewers need to be well-trained persons able to communicate with people with special (unusual) cultural and/or social habits. Understanding cultural norms and culturally/socially sensitive issues will prevent confusion and will help to frame questions so that the answers are correct rather than socially acceptable. Also, in some native tribes and ethnic groups, male interviewers may not be accepted by female community members and vice versa.

Questionnaires. Usually provide detailed information about recent exposure activities and/or about historical exposure of persons involved in the study. Questionnaires typically attempt to characterize sources of exposure in terms of personal habits (e.g. smoking, leisure time activities), occupation, household characteristics and habits (such as ventilation rate, type of heating), travel activities, etc. Questionnaires must also include demographic questions about age, gender, ethnicity, socioeconomic status indicators (household income) or educational level. Answers to these questions may serve as indirect but valuable information about exposure to air pollutants.

In multi-centre international studies, attention should be paid to exact translation of questions and explanatory notes. The text is often developed in one language but used by communities speaking a different language (even different languages). In this case, a series of translations between the original language and the other language(s) must be made to ensure that the original meaning has not been lost in the translation process. Care should be also taken to make sure that local/national idioms are translated correctly to reduce language bias. Questionnaires generally follow more general surveys and frequently serve as an important step in recruitment of participants for personal monitoring studies.

Questionnaires are usually distributed once at the beginning of the study; however, follow ups are also possible. Time allocated to filling-in a questionnaire does not represent a great burden on the participant. Nevertheless, long and/or complicated questionnaires can take more than 1 h to complete even for educated and motivated people. For reluctant participants, however, a few minutes may be too long. For this reason, the approximate time needed to complete the questionnaire should also be included in the questionnaire instructions and explanatory notes.

Time-activity diaries. In exposure studies, time-activity diaries have been used to collect time-resolved quantitative and qualitative information about activities that contribute to an individual's exposure (cooking, smoking, travel) and to identify any important contact with sources of pollutants in microenvironments visited (e.g. at home, at work, in transport). Diaries are time- and attention-demanding tasks. They have to be regularly filled-in throughout the day and frequently for a relatively long period of time. Diaries require literacy and a good sense of time, and usually an extended period of commitment (usually weeks, sometimes months).

The level of temporal detail in a time-activity diary usually corresponds to 15- or 30-min segments over a 24-h period, however, shorter (down to 1 min segments) are also used when changes in activities are rapid and/or the types of environments visited change over a short period of time (Brauer et al. 1999; Quintana et al. 2001; Branis 2006).

Diaries are frequently made up of two parts (e.g. Malhotra et al. 2000): (1) a time line, and (2) a time budget. The time line (or mobility pattern) identifies the person's use of space during the day, typically as a sequence of spaces visited (e.g. a daily mobility pattern can be described as follows : at home, in transit, at work, in transit, in a shopping centre, in transit and at home), while the time budget focuses on activities that occur throughout the day (e.g. cooking in the kitchen, travelling by bus, working with a computer, shopping in a mall, travelling by the underground, working at home in the garden, etc.). Diaries can also contain only simple questions regarding important sources or activities such as "did you spent time in close proximity to a gas oven today?" or "did you clean a stove today?", etc. These questions can also be used together with a simple time-line diaries (Freeman et al. 1999). Time-activity diaries have a long history in recording human activity patterns (Chapin 1974).

Time-activity data are entered in a pre-printed diary according to a standard procedure. In an ideal situation, the monitored person records the activity as soon as it starts and ends using his/her watch or stopwatch. Such intensive registration not only changes the behaviour of the person concerned but also does not allow for performance of prolonged activities without temporary interruptions. The problem of registration of location and activity changes increases when shorter (1 or 5 min) intervals should be monitored together with personal or microenvironmental exposure to air pollutant(s). In this case, sophisticated methods can be used, such as continuous time-stamped voice recording devices or video techniques (Quintana et al. 2001). Scripted time-activity patterns can also be recorded by using other electronic devices, such as hand-held PDA (personal digital assistant) (Cohen and Cotey 1997; Brauer et al. 1999).

Time-activity diaries can also be assessed indirectly by means of surveys and interviews. Within the framework of the National Human Activity Pattern Survey (NHAPS), more than 9,000 respondents were interviewed, with the aim to collect 24-h retrospective time/activity diaries together with answers from each respondent to a number of personal and exposure-related questions. The survey was conducted on a virtually daily basis between September 1992 and September 1994. The resulting database covers ten EPA Regions across the 48 contiguous states. The resulting diary records contained the beginning and ending of each distinct combination of location and activity occurring on the diary day (Klepeis et al. 2001).

It is clear that, in such studies, most of the respondents are unable to remember activities in the more distant past (last week or last month) in detail and can therefore successfully describe only a day or 2 preceding the interview. Consequently, retrospective interviews cannot replace personal time-activity diaries filled out on a daily basis for weeks or months. However, when conducted for longer periods of time and on a large population, even short (24-h data) sequences can provide valuable information about exposure levels for a large population sample.

The EPA National Exposure Research Laboratory Consolidated Human Activity Database (CHAD) is one of the most comprehensive human activity databases that can be used in air pollution exposure estimates. It contains more than 20,000 person-days of activity. All ages and both genders are included in the database, and information regarding every activity undertaken during a day, and lasting for a minute or more, is included in sequential order. The day is defined as a period from midnight to midnight. The Consolidated Human Activity Database uses 144 activity codes and 115 location codes, but data are not available for all of these items. For this reason, aggregated categories (e.g. staying indoor or staying outdoor) have to be used for some exposure estimates instead of more specific information (e.g. staying in the living room, kitchen, etc.). CHAD also includes basic data on meteorology (daily average and maximum temperature, daily accumulated precipitation and duration of the precipitation event [McCurdy et al. 2000]). This allows the assessor to subdivide daily activity patterns into seasonal categories and make more accurate exposure estimates or models.

Observations. Psychologists, sociologists and anthropologists have been using *observations* for decades to describe and understand basic processes and forms of human behaviour, interactions amongst individuals, and culture- ethnicity- or society related specific behavioural patterns. Direct observations are suitable to quantify exposure to air pollutants in situations where time-activity diaries, surveys and/or questionnaires cannot be used. This is mainly related to exposure studies where illiterate, very old, very young or ill individuals are selected for the study.

There are two main types of observations: (1) continuous monitoring, and (2) spot checks. Continuous monitoring involves long term observation and recording of the behaviour of the person of concern in studies where a series or sequence of behaviours needs to be quantified. One of the main disadvantages of continuous monitoring is that few persons can be studied and observer-target person contacts may change the latter's behaviour. Spot-checks are short-time-interval checks (frequently random) performed at various times of the day in various microenvironments. A spot-check begins immediately upon the arrival of the observer at the microevironment of concern. The momentary activities, persons and sources of pollution are registered and the observer moves to another microenvironment. In this way, several microenvironments, activities and persons can be recorded during a day. Spot-checks are useful when a community, population sample or specific individuals do not frequently change the place of activities, for example, in rural areas in developing countries where women spend the majority of their time indoors cooking and looking after children or around the house (Saxena et al. 1992). The advantage of spot-checks is that larger population sample can be observed and that the activities/behaviours of the individual are not disturbed by the permanent presence of the observer in the studied person's vicinity. Clearly, spot-checks do not allow for collection of long-term activity data and have a rather low record detail level.

Observations cannot replace personal time-activity diaries filled in on a daily basis for weeks or months. As longer observational monitoring or extensive spot-checks place a substantial burden on the observer, new techniques have been recently employed. One of the best methods to continuously record the target person's behaviour is to use a video camera (or several cameras if multiple microenvironments have to be monitored). Video-based observations provide a continuous set of data, which can be repeatedly analyzed for various purposes. However, it should be stressed that transcription of long video records can be time-consuming and requires additional laboratory equipment. Video recording is not suitable in communities where photographing and/or videotaping is considered unacceptable (So far, videotaped observations were successfully used to monitor children's dermal exposure and hand-to-mouth activities (e.g. Reed et al. 1999).

Because observations cannot cover either large populations or long periods of time, they are usually combined with other methods aimed at collecting human activity patterns, such as surveys, interviews or (when possible) questionnaires. Using several sources enables the researcher to accumulate more comprehensive time-activity information (Albalak et al. 1999).

Telemetry. Radio-*telemetry* has successfully been used for tracking wild free-ranging animals for decades. Recording of the movement of persons by means of radiotelemetry is nevertheless rather difficult. The participants have to wear a transmitter and its signal is picked up by a receiver to locate the target's position. The main problem lies in the fact that recording may fail when the person of concern moves over a large space or the signal of the transmitter may be obstructed by buildings in a city. Recently, promising new tools have been developed to register the positions of persons whose exposure is studied. The miniaturization of *GPS* (global positioning system) devices allows for *tracking the movement* of target individuals during the whole day in almost unlimited space. Modern GPS systems are not expensive, have sufficient resolution (several meters), and contain dataloggers with high capacity to record movement over a large territory for a long period of time. When combined with other sensors (such as heart or respiration rate monitors), they can provide useful information about the monitored individual's physical activity being performed. The main obstacle in using GPS systems for exposure analysis is that, while they provide excellent information about the person's position, they are not able to provide any exact information about the monitored person's exposure to air pollutants unless combined with other types of monitors. Some problems may also arise when the satellite signals on which the GPS system operates are shielded by construction materials inside building, thus limiting the use of this tracking system in some indoor environments.

4.4.3 Biomonitoring

Many chemical compounds, to which human beings are exposed through inhalation, can be determined either in their original form or as their metabolites in cells, tissues, body fluids, faeces or the exhaled breath of target persons. This approach, using analysis of human biological materials, is called biomonitoring or biological monitoring, and the substances monitored are called biomarkers. Monitoring of biomarkers as means of monitoring personal exposure may be especially valuable when practical obstacles exist in determining exposure to a specific pollutant. Such a situation may arise when it is difficult to assess exposure by personal monitoring techniques (in a large population, in young children, in the elderly, due to unfavourable working conditions of a target person[s], etc.).

The definition of biological monitoring runs parallel to that of personal monitoring: Both assess exposure and health risks, and both ultimately serve to estimate the dose. While personal monitoring ascertains the intake, biological monitoring determines the uptake. (For original definitions, concepts and examples of biomonitoring see, e.g. Berlin et al. 1984; or Tardiff and Goldstein 1991; Leaderer et al. 1993; Grandjean 1995; Ward and Henderson 1996; Dor et al. 1999; Groopman and Kensler 1999; Scherer 2005).

4.4.3.1 Biomarkers

Biomarkers are generally assigned to one of three classes: biomarkers of exposure, effect and susceptibility. *Biomarkers of exposure* (which can be further divided into biomarkers of internal dose and biomarkers of effective dose) should indicate the actual exposure. In other words, a good biomarker of exposure should reflect the concentration of the pollutant that has passed the interface of the human body. The existence of a clear relationship between the pollutant concentration in the inhaled air and the concentration, amount or activity of the biomarker either in the original form of the compound or as its metabolite should be clearly demonstrable (Rappaport et al. 1995). *Biomarkers of effect* (or biomarkers of potential harm) should reflect physiological or morphological changes in cells/tissues as a result of the exposure. In reality, these changes are not necessarily linked to changes in cell/tissue/organ function (symptoms of illness); however, ideal biomarkers of effect can be related to the disease of interest. *Biomarkers of susceptibility* are parameters that can indicate an increase or decrease in the risk of an individual developing adverse effects following exposure to a pollutant. Susceptibility biomarkers to a great extent reflect genetic variability and/or enzyme polymorphism between individuals, which may strongly influence the extent, severity or duration of the response of the organism to the same exposure level, thus modifying the amount or activity of other biomarkers or risk of a disease.

Models can also be used to ascertain the relationship between the concentration of the analyte in the air and in the body. As detailed pharmacokinetic data are necessary as important inputs, these models are still in the phase of development (Georgopoulos et al. 1997; Ren and Stewart 2005; Georgopoulos and Lioy 2006).

Biomarkers in each class should meet several criteria. Schulte and Talaska (1995) and later Dor et al. (1999) suggested validation of biomarkers by their intrinsic qualities and their properties related to the sampling and analytical procedure. The intrinsic qualities are:

- Type of biomarker (biomarker of exposure, effect or susceptibility)
- Time period of exposure reflected by the biomarker (recent/acute or long term/ chronic exposure)
- Sensitivity (low percentage of exposed subjects falsely classified to be not exposed)
- Specificity (low percentage of non-exposed subjects falsely classified as exposed)
- Degree of intra- and inter-individual variation of the biomarker (age, gender, profession specificity)
- Knowledge of background levels of the biomarker in the population
- Existence of a dose-response relationship between the biomarker and the exposure
- Knowledge of confounding factors (endogenous formation, exposure from non-monitored sources)

Properties of a biomarker related to the sampling and successive analytical procedure include:

- Sampling constraints (e.g. timing requirements, specific instrumentation needs)
- Invasiveness (level of disturbing the integrity of the body)
- Stability of the biomarker sample (storage limitations of the biomarker before analysis)

- Possibility of contamination of the sample (during sampling and/or handling and/or analysis)
- Simplicity of the sampling and/or analytical method
- Specificity of the analytical method
- Level of routine use of the method
- Level or possibility of standardization of the method
- Detection and quantification limits
- Temporary constraints (speed of the method)

Biomarkers of exposure from inhaled air. Despite the fact that a number of compound specific biomarkers of exposure have been discovered and used to assess individual and community exposure to air pollutants, recent efforts are aimed at finding and validating biomarkers – surrogates – of mixtures specific for certain sources or environments. In the real world, individuals generally do not inhale single pollutant species. Major sources emit and human activities routinely involve complex mixtures of gases, vapours, and particulate matter that contaminate, more or less, all indoor and outdoor environments. Three main types of complex mixtures known to have adverse health effects are usually mentioned in the literature (Leaderer et al. 1993; Scherer 2005). These are tobacco smoke, diesel/automotive exhaust and polluted ambient air (also called urban smog).

Examples of biomarkers of inhaled complex mixtures. Tobacco smoke represents a single-source complex mixture found and analysed almost exclusively in indoor microenvironments. Environmental tobacco smoke (ETS) is comprised of hundreds of chemical compounds in gaseous and particulate phases, most of which have been identified as having a wide range of acute, chronic and carcinogenic effects. As it is not possible to monitor all the components of ETS, several specific pollutants have been identified as good surrogates and also biomarkers of exposure.

Urban smog (traditionally divided into two major types – photochemical/summer/Los Angeles smog and reduction/winter/London smog) typically includes a set of primary pollutants, such as ultrafine and fine particles, nitrogen oxides, hydrocarbons, carbon monoxide, sulphur dioxide (when fossil fuels are used for heating) and, under specific weather conditions, a mixture of secondary pollutants, such as ozone, volatile organic compounds, polycyclic aromatic hydrocarbons (PAHs) and other photochemical oxidants. The most widely used biomarkers of polluted urban air are carbon monoxide, PAHs and sometimes benzene. As the two former pollutants are products of incomplete combustion, smoking is a strong confounder for exposure to these compounds.

Diesel exhaust consists of a gaseous phase and soot particles (about 100 nm in aerodynamic diameter) containing more than 50% elemental carbon and a large amount of PAHs and nitro-PAHs. 1-Nitropyrene is the major component of nitro-PAHs. Diesel exhaust represents an important risk factor in many workplaces as well as near major roads, road crossings and in urban street canyons. The risk factors in this instance are also confounded by the influence of smoking or exposure to ETS.

One of the principal and widely-used biomarkers of exposure to ETS is *cotinine*. This is a primary metabolite of nicotine with much longer half-life in the body than its precursor. It can be traced in urine, blood and saliva. Its ratio of levels in smokers versus non-smokers can reach values of up to 100 (Scherer and Richter 1997). *Nicotine* (in the blood) and *cadmium* (in the blood or in faeces) can also be used as a biomarker of exposure to ETS.

Inhaled carbon monoxide bonds to hemoglobin by displacing oxygen to form carboxyhemoglobin (COHb). COHb detected in blood (in red blood cells) is one of the oldest biomarkers of ETS and of polluted ambient or indoor air. Carbon monoxide can also be traced in exhaled air. Its major disadvantage for use as a biomarker of exposure to smog and diesel exhaust is its low sensitivity and specificity, namely because of bias from smoking and potential endogenous formation.

The most widely used biomarkers for exposure to benzene are trans, trans-muconic acid and S-phenyl mercapturic acid in urine. Benzene in blood and in exhaled breath has also been used in some studies (Scherer et al. 1995). As benzene is metabolized mainly to phenols, measurement of phenol in urine has been widely used as a biomarker for benzene exposure (Tardiff and Goldstein 1991; Jenkins and Counts 1999; Jo and Pack 1999; Scherer 2005).

Human exposure to polycyclic aromatic hydrocarbons and their derivatives from inhalation of tobacco smoke, diesel exhaust or polluted ambient air varies according to many factors, such as the number of cigarettes smoked, time spent in smoking spaces, degree of urbanisation, industrialisation and traffic density. Of the several 100 PAHs, some have been classified as probable and possible carcinogens and mutagens. Urinary 1-hydroxypyrene is a widely used metabolite of non-carcinogenic pyrene to trace exposure to PAHs (Perico et al. 2001; Hansen et al. 2004). Various types of DNA adducts in white blood cells, hemoglobin and albumin adducts have recently been increasingly used to indicate ambient as well as occupational or smoking exposure to PAHs (Georgiadis et al. 2001; Ruchirava et al. 2002; Scherer 2005). However, further research is needed to identify and employ biomarkers more specifically for diesel exhaust, ETS and polluted ambient air.

At the present time, quite a number of biomarkers are already available for assessing exposure to complex mixtures. Some secondary pollutants can also be traced in the body. Urinary hydroxy-proline has been used as a marker for ETS and NO_2 exposure (Yanagisawa et al. 1988; Perdelli et al. 2002). It is also evident from the number of scientific papers published to date that biomonitoring of active and/or passive exposure to tobacco smoke is more advanced that biomonitoring of exposure to other complex mixtures (ambient air, diesel exhaust). The biological half-life of biomarkers ranges from minutes to months and may thus significantly influence the design of the study and interpretation of the results. Almost all biomarkers also exhibit considerable variability between individuals and most of them indicate only recent exposure. Despite the fact that biomonitoring is being increasingly used for surveillance and spot measurements in environmental and occupational studies, at the present time biomarkers are not suitable as principal or unique proof of exposure either to single airborne contaminants or to complex mixtures. This is mainly because they do not provide exact and adequate information on the magnitude, duration and frequency of exposure. In addition, it is necessary to mention that exposures can also take place through media other than air and by routes of entry other than inhalation. Usually, more than one method for assessing personal or community exposure is needed to estimate the health risk from air pollution. In any case, biomarkers can increase the precision of other methods aimed at estimating total human exposure to air pollutants, such as direct personal monitoring, fixed site monitoring and questionnaires or surveys (Leaderer et al. 1993; Scherer 2005).

4.5 Conclusions

As has been described, personal exposure to airborne pollutants can be assessed using several methods and techniques or their combination. In typical situations, more than one method for assessing exposure is used in epidemiological studies. Complex strategies are needed to integrate the advantages of various exposure assessment methods. The most comprehensive approach lies in the use of "nested exposure assessment" (Leaderer 1990; Leaderer et al. 1993, 1986). This method utilizes a large number of questionnaires to acquire an easily measured indicator of exposure for the whole population under study, simultaneously obtains more detailed information using more sophisticated techniques on a subset of the target population (such as microenvironmental and/or short-term personal monitoring accompanied by time-activity diaries), and finally selects an even smaller subgroup of the population under study to obtain even more detailed data from biomonitoring and/or long term (continuous) personal monitoring. Measurements of exposure obtained from the subgroup of personally monitored individuals can then be used to model the exposure of the whole population to air pollutants or mixtures of pollutants. Outputs from the measurements and models can also serve as a good feedback for analysis of uncertainty associated with the questionnaires.

Today, personal monitoring represents an essential component of comprehensive research in air pollution epidemiology and sophisticated indirect techniques of personal exposure measurements have been established to support it: (1) personal samplers are smaller, pollutant specific, more accurate and able to run for longer time than previously; (2) sophisticated methods of surveying, such as GPS, GIS, allowing tracking of the movement of the target individual are well-established and accessible; (3) an ever-increasing number of specific and sensitive biomarkers can be used together with new methods for their detection and analysis; (4) comprehensive databases of human activity are available for use in epidemiological studies. However, further improvements in techniques and acquisition of new data (namely in developing countries, where air pollution has already reached high levels) are needed to prepare detailed and flexible exposure models, which can assist in assessing potential health risks from inhaled air for various communities, environments and compounds, including areas where limited or no data are available.

Acknowledgments The work was in part supported by the Ministry of Education, Youth and Sports of the Czech republic under the scheme NPV II, No. 2B08077.

References

Adgate, J. L., Church, T. R., Ryan, A. D., Ramachandran, G., Frederickson, A. L., Stock, T. H., et al. (2004). Outdoor, indoor, and personal exposure to VOCs in children. *Environmental Health Perspectives, 112*(14), 1386–1392.

Adgate, J. L., Ramachandran, G., Pratt, G. C., Waller, L. A., & Sexton, K. (2002). Spatial and temporal variability in outdoor, indoor and personal $PM_{2.5}$ exposure. *Atmospheric Environment, 36*, 3255–3265.

Adhikari, A., Martuzevicius, D., Reponen, T., Grinshpun, S. A., Cho, S.-H., Sivasubramani, S. K., et al. (2003). Performance of the button inhalable sampler for the measurement of outdoor aeroallergens. *Atmospheric Environment, 37*, 4723–4733.

Agranovski, I. E., Agranovski, V., Reponen, T., Willeke, K., & Grinshpun, S. A. (2002). Development and evaluation of a new personal sampler for culturable airborne microorganisms. *Atmospheric Environment, 36*, 889–898.

Aizenberg, V., Grinshpun, S. A., Willeke, K., Smith, J., & Baron, P. A. (2000). Performance characteristics of the button personal inhalable aerosol sample. *American Industrial Hygiene Association Journal, 61*(3), 398–404.

Albalak, R., Frisancho, A. R., & Keeler, G. J. (1999). Domestic biomass fuel combustion and chronic bronchitis in two rural Bolivian villages. *Thorax, 54*(11), 1004–1008.

Andresen, P. R., Ramachandran, G., Pai, P., & Maynard, A. (2005). Women´s personal and indoor exposures to PM2.5 in Mysore, India: Imact of domestic fuel usage. *Atmospheric Environment, 39*, 5500–5508.

Balakrishnan, K., Parikh, J., Sankar, S., Padmavathi, R., Srividya, K., Venugopal, V., et al. (2002). Daily average exposures to respirable particulate matter from combustion of biomass fuels in rural households of southern India. *Environmental Health Perspectives, 110*(11), 1069–1075.

Beeson, W. L., Abbey, D. E., & Knutsen, S. F. (1998). Long-term concentrations of ambient air pollutants and incident lung cancer in California adults: Results from the ASHMOG study. *Environmental Health Perspectives, 106*(12), 813–822.

Bell, M. L., & Davis, D. L. (2001). Reassessment of the lethal London fog of 1952: novel indicators of acute and chronic consequences of acute exposure to air pollution. *Environmental Health Perspectives, 109*(Suppl. 3), 389–394.

Berlin, A., Yodaiken, R. E., & Henman, B. A. (1984). *Assessment of toxic agents at the workplace. Roles of ambient and biological monitoring* (684 pp.). The Hague: Martinus Nijhoff.

Boudet, C., Zmirou, D., & Vestri, V. (2001). Can one use ambient air concentration data to estimate personal and population exposures to particles? An approach within the Ezuropean EXPOLIS study. *Science of the Total Environment, 267*, 141–150.

Branis, M. (2006). The contribution of ambient sources to particulate pollution in spaces and trains of the Prague underground transport system. *Atmospheric Environment, 40*(2), 348–356.

Brauer, M. (1995). Assessment of indoor aerosols with an integrating nephelometer. *Journal of Exposure Analysis & Environmental Epidemiology, 5*, 45–56.

Brauer, M., Hirtle, R. D., Hall, A. C., & Yip, T. R. (1999). Monitoring personal particle exposure with a particle counter. *Journal of Exposure Analysis & Environmental Epidemiology, 9*, 228–236.

Brouwer, D. H., Gijsbers, J. H. J., & Lurvink, M. W. M. (2004). Personal exposure to ultrafine particles in the workplace: Exploring sampling techniques and strategies. *Annals of Occupational Hygiene, 48*(5), 439–453.

Brown, R. H. (Ed.). (1993). The use of diffusive samplers for monitoring of ambient air (IUPAC report). *Pure & Applied Chemistry, 65*(8), 1859–1874.

Buckley, T. J., Waldman, J. M., Freeman, N. C. G., Marple, V. A., Turner, W. A., & Lioy, P. J. (1991). Calibration, intersampler comparison, and field application of a new PM-10 personal air-sampling impactor. *Aerosol Science and Technology, 14*, 380–387.
Chakrabarti, B., Fine, P. M., Delfino, R., & Sioutas, C. (2004). Performance evaluation of the active flow personal DataRAM PM2.5 mass monitor (Thermo Anderson pDR – 1200) designed for continuous personal exposure measurements. *Atmospheric Environment, 38*, 3329–3340.
Chang, Li-Te, Koutrakis, P., Catalano, P. J., & Suh, H. H. (2000). Hourly personal exposures to fine particles and gaseous pollutants – results from Baltimore, Maryland. *Journal of the Air & Waste Management Association, 50*(7), 1223–1235.
Chang, Li-Te, Suh, H. H., Wolfson, J. M., Misra, K., Allen, G. A., Catalano, P. J., et al. (2001). Laboratory and field evaluation of measurement methods for one-hour exposures to O_3, $PM_{2.5}$, and CO. *Journal of the Air & Waste Management Association, 51*(10), 1414–1422.
Chapin, F. S., Jr. (1974). *Human activity patterns in the city: Things people do in time and space* (272 pp.). New York: Wiley Intersciences.
Chen, B. T., Feather, G. A., Maynard, A., & Rao, C. Y. (2004). Development of a personal sampler for collecting fungal spores. *Aerosol Science and Technology, 38*, 926–937.
Chow, J. C., Engelbrecht, J. P., Freeman, N. C. G., Hashim, J. H., Jantunen, M., Michaud, J.-P., et al. (2002). Chapter one: Exposure measurement. *Chemosphere, 49*, 875–901.
Cohen, M. A., & Cotey, M. R. (1997). The use of hand held pen computer for field data entry. *Applied Occupational and Environmental Hygiene, 12*, 792–795.
Demokritou, P., Kavouras, I. G., Ferguson, S. T., & Koutrakis, P. (2001). Development and laboratory performance evaluation of a personal multipollutant sampler for simultaneous measurement of particulate and gaseous pollutants. *Aerosol Science and Technology, 35*, 741–752.
Dockery, D. W., & Pope, C. A., III. (1994). Acute respiratory effects of particulate air pollution. *Annual Review of Public Health, 15*, 107–132.
Dockery, D. W., Pope, C. A., III, Xu, X., Spengler, J. D., Ware, J. H., Fay, M. E., et al. (1993). An association between air pollution and mortality in six US cities. *New England Journal of Medicine, 329*, 1753–1759.
Dockery, D. W., & Spengler, J. D. (1981). Personal exposure to respirable particulates and sulphates. *Journal of the Air Pollution Control Association, 31*, 153–159.
Dor, F., Dab, W., Empereur-Bissonet, P., & Zmirou, D. (1999). Validity of biomarkers in environmental health studies: The case of PAHs and benzene. *Critical Reviews in Toxicology, 29*(2), 129–168.
Duan, N. (1982). Model for human exposure to air pollution. *Environment International, 8*, 305–309.
Freeman, N. C. G., Lioy, P. J., Pellizzari, E., Zelon, H., Thomas, K., Clayton, A., et al. (1999). Responses to the Region 5 NHEXAS time/activity diary. *Journal of Exposure Analysis & Environmental Epidemiology, 9*, 414–426.
Freeman, N. C. G., & Saenz de Tejada, S. (2002). Methods for collecting time-activity information related to exposure to combustion products. *Chemosphere, 49*, 979–992.
Gauvin, S., Reungoat, P., Cassadou, S., Déchenaux, J., Momas, I., Just, J., et al. (2002). Contribution of indoor and outdoor environments to $PM_{2.5}$ personal exposure of children – VESTA study. *Science of the Total Environment, 297*, 175–181.
Georgiadis, P., Topinka, J., Stoikidou, M., Kaila, S., Gioka, M., Katsouyanni, K., et al. (2001). Biomarkers of genotoxicity of air pollution (the AULIS project): bulky DNA adducts in subjects with moderate to low exposures to airborne polycyclic aromatic hydrocarbons and their relationship to environmental tobacco smoke and other parameters. *Carcinogenesis, 22*(9), 1447–1457.
Georgopoulos, P. G., & Lioy, P. J. (2006). From a theoretical framework of human exposure and dose assessment to computational system implementation: The Modeling Environment for Total Risk studies (MENTOR). *Journal of Toxicology & Environmental Health – Part B, 9*(6), 457–483.
Georgopoulos, P. G., Wallace, L. A., Roy, A., & Lioy, P. J. (1997). Integrated exposure and dose modelling and analytical system. 1. Formulation and testing of microenvironmental and pharmacokinetic components. *Environmental Science and Technology, 31*(1), 17–27.

Geyh, A. S., Roberts, P. T., Lurmann, F. W., Schoell, B. M., & Avol, E. L. (1999). Initial field evaluation of the Harvard active ozone sampler for personal ozone monitoring. *Journal of Exposure Analysis & Environmental Epidemiology, 2*, 143–149.

Gore, R. B., Curbishley, L., Truman, N., Hadley, E., Woodcock, A., Langley, S. J., et al. (2006). Intranasal air sampling in homes: relationships among reservoir allergen concentrations and asthma severity. *Allergy & Clinical Immunology, 117*(3), 649–655.

Gore, R. B., Hadi, E. A., Craven, M., Smillie, F. I., O'Meara, T. J., Tovey, E. R., et al. (2002). Personal exposure to house dust mite allergen in bed: Nasal air sampling and reservoir allergen levels. *Clinical & Experimental Allergy, 32*(6), 856–859.

Gorecki, T., & Namiesnik, J. (2002). Passive sampling. *Trends in Analytical Chemistry, 21*(4), 276–291.

Görner, P., Bemer, D., & Fabriés, J. F. (1995). Photometer measurements of polydisperse aerosols. *Journal of Aersol Science, 8*, 1281–1302.

Graham, J. A. H., Pavlicek, P. K., Sercombe, J. K., Xavier, M. L., & Tovey, E. R. (2000). The nasal air sampler: A device for sampling inhaled aeroallergens. *Annals of Allergy, Asthma & Immunology, 84*(6), 599–604.

Grandjean, P. (1995). Biomarkers in epidemiology. *Clinical Chemistry, 41*(12), 1800–1803.

Groopman, J. D., & Kensler, T. W. (1999). The light at the end of the tunnel for chemical-specific biomarkers: Daylight or headlight? *Carcinogenesis, 20*, 1–11.

Hagenbjörk-Gustafsson, A., Lindahl, R., Levin, J.-O., & Karlsson, D. (2002). Validation of the Willems badge diffusive sampler for nitrogen dioxide determinations in occupational environments. *Analyst, 127*, 163–168.

Hansen, A. M., Wallin, H., Binderup, M. L., Dybdahl, M., Autrup, H., Loft, S., et al. (2004). Urinary 1-hydroxypyrene and mutagenicity in bus drivers and mail carriers exposed to urban air pollution in Denmark. *Mutation Research, 557*(1), 7–17.

Harrison, R. M., Thornton, C. A., Lawrence, R. G., Mark, D., Kinerslay, R. P., & Ayers, J. G. (2002). Personal exposure monitoring of particulate matter, nitrogen dioxide, and carbon monoxide, including susceptible groups. *Occupational and Environmental Medicine, 59*, 671–679.

Herber, R. F. M., Duffus, J. H., Christensen, J. M., Olsen, E., Park, M. V. (2001). Risk assessment for occupational exposure to chemicals. Review of current methodology (IUPAC Technical Report). *Pure and Applied Chemistry, 73*(6), 993–1031.

Hood, L., Heath, J. R., Phelps, M. E., & Lin, B. (2004). Systems biology technologies enable predictive and preventive medicine. *Science, 306*(5696), 640–643.

Howard-Reed, C., Rea, A. W., Zufall, M. J., Burke, J. M., Williams, R. W., Suggs, J. C., et al. (2000). Use of a continuous nephelometer to measure personal exposure to particles during the US Environmental Protection Agency Baltimore and Fresno panel studies. *Journal of the Air & Waste Management Association, 50*, 1125–1132.

Janssen, N. A. H., de Hartog, J. J., Hoek, G., & Brunekreef, B. (2000). Personal exposure to fine particulate matter in elderly subjects: relation between personal, indoor and outdoor concentrations. *Journal of the Air & Waste Management Association, 50*, 1133–1143.

Janssen, N. A. H., Hoek, G., Harssema, H., & Brunekreef, B. (1999). Personal exposure to fine particles in children correlates closely with the ambient fine particles. *Archives of Environmental Health, 54*(2), 95–101.

Jantunen, M., Hanninen, O., Katsouyanni, K., Knoppel, H., Kuenzli, N., Lebret, E., et al. (1998). Air pollution exposure in European cities: The EXPOLIS study. *Journal of Exposure Analysis & Environment Epidemiology, 8*(4), 495–518.

Jantunen, M., Hänninen, O., Koistinen, K., & Hashim, J. H. (2002). Fine PM measurements: personal and indoor monitoring. *Chemosphere, 49*, 993–1007.

Jenkins, R. A., & Counts, R. W. (1999). Personal exposure to environmental tobacco smoke: salivary cotinine, airborne nicotine, and nonsmoker misclassification. *Journal of Exposure Analysis & Environment Epidemiology, 9*, 352–363.

Jenkins, R. A., Ilgner, R. H., & Tomkins, B. A. (2004). Development and application of protocols for the determination of response of real-time particle monitors to common indoor aerosols. *Journal of the Air & Waste Management Association, 54*, 229–241.

Jo, W.-K., & Pack, K.-W. (1999). Utilization of breath analysis for exposure estimates of benzene associated with active smoking. *Environmental Research, 83*, 180–187.

Jones, A. Y. M., Lam, P. K. W., & Dean, E. (2006). Respiratory health of bus drivers in Hong Kong. *International Archives of Occupational and Environmental Health, 79*, 414–418.

Katsouyani, K., Zmirou, D., Spix, C., Schwartz, J., Balducci, F., Medina, S., et al. (1997). Short term effect of ambient sulphur dioxide and particulate matter on mortality in 12 European cities: results from time-series data from the APHEA project. *British Medical Journal, 314*, 1658–1663.

Kim, D., Sass-Kortsak, A., Purdham, J. T., Dales, R. E., & Brook, J. R. (2006). Association between personal exposures and fixed-site ambient measurements of fine particulate matter, nitrogen dioxide, and carbon monoxide in Toronto, Canada. *Journal of Exposure Analysis & Environmental Epidemiology, 16*, 172–183.

Klepeis, N. E., Nelson, W. C., Ott, W. R., Robinson, J. P., Tsang, A. M., Switzer, P., et al. (2001). The National Human Activity Pattern Survey (NHAPS): A resource for assessing exposure to environmental pollutants. *Journal of Exposure Analysis & Environmental Epidemiology, 11*, 231–252.

Kousa, A., Oglesby, L., Koistinen, K., Künzli, N., & Jantunen, M. (2002). Exposure chain of urban air $PM_{2.5}$ – associations between ambient fixed site, residential outdoor, indoor, workplace and personal exposures in four European cities in the EXPOLIS-study. *Atmospheric Environment, 36*, 3031–3039.

Kruppa, S. V. (1998). Passive sampling of ambient gaseous air pollutants: A review. In A. H. Legge & L. L. Jones (Eds.), *Proceedings of the international speciality conference, engineering air issues for the 21st century: the need for multidisciplinary management, VIP-78* (pp. 485–505). Pittsburgh, PA: Air and Waste Management Association.

Kruppa, S. V., & Legge, A. H. (2000). Passive sampling of ambient, gaseous air pollutants: An assessment from an ecological perspective. *Environmental Pollution, 107*, 31–45.

Kwok, R. (2004). *Use of 1) Sensors and 2) Radio Frequency ID (RFID) for the national children's study* (EPA 600/R-05/018, 85pp.). Final Report.

Lanki, T., Alm, S., Ruuskanen, J., Janssen, N. A. H., Jantunen, M., & Pekkanen, J. (2002). Photometrically measured continuous personal $PM2_{2.5}$ exposure and correlation to a gravimetric method. *Journal of Exposure Analysis & Environmental Epidemiology, 12*, 172–178.

Leaderer, B. P. (1990). Assessing exposures to environmental tobacco smoke. *Risk Analysis, 10*, 19–26.

Leaderer, B. P., Lioy, P. J., & Spengler, J. D. (1993). Assessing exposures to inhaled complex mixtures. *Environmental Health Perspectives, 101*(Suppl. 4), 167–177.

Leaderer, B. P., Zagraniski, R. T., Berwick, M., & Stolwijk, J. A. (1986). Assessment of exposure to indoor air contaminants from combustion sources: methodology and application. *American Journal of Epidemiology, 124*, 275–289.

Lioy, P. J. (1990). Assessing total human exposure to contaminants: A multidisciplinary approach. *Environmental Science and Technology, 24*(7), 938–945.

Lioy, P. J. (1993). Measurements of personal exposure to air pollution: Status and needs. In L. Newan (Ed.), *Measurement challenges in atmospheric chemistry* (Chapter 13, pp. 373–390). Advances in Chemistry Series 232. Washington, DC: American Chemical Society.

Lioy, P. J., Waldman, J. M., Buckley, T., Butler, J., & Pietarinen, C. (1990). The personal, indoor and outdoor concentration of PM-10 measured in an industrial community during the winter. *Atmosperic Environment, 24B*, 57–66.

Liu, L.-J. S., Box, M., Kalman, D., Kaufman, J., Koenig, J., Larson, T., et al. (2003). Exposure assessment of particulate matter for susceptible populations in Seattle. *Environmental Health Perspectives, 111*(7), 909–918.

Liu, L.-J. S., Olson, M. P., Allen, G. A., Koutrakis, P., McDonell, W. F., & Gerrity, T. R. (1994). Evaluation of the Harvard ozone passive sampler on human subjects indoors. *Environmental Science and Technology, 28*(5), 915–923.

Liu, L.-J. S., Slaughter, J. C., & Larson, T. V. (2002). Comparison of light scattering devices and impactors for particulate measurements in indoor, outdoor, and personal environments. *Environmental Science and Technology, 36*, 2977–2986.

Logan, W. P. D. (1953). Mortality in the London fog incident. *Lancet, 261*(6755), 336–338.
Malhotra, P., Saksena, S., & Joshi, V. (2000). Time budgets of infants for exposure assessment: A methodological study. *Journal of Exposure Analysis & Environmental Epidemiology, 10*, 267–284.
Martos, P. A., & Pawlyszyn, J. (1999). Time-weighted average sampling with solid-phase microextraction devise: Implications for enhanced personal exposure monitoring to airborne pollutants. *Analytical Chemistry, 71*, 1513–1520.
McCurdy, T., Glen, G., Smith, L., & Lakkadi, Y. (2000). The National Exposure Research Laboratory's consolidated human activity database. *Journal of Exposure Analysis & Environmental Epidemiology, 10*, 566–578.
Meng, Q. Y., Turpi, B. J., Polidori, A., Lee, J. H., Weisel, C., Morandi, M., et al. (2005b). $PM_{2.5}$ of ambient origin: Estimates and exposure errors relevant to PM epidemilogy. *Environmental Science and Technology, 39*(14), 5105–5112.
Meng, Q. Y., Turpin, B. J., Korn, L., Weisel, C. P., Morandi, M., Colome, S., et al. (2005a). Influence of ambient (outdoor) sources on residential indoor and personal $PM_{2.5}$ concentrations: analyses of RIOPA data. *Journal of Exposure Analysis & Environmental Epidemiology, 15*, 17–28.
Misra, Ch, Singh, M., Shen, S., Sioutas, C., & Hall, P. (2002). Development and evaluation of a personal cascade impactor sampler. *Journal of Aerosol Science, 33*, 1027–1047.
Mitakakis, T. Z., Tovey, E. R., Xuan, W., & Marks, G. B. (2000). Personal exposure to allergenic pollen and mould spores in inland New South Wales, Australia. *Clinical & Experimental Allergy, 30*, 1733–1739.
Monn, C. H. (2001). Exposure assessment of air pollutants: A review on spatial heterogeneity and indoor/outdoor/personal exposure to suspended particulate matter, nitrogen dioxide and ozone. *Atmospheric Environment, 35*(1), 1–32.
Monn, Ch., Fuchs, A., Högger, D., Junker, M., Kogelschatz, D., Roth, N., et al. (1997). Particulate matter less than 10μm (PM_{10}) and fine particles less than 2.5μm ($PM_{2.5}$): relationships between indoor, outdoor and personal concentrations. *Science of the Total Environment, 208*, 15–21.
Morawska, L., He, C., Hitchins, J., Mengersen, K., & Gilbert, D. (2003). Characteristics of particulate and mass concentrations in residential houses in Brisbane, Australia. *Atmospheric Environment, 37*, 4195–4203.
Namiesnik, J., Zabiegala, B., Kot-Wasik, A., Partyka, M., & Wasik, A. (2005). Passive sampling and/or extraction techniques in environmental analysis: A review. *Analytical and Bioanalytical Chemistry, 381*, 279–301.
O'Neill, M. S., Ramirez-Aguilar, M., Meneses-Gonzales, F., Hernández-Avila, M., Geyh, A. S., Sienra-Monge, J. J., et al. (2003). Ozone exposure among Mexico City outdoor workers. *Journal of the Air & Waste Management Association, 53*(3), 339–346.
Oglesby, L., Künzli, N., Röösli, M., Braun-Fahrländer, Ch., Mathys, P., Stern, W., et al. (2000a). Validity of ambient levels of fine particles as surrogate for personal exposure to outdoor air pollution – results of the European EXPOLIS-EAS study (Swiss Center Basel). *Journal of the Air & Waste Management Association, 50*, 1251–1261.
Oglesby, L., Rotko, T., Krütli, P., Boudet, C., Kruize, H., Jantunen, M. J., et al. (2000b). Personal exposure assessment studies may suffer from exposure-relevant selection bias. *Journal of Exposure Analysis & Environmental Epidemiology, 10*, 251–266.
Ott, W. R. (1982). Concepts of human exposure to air pollution. *Environment International, 7*, 179–186.
Ott, W. R. (1985). Total human exposure. *Environmental Science and Technology, 19*(10), 880–886.
Ouyang, G., & Pawliszyn, J. (2006). Recent developments in SPME for on-site analysis and monitoring. *Trends in Analytical Chemistry, 25*, 602–703.
Palmes, E. D., & Gunnison, A. F. (1973). Personal monitoring device for gaseous contaminants. *American Industrial Hygiene Association Journal, 34*, 78–81.
Palmes, E. D., Gunnison, A. F., DiMattio, J., & Tomczyk, C. (1976). Personal sampler for nitrogen dioxide. *American Industrial Hygiene Association Journal, 37*, 570–577.

Pawliszyn, J. (1997). *Solid phase microextraction. Theory and practice* (264 pp.). New York: Willey WCH.

Pellizzari, E., Perritt, R. L., & Clayton, C. A. (1999). National human exposure assessment survey (NHEXAS): exploratory survey of exposure among population subgroups in EPA Region 5. *Journal of Exposure Analysis & Environmental Epidemiology, 9*, 49–55.

Perdelli, F., Cristina, M. L., Sartini, M., & Orlando, P. (2002). Urinary hydroxyproline as a biomarker of effect after exposure to nitrogen dioxide. *Toxicology Letters, 134*(1–3), 319–323.

Perico, A., Gottardi, M., Boddi, V., Bavazzano, P., & Lanciotti, E. (2001). Assessment of exposure to polycyclic aromatic hydrocarbons in police in Florence, Italy, through personal air sampling and biological monitoring of the urinary metabolite 1-hydroxypyrene. *Archives of Environmental Health, 56*, 506–512.

Perlin, S. A., Sexton, K., & Wong, D. W. S. (1999). An examination of race and poverty for populations living near industrial sources of air pollution. *Journal of Exposure Analysis and Environmental Epidemiology, 9*(1), 29–48.

Pope, C. A., III, & Dockery, D. W. (2006). Health effects of fine particulate air pollution: Lines that connect. *Journal of the Air & Waste Management Association, 56*, 709–742.

Quintana, P. J. E., Valenzia, J. R., Delfino, R. J., & Liu, L.-J. S. (2001). Monitoring of 1min personal particulate matter exposures in relation to voice recorded time activity data. *Environmental Research Section A, 87*, 199–213.

Ramachandran, G., Adgate, J. L., Pratt, G. C., & Sexton, K. (2003). Characterizing indoor and outdoor 15 minute average $PM_{2.5}$ concentrations in urban neighborhoods. *Aerosol Science and Technology, 37*, 33–45.

Rappaport, S. M., Symanski, E., Yager, J. W., & Kupper, L. L. (1995). The relationship between environmental monitoring and biological markers in exposure assessment. *Environmental Health Perspectives, 103*(Suppl. 3), 49–53.

Raynie, D. E. (2000). Modern extraction techniques. *Analytical Chemistry, 78*, 3997–4003.

Reed, K. J., Jimenez, M., Lioy, P. J., & Freeman, N. C. G. (1999). Quantification of children's hand and mouthing activities. *Journal of Exposure Analysis & Environmental Epidemiology, 9*, 513–520.

Ren, Z. G., & Stewart, J. (2005). Prediction of personal exposure to contaminant sources in industrial buildings using a sub-zonal model. *Environmental Modelling & Software, 20*(5), 623–638.

Rojas-Bracho, L., Suh, H., & Koutrakis, P. (2000). Relationships among personal, indoor, and outdoor fine and coarse particle concentrations for individuals with COPD. *Journal of Exposure Analysis & Environmental Epidemiology, 10*, 294–306.

Rubow, K. L., Marple, V. A., Olin, J., & McCawley, M. A. (1987). A personal cascade impactor: design, evaluation and calibration. *American Industrial Hygiene Association Journal, 48*(6), 532–538.

Ruchirava, M., Mahidol, C., Tangjarukij, C., Pui-Ock, S., Jensen, O., Kampeerawipakorn, O., et al. (2002). Exposure to genotoxins present in ambient air in Bangkok, Thailand – particle associated polycyclic aromatic hydrocarbons and biomarkers. *Science of the Total Environment, 287*, 121–132.

Sarnat, S. E., Coull, B. A., Schwartz, J., Gold, D. R., & Suh, H. H. (2006). Factors affecting the association between ambient concentrations and personal exposures to particles and gases. *Environmental Health Perspectives, 114*, 649–654.

Sarnat, J. A., Koutrakis, P., & Suh, H. (2000). Assessing the relationship between personal particulate and gaseous exposures of senior citizens living in Baltimore. *Journal of the Air & Waste Management Association, 50*, 1184–1198.

Sarnat, J. A., Schwartz, J., Catalano, P. J., & Suh, H. H. (2001). Gaseous pollutants in particulate matter epidemiology: Confounders or surrogates? *Environmental Health Perspectives, 109*(10), 1053–1061.

Saxena, S., Prasad, R., Pal, R. S., & Joshi, V. (1992). Patterns of daily exposure to TSP and CO in the Garhwal Himalaya. *Atmospheric Environment, 26A*(11), 2125–2134.

Scherer, G. (2005). Biomonitoring of inhaled complex mixtures – ambient air, diesel exhaust and cigarette smoke. *Experimental and Toxicologic Pathology, 57*, 75–110.

Scherer, G., & Richter, E. (1997). Biomonitoring exposure to environmental tobacco smoke (ETS): A critical reappraisal. *Human & Experimental Toxicology, 16*(8), 449–459.

Scherer, G., Ruppert, T., Daube, H., Kossien, I., Riedel, K., Tricker, A. R., et al. (1995). Contribution of tobacco smoke to environmental benzene exposure in Germany. *Environment International, 21*, 779–789.

Schulte, P. A., & Talaska, G. (1995). Validity criteria for the use of biological markers of exposure to chemical agents in environmental epidemiology. *Toxicology, 101*, 73–78.

Sexton, K., Gong, H., Ailar, J. C., Ford, J. G., Gold, D. R., Lambert, W. E., et al. (1993). Air pollution health risks. Do class and race matter? *Toxicology & Industrial Health, 9*(5), 843–878.

Sexton, K., & Ryan, P. B. (1988). Assessment of human exposure to air pollution: methods, measurements and models. In A. Y. Watson, R. R. Bates, & D. Kennedy (Eds.), *Air pollution, the automobile and public health* (pp. 207–238). Washington, DC: National Academic Press.

Sexton, K., Selevan, S., Wagne, D., & Lybarger, J. (1992). Estimating human exposures to environmental pollutants: The availability of existing databases. *Archives of Environmental Health, 47*(6), 398–407.

Sexton, K., Spengler, J. D., & Treitman, R. D. (1984). Personal exposure to respirable particles: A case study in Waterbury, Vermont. *Atmospheric Environment, 18*, 1385–1398.

Singh, M., Misra, Ch, & Sioutas, C. (2003). Field evaluation of a personal cascade impactor sampler (PCIS). *Atmospheric Environment, 37*, 4781–4793.

Sioutas, S., Kim, S., Chang, M., Terrel, L. L., & Gong, H., Jr. (2000). Field evaluation of a modified DataRAM MIE scattering monitor for real-time $PM_{2.5}$ mass concentration measurements. *Atmospheric Environment, 34*, 4829–4838.

Smith, K. R. (2002). Indoor air pollution in developing countries: Recommendations for research. *Indoor Air, 12*(3), 198–207.

Staimer, N., Delfino, R. J., Buffalino, Ch, Fine, P. M., Sioutas, C., & Kleinman, M. T. (2005). A miniaturized active sampler for the assessment of personal exposure to nitrogen dioxide. *Analytical and Bioanalytical Chemistry, 383*, 955–962.

Tardiff, R. G., & Goldstein, B. D. (Eds.). (1991). *Methods for assessing exposure of human and non-human biota* (417 pp.). SCOPE 46, IPCS Joint Symposia. Chichester, Wiley.

Thomas, K. W., Pellizzari, E. D., Clayton, A., Whitaker, D. A., Shores, R. C., Spengler, J., et al. (1993). Particle total exposure assessment methodology (PTEAM) 1990 study: Method performance and data quality for personal, indoor, and outdoor monitoring. *Journal of Exposure Analysis & Environmental Epidemiology, 3*, 203–226.

Vinzents, P. S. (1996). A passive personal dust monitor. *Annals of Occupational Hygiene, 40*(3), 261–280.

Wallace, L. (2000). Correlations of personal exposure to particles with outdoor air measurements: a review of recent studies. *Aerosol Science and Technology, 32*, 15–25.

Wallace, L. A., Mitchell, H., O'Connor, G. T., Neas, L., Lippmann, M., Kattan, M., et al. (2003). Particle concentrations in inner city homes of children with asthma: the effect of smoking, cooking, and outdoor pollution. *Environmental Health Perspectives, 111*(9), 1265–1272.

Ward, J. B., Jr., & Henderson, R. E. (1996). Identification of needs in biomarker research. *Environmental Health Perspectives, 104*, 895–900.

Watts, R., Lewtas, J., Stevens, R., Hartlage, T., Pinto, J., Williams, R., et al. (1994). Czech-United-States EPA health study – assessment of personal and ambient air exposures to pah and organic mutagens in the Teplice district of Northern Bohemia. *International Journal of Environmental Analytical Chemistry, 56*(4), 271–287.

Weis, B. K., Balshawl, D., Barr, J. R., Brown, D., Ellisman, M., Liov, P., et al. (2005). Personalized exposure assessment: Promising approaches for human environmental health research. *Environmental Health Perspectives, 113*(7), 840–848.

Wheeler, A. J., Williamd, I., Beaumont, R. A., & Hamilton, R. S. (2000). Characterization of particulate matter sampled during a study of children's personal exposure to airborne particulate matter in a UK urban environment. *Environmental Monitoring and Assessment, 65*, 69–77.

Whitmore, R. W., Byron, M. Z., Clayton, C. A., Thomas, K. W., Zelon, H. S., Pellizzari, E., et al. (1999). Sampling design, response rate and analysis weights for the National Human Exposure Assessment Survey (NHEXAS) in EPA region 5. *Journal of Exposure Analysis & Environmental Epidemiology, 9*, 369–380.

Williams, R., Suggs, J., Creason, J., Rodes, C. H., Lawless, P., Kwok, R., et al. (2000). The 1998 Baltimore particulate matter epidemiology–exposure study: Part 2. Personal exposure assessment associated with an elderly study population. *Journal of Exposure Analysis & Environmental Epidemiology, 10*, 533–543.

Wilson, W. E., Mage, D. T., & Grant, L. D. (2000). Estimating separately personal exposure to ambient and nonambient particulate matter for epidemiology and risk assessment: Why and how. *Journal of the Air & Waste Management Association, 50*, 1167–1183.

Wu, Ch.-F., Delfino, R. J., Floro, J. N., Quintana, P. J. E., Samimi, B. S., Kleinman, M. T., et al. (2005b). Exposure assessment and modelling of particulate matter for asthmatic children using personal nephelometers. *Atmospheric Environment, 39*, 3457–3469.

Wu, Ch.-F., Delfino, R. J., Floro, J. N., Samimi, B. S.,Quintana, P. J. E., Kleinman, M. T., et al. (2005a). Evaluation and quality control of personal nephelometers in indoor, outdoor and personal environments. *Journal of Exposure Analysis & Environmental Epidemiology, 15*, 99–110.

Yanagisawa, Y., Nishimura, H., Matsuki, H., Osaka, F., & Ksuga, H. (1988). Urinary hydroxyproline to creatinine ratio as a biological effect marker of exposure to NO_2 and tobacco smoke. *Atmospheric Environment, 22*(10), 2195–2203.

Yanoski, J. D., Williams, P. L., & MacIntosh, D. L. (2002). A comparison of two direct-reading aerosol monitors with the federal reference method for $PM_{2.5}$ in indoor air. *Atmospheric Environment, 36*, 107–113.

Zagury, E., Le Moulec, Y., & Momas, I. (2000). Exposure of Paris taxi drivers to automobile air pollutants within their vehicles. *Occupational and Environmental Medicine, 57*, 406–410.

Chapter 5
Health Effects of Air Pollutants

Sotiris Vardoulakis, Xiyu Phoon, and Caroline Ochieng

Abstract Although air quality in large cities and industrial areas of Western Europe was deteriorating since the beginning of the industrial revolution in nineteenth century, it was not until the first half of the twentieth century that the adverse health effects of air pollution started attracting scientific attention. Two widely quoted air pollution episodes, the Great London Smog of 1952 and the Meuse Valley (Belgium) episode of 1930, raised public awareness of the potential health effects of air pollution. The London Smog, attributed to the widespread use of coal for domestic heating in London and the stagnant atmospheric conditions prevailing in the capital in December 1952, caused approximately 4,000 premature deaths in a period of around 1 week (representing a 200–300% increase in mortality), with mortality remaining above normal levels for several months after the smog episode. Although air quality has improved a lot since the 1950s due to cleaner fuels, better industrial processes and control technologies, road traffic has emerged as the dominant source of air pollution in developed countries. In December 1991, another winter smog episode induced by similar weather conditions as in 1952 occurred in London causing a 10–25% increase in mortality. In this event, road transport and to a lesser extent domestic heating were the main sources of air pollution. In this Chapter, the health effects of a wide range of outdoor and indoor air pollutants are reviewed based on current toxicological and epidemiological evidence.

5.1 Introduction

In the last 2 decades, substantial improvements in urban air quality have been achieved in most developed countries mainly with the introduction of unleaded petrol, three-way catalytic converters and enhanced engine technology in vehicles.

S. Vardoulakis (✉), X. Phoon, and C. Ochieng
Public & Environmental Health Research Unit,
London School of Hygiene & Tropical Medicine, Keppel Street, London WC1E 7HT, UK
e-mail: sotiris.vardoulakis@lshtm.ac.uk

M. Lazaridis and I. Colbeck (eds.), *Human Exposure to Pollutants via Dermal Absorption and Inhalation*, Environmental Pollution 17, DOI 10.1007/978-90-481-8663-1_5,

Nevertheless, the ever increasing volume of road traffic has offset some of these environmental gains. In developing countries, the rapid industrialisation and increased car ownership have resulted in deteriorating air quality conditions, especially in East Asian, Latin American and African megacities.

In this period of improving ambient air quality (at least in Europe and North America), indoor air quality has attracted public concern and increasing scientific interest. Public concern has been fuelled by widely publicised acute poisoning events such as CO poisoning (e.g. caused by faulty gas appliances in houses), chronic health effects such as leukaemia caused by long-term exposure to certain solvents, allergies and other symptoms associated with the so called sick building syndrome.

It should be emphasised that in the developing world, the greater air pollution related health risks are posed by the use of biomass fuels indoors mainly in rural areas. Globally indoor air pollution from biomass fuels use is responsible for around two million excess deaths in developing countries and for 4% of the global burden of disease. This makes it one of the top ten global health risks, responsible for a similar proportion of the global burden of disease as tobacco smoking and unsafe sex (WHO 2002).

This chapter discusses the health effects of the main outdoor and indoor air pollutants, including pollutant mixtures, mainly associated with non-occupational exposures. A brief discussion of the epidemiology, risk analysis and health impact assessment of air pollution is also included. The paper does not deal with the indirect health effects of air pollution, including knock-on effects of pollutants on vegetation, building materials, visibility and global/regional climate. Physical and human health risks from explosion of flammable gases (e.g. methane), radon and asbestos are not discussed either.

5.2 Epidemiological Evidence and Mechanisms of Toxicity

Several air pollutants have been associated with acute effects (e.g. asthma attacks, respiratory infections, etc.) as well as chronic health effects (such as chronic obstructive pulmonary disease and lung cancer). For example, there is convincing evidence linking short-term exposure to particulate matter (PM_{10}) with cardiovascular deaths (COMEAP 2006). Long-term exposure to polycyclic aromatic hydrocarbons (PAH) and especially benzo[a]pyrene has been shown to increase the risk of lung cancer (Armstrong et al. 2004). However, only a small proportion of the population is expected to suffer severe health effects including premature death due to air pollution, while larger numbers of individuals may experience relatively mild symptoms such as impaired pulmonary function (Fig. 5.1). It should be noted that some individuals are more susceptible than others to air pollution, with the young being more sensitive in general. For example, the high bronchial lability and responsiveness in the very young seems to disappear with growth (Lebowitz 1991).

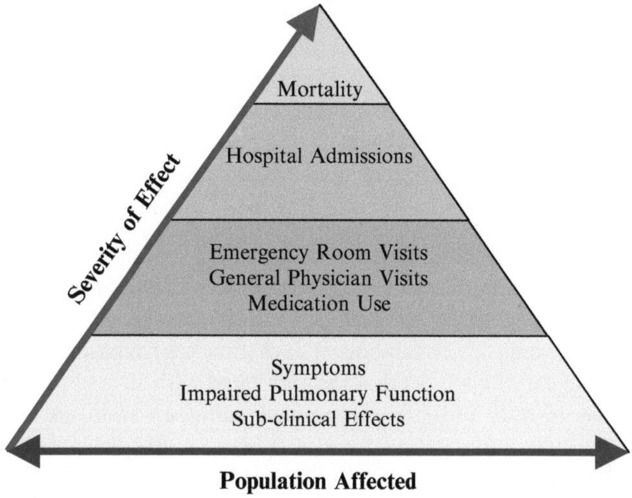

Fig. 5.1 Severity of health effects and size of affected population

Susceptibility may be genetic and/or due to pre-existing illness, such as asthma. These individuals require special attention and air quality regulations should be stringent enough to protect them.

Air pollution epidemiology relies on advance statistical techniques such as multivariable regression analysis, to detect even small effects of air pollutants on human health. These techniques allow for adjustment for the many potential confounding factors, like smoking, socio-economic status and weather, that may affect the strength of the statistical association between air pollution concentrations and incidence of certain health outcomes.

The evidence of the chronic and acute health effects of air pollution comes mainly from two types of epidemiological studies, *cohort* and *time-series* studies, based on personal and population exposure information respectively. Cohort studies follow a group of individuals over a long period of time, estimating individual levels of exposure to air pollutants, morbidity and mortality (e.g. cancer incidence), and other factors such as smoking and family history of disease. Time-series studies correlate daily pollutant concentrations obtained from one or a few fixed air quality monitoring sites with daily morbidity (e.g. respiratory hospital admissions) or mortality (e.g. cardiovascular deaths) within the same urban area. It should be remembered that the health impacts of air pollution derived from epidemiological time-series and cohort studies are not directly comparable as they rely on different statistical methods and assumptions, but it is likely that cohort studies can predict more closely the total impact on human health (Künzli et al. 2000).

Epidemiological evidence, strictly speaking, does not prove causality between air pollution and health effects as it can be argued that demonstrated associations are "merely statistical" (Maynard 2004). In order to prove causality, researchers have developed mechanistic hypothesis to explain the epidemiological evidence. For example, Seaton and coworkers (1995) argued that ultrafine particles can pass through the lung epithelium affecting blood clotting. Experimental animal models have been used to test such hypotheses.

A brief summary of the different methods that can be used to investigate the health effects of air pollutants is given below:

1. Time-series studies, examining the relationship between daily air pollution levels and daily health outcomes, such as mortality (total or cause specific) and hospital admissions. Daily air pollution can be associated with health effects observed on the same day or subsequent days (time-lag), adjusting for meteorology and other time varying factors that are correlated with air pollution. Exposure to air pollution is only indirectly estimated using an exposure surrogate, for example ambient air quality measurements from an urban centre monitoring station. Time-series and other epidemiological studies may suffer from lack of specificity as people are exposed to mixtures of pollutants in their everyday lives.
2. Cohort (or panel) studies, including occupational and non-occupational cohorts of individuals followed up for a period of several years or decades. In this case, health outcomes are associated with long-term exposure to air pollution.
3. Case-control studies, which compare *cases* of individuals who have developed an air pollution related disease against a representative *control* population. Case-control studies are distinguished from cohort studies by their sampling strategy, which allows determining whether an association exists between environmental exposure and relatively rare health outcomes.
4. Cross-sectional studies, in which prevalence of disease rates are compared across different geographical areas at the same point in time. These studies can be used to examine chronic health effects, but they are less powerful (and less expensive) than cohort studies as exposure to air pollutants is not reconstructed over time.
5. Chamber (or controlled challenge) studies, in which the effects of single pollutants or pollutant mixtures are examined on healthy adults or adults with mild asthma in a controlled environment. The advantage of these experiments is that the dose administered and the environmental conditions are strictly controlled. On the other hand the participants can only be relatively healthy adults (children and other susceptible groups are excluded for ethical reasons).
6. Laboratory animal experiments (in vivo), which can be used to improve our understanding of relationships between exposure to air pollutants and health effect.
7. Laboratory in vitro experiments investigating the various changes that may occur when a pollutant encounters a living cell.

The above methods can be broadly classified into epidemiological (1–4) and toxicological (5–7) studies. More information about study design in environmental epidemiology can be found elsewhere (Morgenstern and Thomas 1993).

5.3 Outdoor Air Pollutants

This section summarises the sources and health effects of key outdoor air pollutants, including inorganic gases, hydrocarbons and atmospheric particles. Epidemiological evidence on health effects of air pollutants can be divided according to the investigated period of exposure, that is short-term and long-term effects. Short-term effects relate to changes in health outcomes associated with assessed hourly, daily or weekly pollution exposures (as shown by time-series studies). Long-term effects are associated with exposure extending for a longer time-period (e.g. 1 year or more), and are able to capture effects of the pollutants on disease induction (i.e. incidence).

5.3.1 Inorganic Gases

5.3.1.1 Sulphur Dioxide

Sulphur dioxide (SO_2) is among the most common air pollutants. Man-made sources include fossil fuel combustion, smelting, manufacture of sulphuric acid, conversion of wood pulp to paper, incineration of refuse and production of elemental sulphur. Coal burning is the single largest man-made source of SO_2 accounting for about 50% of annual global emissions, with oil burning accounting for a further 25–30%. Ambient levels of SO_2 have declined greatly in recent decades in developed countries due to the use of cleaner fuels and flue-gas desulphurisation end-of-pipe techniques in industry. Natural sources of SO_2 include oceans, biological decay and forest fires. It also contributes to particulate aerosol formation (secondary particles).

Mechanisms of Action

SO_2 is thought to produce its effects by initiation of bronchoconstriction which occurs when the inhaled SO_2 is detoxified by the liver into sulphates, producing bisulfite ion. The ion produced reacts with water to form sulphurous acid, an irritant and inhibitor of mucociliary transport.

The acute effects of SO_2 have mainly been demonstrated through controlled chamber studies. SO_2 is a fast acting lung irritant and bronchoconstrictor (Katsouyanni et al. 1997). Exposures of between 1 and 6 h to concentrations as low as 2,860 $\mu g/m^3$ (1 ppm) have been shown to produce a reversible decrease in lung function ATSDR 1998. An exposure for 10–30 min at concentrations of 1.43×10^4 $\mu g/m^3$ (5 ppm) causes constriction of the bronchiole tubes. A 20 min exposure to 2.29×10^4 $\mu g/m^3$ (8 ppm) results in reddening of the throat and mild nose and throat irritation. Exposure to 5.72×10^4 $\mu g/m^3$ (20 ppm) is very irritating, while very high concentrations of 14.3×10^5 $\mu g/m^3$ (500 ppm) makes it impossible to breathe. One minute exposure to these very high concentrations results in severe airways obstruction, hypoxemia (insufficient oxygenation of the blood), pulmonary edema (fluid accumulation in the lungs) and death. The irritant effects are also observed in the eyes (Linn et al. 1997).

Short-term Effects

Several time-series studies on the health effects of SO_2 have been carried out. COMEAP gives a summary of nearly 90 studies exploring the relationship between SO_2 and cardiovascular hospital admissions and mortality (COMEAP 2006). The majority of the studies showed significant associations, although the estimates of total cardiovascular mortality were highly heterogeneous. The pooled estimate showed a 0.8% (95% CI, 0.6–1.0%) increase in total cardiovascular mortality for an increase of 10 μg/m^3 in SO_2. A study on SO_2 concentrations and hospital admissions in Hong Kong and London have also shown positive associations, with no evidence of a concentration threshold within the range of 5–40 μg/m^3 (Wong et al. 2002). Epidemiological studies of ambient SO_2 concentrations and childhood asthma hospitalisation and/or emergency room visits have however yielded mixed results, with some finding statistically significant associations while others reporting no association or findings of marginal statistical significance (Galan et al. 2003).

In the APHEA study which investigated the effects of several air pollutants in 12 European cities in ten countries (Katsouyanni et al. 1997), an increase of 50 μg/m^3 in SO_2 was associated with a 3% and 0.8% increase in daily mortality in Western and Central Europe respectively. The effect was significant independent of the level and composition of PM_{10}, suggesting that SO_2 may have had a direct effect on mortality (Katsouyanni et al. 1997). In the US National Morbidity, Mortality, and Air Pollution Study (NMMAPS) the effects of SO_2 on mortality was observed only when it was analysed independently. When analysed in combination with airborne particles, the effect was no longer significant (Samet et al. 2000). In Germany, a strong reduction of SO_2 concentrations occurred over a 10 year period following implementation of pollution control measures. Although there was a decline in mortality, it was attributed to reduction in concentrations of ultrafine particles and not SO_2 (Susanne Breitner et al. 2009).

A study of air pollution in Asian cities (PAPA) has further demonstrated associations between SO_2 (and NO_2, PM and O_3) and cardiovascular and respiratory mortality at rates greater than those in industrialised nations (Wong et al. 2008). In a study of 12 Canadian cities which examined the effect of NO_2 on mortality, SO_2 was also significantly associated with daily mortality at an average concentration of only 5 μg/m^3 (Burnett et al. 2004). In a study in Hong Kong (Peters et al. 1996), a major reduction in sulphur content in fuels over a very short period of time was associated with a significant decrease in mortality.

An epidemiologic study in Japan following a volcanic eruption demonstrated the effect of SO_2 on acute respiratory symptoms in healthy subjects (Ishigami et al. 2008). There was significant association between hourly incidence of respiratory symptoms and SO_2 concentrations, clearly demonstrating an exposure-response relationship. The effects were considered specific to SO_2.

Long-Term Effects

Long-term exposure to low level SO_2 has been shown to lead to the development of chronic obstructive lung disorders in both general and occupational populations (Schwela 2000) at concentrations as low as 2,860 $\mu g/m^3$ (1 ppm). The mechanism is explained by repeated episodes of bronchoconstriction due to repeated exposure, leading to permanent pulmonary impairment. It can also result in an altered sense of smell and increased susceptibility to respiratory infections (ASTDR 1998).

The American Cancer Society (ACS) cohort study that followed 1.2 million adults for 17 years reported associations between SO_2 and health effects at mean concentrations of 18 $\mu g/m^3$.

Other epidemiological studies have focused on cancers such as lung cancer, stomach cancer or brain tumours (Pope III et al. 2002). In most of the studies however, the role of confounding factors including exposures to other chemicals could not be overruled (WHO 2003). The International Agency for Cancer (IARC) considers the evidence of SO_2 carcinogenicity as inadequate.

5.3.1.2 Nitrogen Dioxide

Nitrogen dioxide (NO_2) is a ubiquitous product of combustion, occurring in motor vehicle exhausts, emissions from the combustion of stationary fuel sources such as coal, oil or natural gas and various industrial sources. In ambient conditions, NO_2 is generated through oxidation of nitric oxide by atmospheric oxidants such as ozone.

The major source of anthropogenic emissions of nitrogen oxides into the atmosphere is the combustion of fossil fuels from stationary sources (heating, power generation) and in motor vehicles, which is responsible for half the total NO_2 emissions. NO_2 production and use as an intermediate in the manufacture of nitric acid and as an oxidizing agent may also lead to its release to the environment through various waste streams. Natural sources, such as forest fires, atmospheric lightning discharges and the biogenic oxidation of nitrogen containing compounds in soils may also release quantities of NO_2 directly to the environment. NO_2 is a major component of the photochemical smog and thus a contributor to the formation of ozone.

Mechanisms of Action and Chamber Studies

NO_2 has the capacity to impair the function of alveolar macrophages and epithelial cells, hence can contribute to exacerbations of respiratory disease and increase the risk of lung infection. Two biochemical mechanisms of NO_2 cellular injury have been identified. One emphasizes lipid peroxidation, while the other the oxidation of low molecular weight reducing substances and proteins. Other studies have investigated the possibility that inhalation of NO_2 at moderate concentrations may cause an inflammatory reaction in the lungs or may increase the susceptibility of individuals to subsequent inhalation of allergens, such as those from house dust mite or grass pollen. The inflammatory reaction, if repeated frequently, might act to decrease the resistance of individuals to infection.

Around 80–90% of NO_2 is absorbed by the respiratory system, irritating the mucosal membranes and can reach the deeper regions of the respiratory tract acting on the bronchotracheal and alveolar regions (Chitano et al. 1995). Exposure to high levels of the gas (>18.8 mg/m^3 [>10 ppm]) can be fatal with the cause of death being pulmonary oedema (Kraft et al. 2005). It has been demonstrated that NO_2 odour can be detected at 0.2 mg/m^3 (0.11 ppm). Acute exposure to NO_2 (3–3.8 mg/m^3 [1.6–2.0 ppm] for 15 min) resulted in increased respiratory resistance which progressively increased above 3.8 mg/m^3 (2.0 ppm). Exposures to 7.5–9.4 mg/m^3 (4–5 ppm) NO_2 can also decrease arterial partial pressure of oxygen and carbon monoxide diffusion (WHO 1977).

In asthmatics, changes in lung function have been observed in some of the subjects at exposures of around 300 ppb when exercising (Department-of-the-Environment 1996). Inhalation of dust mite following an exposure to concentrations of 200–400 ppb for an hour resulted in great reduction in lung function, suggesting a negative effect of NO_2 on immunological response.

It should be noted however that the above levels are not normally reached in ambient air, even during high pollution events. Furthermore, the effects are very minimal in healthy adults even at exposures well above those observed in ambient air. Small changes are observed in lung function at concentrations above 500 μg/m^3 (260 ppb) which do not occur in ambient air.

Short-Term Effects

Results of time-series studies have shown a positive association between NO_2 levels and increased hospital admissions and mortality (WHO 2003). A review has also reported an association between NO_2 and cardiovascular mortality with a pooled risk estimate of 1% (95% CI, 0.8–1.3%) increase in hospital admissions for an increase of 10 μg/m^3 in NO_2 (COMEAP 2006).

A number of time-series studies have shown associations between hourly and daily fluctuations in NO_2 concentrations and mortality. However, the NMMAPS study that assessed mortality and NO_2 concentrations in 19 cities did not report any association (Samet et al. 2000). Similarly, the APHEA study did not report significant association between NO_2 levels and mortality (Katsouyanni et al. 1997). In a study of seven cities in Spain (EMECAM) (Ballester et al. 2006), positive associations were found between daily mortality (all causes and cardiovascular) and NO_2, once the rest of air pollutants were taken into account. A 10 μg/m^3

increase in the 24-h average 1-day NO_2 level was associated with an increase in the daily number of deaths of 0.43% (95% CI, –0.003–0.86%) for all causes. In a meta analysis of 109 studies published between 1982 and 2000, a random effects model showed consistent association between NO_2 levels and mortality from respiratory diseases (Stieb et al. 2002).

Questions have been raised as to whether NO_2 *per se* has adverse respiratory effects at ambient levels, since NO_2 may also be an indicator of traffic emissions (Brunekreef 2007). Recent studies have however shown NO_2 to have a higher spatial variation than particulate matter (Pattenden et al. 2006). In addition, adverse health effects of NO_2 have been observed in areas with high NO_2 concentrations but relatively low concentrations of other gaseous pollutants. Recent studies showing association between indoor NO_2 concentrations and respiratory symptoms in children further add strength to this argument (WHO 2003).

Long-Term Effects

Compared to time-series studies, the number of panel studies investigating associations of long-term exposure to NO_2 and health effects are small. Large cohort studies on air pollution and health, that is the Six Cities Study, the American Cancer Society Study and the Adventist Health SMOG (AHSMOG) study, reported significant associations between NO_2 concentrations and mortality (WHO 2003). In a cohort study of lung cancer and traffic related air pollution in the Netherlands, exposure to NO_2 was not associated with all causes of mortality after adjustment for confounding factors. The association however was significant for those living near a major road (Hoek et al. 2002). A case-control study on air pollution and respiratory mortality in infants showed positive associations with NO_2 (Bobak and Leon 1999). Finally, annual variations in NO_2 concentrations have been associated with symptoms of bronchitis in asthmatic children (WHO 2006).

5.3.1.3 Ozone

Ozone (O_3) is mainly a secondary pollutant generated by photochemical reactions in the atmosphere involving the action of sunlight on NO_2 and hydrocarbons. Its abundance in the atmosphere is linked to the levels of its precursors, that is the ratio of NO and NO_2 and presence of volatile organic hydrocarbons (VOC) in the atmosphere. Thus high levels tend to occur during hot summer days in or downwind of heavily populated areas, where sources emit the necessary VOC and NO_X to produce ozone. In Northern Europe, typical atmospheric concentrations are in the order of a few tens of μg/m^3, while substantially higher concentrations are routinely observed in Southern Europe especially during spring and summer. Concentrations in city centres tend to be lower than those in suburbs, mainly as a result of the scavenging of O_3 by NO originating from traffic.

Ground-level O_3 greatly differs from the protective stratospheric O_3 (the *ozone layer*), which shields the Earth's surface from intense ultraviolet (UV) radiation from the sun.

Mechanisms of Action and Chamber Studies

Ozone is a powerful oxidant as demonstrated by animal and human in vitro and in vivo studies. Upon inhalation, O_3 activates the stress signalling pathways in epithelial cells and alveolar inflammatory cells (Brunekreef and Holgate 2002). This induces an inflammatory response in the lung and releases neuropeptides from sensory neurons. This mechanism explains the acute bronchoconstrictor response and hyper-responsiveness seen in asthma patients upon exposure to O_3, as well as to the tolerance observed in subjects who have been repeatedly exposed to O_3 (Schelegle et al. 2001).

Aside from triggering asthmatic attacks, it has been suggested that in some cases O_3 might cause asthma (WHO 2003). In healthy individuals, chamber studies have shown that exposure for even short periods at relatively low concentrations significantly reduce lung function during periods of exercise (Thurston et al. 1997). This decrease in lung function is generally accompanied by other symptoms, including tightness of the chest, pain and difficulty breathing, coughing, and wheezing. Short-term inhalation at concentrations that occur in the urban environment have been shown to induce acute coronary artery vasoconstriction (WHO 2003).

Short-Term Effects

Like SO_2 and NO_2, hourly and daily fluctuations in O_3 concentrations have been associated with mortality and hospital admissions. The effect on hospital admissions has however been variable compared to the effect of other pollutants.

Several epidemiological studies have assessed the short-term effects of O_3 on various health outcomes. Most show significant increase of mortality risk (between 0.2% and 0.6% per increase of 10 $\mu g/m^3$ in O_3) (WHO 2003). The NMMAPS study reported a significant effect of O_3 during the summer season, which was a 0.4% increase in mortality associated with an increase of 20 $\mu g/m^3$ in daily O_3 concentrations at lag 0 (Samet et al. 2000). A larger effect was found at lag 2 (O_3 levels 2 days earlier), independent of other pollutants. Ozone daily levels have been associated with hospital respiratory admissions at all ages in most of the studies using 8-h mean concentrations as well as other averaging periods. The magnitude of the association was slightly larger than that obtained for mortality (0.5–0.7% increase in admissions per increase of 10 $\mu g/m^3$ in O_3) (WHO 2003). Studies on admissions for asthma in children have not found conclusive associations with O_3 measurements. However, there is evidence that during days when ozone levels are high, asthmatic subjects increase their use of medication that may mask any adverse O_3 effect

(Thurston et al. 1997). A study in Southern California has recently suggested that current levels of O_3 contribute to an increased risk of hospitalisation for children with asthma (Moore et al. 2008). A review of studies conducted from 1996 and 2001 has also shown a negative impact of O_3 on lung function (WHO 2003).

Estimates for the effect of 1-h O_3 concentrations on cardiovascular mortality have mostly reported significant and positive associations (COMEAP 2006). In a combined analysis of studies based on measurements of 8-h O_3 concentrations, COMEAP reported an overall increase of 0.4% (95% CI, 0.3–0.5%) in cardiovascular mortality for an increase of 10 μg/m³ in O_3. In contrast, estimates for 24-h O_3 were more variable and fewer were statistically significant.

Long-Term Effects

There are few epidemiological studies on the chronic effects of O_3 on human health. Incidence of asthma, decreased lung function growth, lung cancer and total mortality are the main health outcomes studied (Brunekreef and Holgate 2002). At levels currently observed in Europe, the evidence linking O_3 exposure to asthma incidence and prevalence in children and adults is not consistent.

Available evidence however suggests that long-term O_3 exposure reduces lung function growth in children. There is little evidence for an independent long-term O_3 effect on lung cancer or total mortality (WHO 2003). In the AHSMOG study where a cohort of 3,091 non-smokers was followed over 15 years, a significant association was found between asthma and 20 year mean 8-h average ambient O_3 concentration. In a cohort study of 3,535 children, aged 10–16 years with no history of asthma recruited from 12 communities in the Southern California study and followed during 5 years, the relative risk of developing asthma among children playing three or more sports (8% of the children) was 3.3 (95% CI, 1.9–5.8%) compared with children playing no sports in communities with high O_3 concentrations, but not in communities of low O_3 (McConnell et al. 2002).

5.3.1.4 Carbon Monoxide

The principal anthropogenic source of carbon monoxide (CO) is from incomplete combustion of fossil fuels, primarily from motor vehicles. Most ambient CO (75% in the UK) is due to road traffic. Of this, 97% comes from petrol engines, and only 3% from diesel engines. Typical ambient levels have substantially declined in Europe to a few ppm mainly due to the introduction of three-way catalytic converters.

Other persistent sources of CO are tobacco smoking and combustion of wood and coal, which can produce high concentrations close to the source. A significantly larger portion of an individual's total CO exposure may come from these sources. Natural sources include volcanoes and forest fires.

Mechanisms of Action

The most prominent pathophysiological effect of CO is hypoxemia caused by binding of CO to haemoglobin (Hb) (EPA 2000). Formation of carboxyhaemoglobin (COHb) reduces the O_2 carrying capacity of blood and impairs release of O_2 from O_2Hb to tissues. Failure of vasodilation to compensate causes tissue hypoxia. In addition to tissue hypoxia, ultimate diffusion of CO to cells may affect adversely their function. The brain and heart tissues are particularly sensitive to CO induced hypoxia and cytotoxicity.

Apart from impaired O_2 delivery to the tissues because of COHb formation, recent studies of CO pathophysiology suggest cytotoxic effects independent of O_2. There is a growing recognition of the role that CO may play in normal neurophysiology and in microvascular vasomotor control. The impact of CO from ambient air on these processes has however not been investigated adequately.

Unlike other air pollutants, CO does not directly affect eyes, nasal passages or lungs. There is evidence that CO even at low levels of concentration is associated with neurological damage especially in young children, and with adverse developmental effects in unborn children (EPA 2000).

Short-Term Effects

Short-term fluctuations in CO concentrations have been linked to exacerbations of pre-existing cardiovascular disease, mortality and respiratory illness in several time-series studies. The associations tend to be stronger during cold weather.

Several studies on CO levels and hospital admissions have been conducted in USA, UK and Europe (Ballester et al. 2002). Most of the studies have shown significant associations between ambient CO concentration and hospital admissions for heart disease. In North America studies, the relative risk of admissions associated with a 11.45 mg/m^3 (10 ppm) increase in ambient CO ranged from 1.10 in New York City to 1.37 in Los Angeles.

Associations of short-term fluctuations in CO levels and respiratory illness have been reported in some studies (EPA 2000). However, most of the studies report exacerbation of pre-existing respiratory illness rather than disease induction. Two studies have reported positive significant associations between CO levels (averaged over a 3 months period) and increased incidence of low birth weight (Pereira et al. 1998; Ritz and Yu 1999), while one study has demonstrated a protective effect (Alderman et al. 1987).

Studies of short-term ambient CO levels and daily mortality have yielded mixed results. In North and South America, four studies did not find association after adjusting for ambient particles, time trends, and weather (EPA 2000). One study in Europe (Verhoeff et al. 1996) found no relationship between 24-h mean CO concentrations and daily mortality with or without adjustment for PM_{10} and other gaseous pollutants. Three other studies (Touloumi et al. 1994; Salinas and Vega 1995; Wietlisbach et al. 1996) showed small, statistically significant relationships between CO and daily mortality. However, the effects of other pollutants (e.g., Total Suspended Particles, SO_2, NO_2, Black Smoke) and of meteorological variables were also significant. One study observed a distinct positive association of ambient CO levels with daily mortality (Touloumi et al. 1996).

Questions have been raised on whether low-level CO exposure actually increases mortality, particularly in the elderly population. It is also unclear whether CO is a surrogate for other combustion-related pollutants, or whether CO is a surrogate for the total combustion-related or traffic-related pollutant mixture. It should be also noted that outdoor CO levels may correlate with indoor levels of CO and other pollutants (which could be higher than outdoor levels) that are not usually accounted for in these studies.

Long-Term Effects

There are few epidemiologic studies that have investigated the long-term effects of CO exposure (EPA 2000). Health effects of long-term exposure to CO at present-day low ambient pollutant concentrations in developed countries are particularly uncertain. Also, it is not known whether long-term exposure to ambient CO plays a role in the induction of new cases of illness. However, in the Six Cities study (Laden et al. 2000), patterns of increasing hospital admissions with increasing ambient CO concentrations were observed across all the cities.

5.3.2 *Particles*

Airborne particulate matter represents a complex mixture of organic and inorganic substances which vary widely in origin, size, shape and chemical composition. The size of suspended particles in the atmosphere varies over four orders of magnitude, from a few nanometres to tens of micrometres.

Particles are classified by their aerodynamic diameter, which is the size of a unit density sphere with the same aerodynamic characteristics. This is convenient because these properties govern their transport and removal from the air, their deposition within the respiratory system and their chemical composition and sources (WHO 2003). In practical terms therefore, a distinction is made between PM_{10} or inhalable particles (<10 μm and can penetrate into the lower respiratory system), $PM_{2.5}$ or respirable particles (< 2.5 μm and can penetrate into the gas-exchange region of the lung), and PM_1 or small particles (<1 μm), the smallest of which are <0.1 μm (ultrafine particles) and have higher degrees of lung penetration (Halonen et al. 2008).

Small particles are largely formed from gases. Ultrafine particles are formed by nucleation resulting from condensation or chemical reactions that form new particles. Even though they contribute little to the total particulate mass, ultrafine particles are very high in number, reaching several 100,000 particles per cubic centimeter.

Due to their small size, they have a large total surface area and may become coated or "enriched" with contaminants in combustion flue gas or the atmosphere including toxic metals, sulphur, PAH and partially oxidized hydrocarbons, polychlorinated dibenzo-p-dioxins and polychlorinated dibenzofurans (PCDD/F). They are however relatively short lived and combine into accumulation mode particles (Forastiere and Faustini 2008).

Particles can also be broadly subdivided into primary and secondary. Both primary particles and precursor gases can be emitted from natural and anthropogenic sources. A major source of primary particles is incomplete combustion (diesel cars, power generation, domestic heating, etc.) leading to particles of elemental carbon, usually with a coating of semi-volatile organic compounds. An example is black smoke emitted from diesel sources. Other primary sources of particles are construction, agriculture, mining, quarrying, cement and brickworks. Natural sources include sea sprays and wind-blown dust. Secondary particles, mainly nitrates and sulphates, are produced through a number of chemical processes in the atmosphere.

The largest particles (coarse fraction) are mechanically produced by attrition of larger particles. The suspended portion (total suspended particulates or TSP) is generally less than 40 μm.

5.3.2.1 Mechanism of Action

The following mechanisms have been advanced to explain how PM affects health (Brunekreef and Holgate 2002; COMEAP 2006):

1. Particles cause pro-inflammation of lung epithelial cells, leading to acute tissue damage and organ-repair responses. A repeated cycle of tissue damage and repair contributes to airway inflammation.
2. Inhaled particles produce an effect on alveolar macrophages (through activation of oxidant pathways), thereby increasing susceptibility to pathogens.
3. Particles set up an inflammatory response in the interstitium of the lung, which eventually provokes an increase in the likelihood of blood to clots, which can cause heart attacks.
4. Inhaled particles causes a reflex effect on the heart. The effect is provoked by the interaction of pollutants, or secondary factors produced by inflammation, with receptors in the airways and lung.

5.3.2.2 Short-Term Effects

Several studies have reported increased mortality and hospitalisation for cardiopulmonary causes with increase in airborne concentrations of PM_{10} and $PM_{2.5}$ (Schwartz and Morris 1995; Katsouyanni et al. 1997; Atkinson et al. 1999;

Schwartz 1999; Laden et al. 2000). Examples of associated illnesses include respiratory problems, changes in heart rhythms, heart attacks, and severe respiratory and heart malfunctions leading to death.

The APHEA-2 mortality study covered a population of more than 43 million people living in 29 European cities (Katsouyanni et al. 2001). The combined effect estimate showed that all cause daily mortality increased by 0.6% (95% CI, 0.4–0.8%) for each 10 μg/m^3 increase in PM_{10}. Hospital admissions for asthma and chronic obstructive pulmonary disease (COPD) among people older than 65 years were increased by 1.0% (95% CI, 0.4–1.5%) for each 10 μg/m^3 increase in PM_{10}, and admissions for cardiovascular disease (CVD) were increased by about 0.5% (95% CI, 0.2–0.8%).

In the USA, the NMMAPS studies reported an increase of 0.5% (95% CI, 0.1–0.9%) in all causes of mortality for each 10 μg/m^3 of PM_{10}, a value close to the European results. Effects on hospital admissions for people aged 65 years and above were 1.5% (95% CI, 1.0–1.9%) for COPD and 1.1% (95% CI, 0.9–1.3%) for CVD, for each 10 μg/m^3 increase in PM_{10} levels.

In the Netherlands, increases in mortality with increasing airborne concentrations of black smoke (Hoek et al. 2002) were reported. In Dublin, where a law prohibiting sale of coal in the city was passed in 1990, a study conducted using data obtained prior to and following implementation of the law reported decreases in concentrations of black smoke along with decreases in mortality from lung disease. PAPA study in Asia (Wong et al. 2008) reported associations between PM_{10} and mortality from various health outcomes at an effect greater than that observed in US and European studies.

A number of studies have been conducted on respirable particles ($PM_{2.5}$), some of which suggest associations with mortality and asthma exacerbations (Avakian et al. 2002). A study in Helsinki, Finland, reported a 7.8% (95% CI, 3.5–12.3%) increase in emergency room visits of children for asthma for each 1.1 μg/m^3 increase in traffic-related $PM_{2.5}$ at lag 4. An increase of 3.9% (95% CI, 0.28–7.7%) in pooled asthma and COPD visits among elderly people was reported for a similar increase in traffic-related $PM_{2.5}$ at lag 0 (Halonen et al. 2008). A study in 24 US and Canadian communities showed significant associations between exposure to $PM_{2.5}$ and children's acidity and lung function as well as symptoms of bronchitis, but not with asthma (Dockery et al. 1996; Raizenne et al. 1996). Other studies in USA (Laden et al. 2000) showed that $PM_{2.5}$ was associated with life-threatening arrhythmia leading to therapeutic interventions by an implanted cardioverter defibrillator, and that $PM_{2.5}$ concentrations were higher in the hours and days before onset of myocardial infarction.

COMEAP analysed 117 studies for PM_{10}, $PM_{2.5}$ and TSP and mortality from a cardiovascular diagnosis (COMEAP 2006). The vast majority of these studies have produced positive associations, with only six producing negative associations. A retrospective study of an air pollution episode in Augsburg, Germany (the MONICA project), found that in 1985 plasma viscosity, as well as heart rate and concentrations of C-reactive protein, were increased during the episode (Peters et al. 1997), which can contribute to an increased risk of cardiovascular events. Increased plasma sCD40L levels associated with particulate pollution have been reported in another study (Rücker et al. 2007), further lending support to the hypothesis that higher levels of ambient air pollution lead to an inflammatory response in patients with coronary heart disease.

Studies on certain health effects have not been as conclusive as mortality studies. One large-scale collaborative European study, the PEACE (Pollution Effects on Asthmatic Children in Europe) study, failed to show effects of PM (and NO_2) on lung function and acute symptoms (Roemer et al. 1998).

5.3.2.3 Long-Term Effects

Although most of the epidemiological research has focused on effects of short-term exposures, several studies suggest that long-term exposure may be more important in terms of public health (Brunekreef and Holgate 2002). In parallel with the development of time-series studies, a number of cohort studies have been reported. These studies have shown that inhabiting a relatively polluted city for a prolonged period leads to a shortening of life expectancy.

The analysis of long-term prospective cohorts from the Six Cities study have shown an association between levels of fine particulate matter ($PM_{2.5}$) and an elevated risk of death from all causes and from cardiovascular disease (Dockery et al. 1993). More recent data have shown that non-fatal ischemic events are also associated with an increase in fine PM concentrations in the community (Miller et al. 2007).

Evidence has also been provided by the American Cancer Society (ACS) cohort study of approximately 1.2 million adults (Pope III et al. 2002). This study demonstrated associations between ambient $PM_{2.5}$ and elevated risks of both cardiopulmonary and lung cancer mortality. Each 10 $\mu g/m^3$ elevation in long-term average $PM_{2.5}$ ambient concentrations was associated with approximately a 4%, 6%, and 8% increased risk of all-cause, cardiopulmonary, and lung cancer mortality, respectively. This evidence was further supported by a thorough re-analysis of the original Six Cities and ACS cohort studies by the Health Effects Institute (HEI 2000).

Other studies have been cross-sectional, and assumed that current air pollution exposure is sufficiently representative of long-term, previous exposure to make a plausible link with current health status (Brunekreef and Holgate 2002). An example is a series of Swiss studies on the association between air pollution and health among children and adults. In eight different communities, decreased lung function as well as symptoms of bronchitis in adults were associated with PM_{10} (and NO_2 and SO_2) (Ackermann-Liebrich et al. 1997). There was however no association with asthma. In children from ten Swiss communities, the same pollutants were found to be associated with symptoms of bronchitis but not asthma or allergy (Braun-Fahrlander et al. 1997). The associations were seen at a range of PM_{10} concentrations of 10–33 $\mu g/m^3$ only, which is well below ambient concentrations in many European countries. However, the high correlation between pollutants prevented separation of their individual effects.

Exposure to particles has been suggested to be related to reduced lung function growth in children, and even children relocating from high to low pollution areas (or vice versa) were shown to experience changes in lung function growth that mirrored changes in exposure to particles.

5.3.2.4 Health Effects of Ultrafine Particles

There has been recent recognition that ultrafine particles (<0.1 μm) might be more toxic due to their physical and chemical characteristics (Avakian et al. 2002). Most of the available evidence is however mainly based on animal experiments and studies in occupational environments.

Ultrafine particles of <0.02 μm have been shown in animal experiments to have the highest deposition efficiency in the alveolar region of the lungs. They can rapidly penetrate the epithelium so that they can be detected in the interstitium of the lung shortly after exposure (Avakian et al. 2002). Based on this mechanism, it is hypothesised that ultrafine particles in ambient air may play a role in causing acute lung injury in sensitive individuals.

Research with rodent models indicates that, per given mass, ultrafine particles administered to the lung causes a greater inflammatory response than larger particles and that coagulation of ultrafine particles to larger particles significantly decreases their toxicity. The pulmonary inflammatory potential of ultrafine carbonaceous particles can be significantly enhanced in rodents by gaseous co-pollutants such as O_3 (Avakian et al. 2002). In addition, surface chemistry may play an important role in ultrafine particle toxicity. Susceptibility to the inflammatory effects of ultrafine particles in rats and mice increases significantly with age and/or a compromised/sensitized respiratory tract. It also appears that after deposition in the lung, ultrafine particles are not efficiently phagocytized by alveolar macrophages (Oberdorster 2000). Translocation of ultrafine particles to the liver has been documented, suggesting that organ systems other than the lungs may be impacted by ultrafine particles (Avakian et al. 2002).

In a population study, associations of ultrafine and fine particles have been demonstrated to be largely independent of each other, and the risk was enhanced if both were considered at the same time. Furthermore, when the associations were summed for the six size classes between 0.01 and 2.5 μm, the overall association was clearly stronger than the associations of the individual size classes alone (Wichmann et al. 2000).

5.3.2.5 Lead

Lead (Pb), associated in the atmosphere with airborne particulate matter, has been extensively used as a fuel additive to improve the running of petrol-fuelled car engines. Its use has now been banned in many regions, including North America and the European Union, and consequently lead levels have declined greatly in recent years. Non-ferrous metal smelting remains a source of atmospheric lead, and historical deposits from vehicle exhaust remain in road dust and some soils. Lead is acutely toxic to humans in high doses. At lower concentrations, such as those occurring in the vicinity of busy roads prior to the ban on lead in fuel, there is evidence that it impairs the neurological development of children, leading to behavioural problems and a reduction in intelligence as measured by IQ.

5.3.3 *Organic Gases*

5.3.3.1 Benzene

Benzene has been extensively used as a solvent and industrial raw material. It is also a component of petrol and its most widespread source in urban areas is evaporation and combustion from petrol car engines (Suh et al. 2000). Levels of benzene in vehicle exhaust have been greatly reduced since the introduction of three-way catalytic converters. Low aromatic content fuels and solvents are now widely used in Europe and North America.

Current ambient levels of benzene in most urban areas in Europe are low (typically <1 ppb) with substantially higher concentrations occurring near busy streets and petrol stations (Gonzalez-Flesca et al. 2002; Vardoulakis et al. 2005; Karakitsios et al. 2007). Tobacco smoking is also a major source of personal exposure to benzene. Benzene is a known carcinogen, with chronic exposure being particularly linked to leukaemia.

Mechanisms of Action

The adverse effects of benzene can occur from the production of harmful metabolites, such as semiquinone, and from autoxidation of benzene polyphenol metabolites, which in turn can form quinone oxidation products and superoxide radicals (Greenlee et al. 1981). Bone marrow damage as a result of benzene exposure is thought to be due to the presence of benzene metabolites in the liver. However, the liver itself does not suffer from hepatotoxicity and is protected from damage by the enzymes, DT-diaphorase and carbonyl reductase (WHO 1993).

The presence of benzene or benzene metabolites in blood or urine may be used to assess exposure. This can be done by measuring the presence of benzene itself circulating in the blood; hydroxylated aromatic urinary metabolites (phenol, catechol, hydroquinone and 1,2,4-trihydroxybenzene); trans,trans-muconic acid (urinary ring-open metabolite); N-acetyl-S-(2,5-dihydroxyphenyl)-L-cystine (urinary metabolite of benzene, phenol and hydroquinone); S-phenylmercapturic acid (glutathione-derived adduct); N-phenylguanine (DNA adduct); and S-phenylcystine and N-phenylvaline, haemoglobin/protein-derived adducts (Medeiros et al. 1997). The use of biomarkers to assess benzene exposures as been conducted by Qu and co-workers (Qu et al. 2005) showed that urinary levels of S-phenylmercapturic acid and trans,trans-muconic acid had a significant exposure-response pattern for benzene levels between 0 and 3.19 mg/m^3 (0–1 ppm).

Carcinogenicity

A study of around 1,800 workers from a rubber hydrochloride factory revealed that there was a strong association between benzene exposure and leukaemia. Daily benzene exposures for each member of the cohort were calculated using data from past air sampling surveys that stretched over 35 years. An overall standardized mortality ratio of 337 was calculated using expected and observed leukaemia deaths (Rinsky 1989). Workers exposed to 478.5–2,073.5 mg/m^3 (150–650 ppm) benzene over 4 months–15 years were reported to have abnormal bone marrow, eight of whom also had thrombocytopenia leading to haemorrhage and infection (Aksoy et al. 1971). Most notably, long-term exposure to benzene causes nonlymphocytic leukaemia in particularly acute myelogenous leukaemia which has been found to occur from cumulative benzene doses between 127.6–638 mg/m^3 (40–200 ppm) (Rinsky et al. 2002).

Other Health Effects

Chronic exposure to benzene can cause significant haematological, immunological and neurological effects. A number of blood abnormalities including pancytopenia, aplastic anaemia, thrombocytopenia, granulocytopenia, and lymphocytopenia have been associated with benzene exposure (WHO 1993). A significant decrease in total red blood cells, white blood cells, lymphocyte count, platelets and haematocrit levels were observed in a group of 44 healthy individuals exposed to 98.9 mg/m^3 (31 ppm) (8-h Time Weighted Average) for around 6.3 years (Rothman et al. 1996a, b). More recently, a comparison between chronically exposed workers with age and gender matched persons without occupational benzene exposure showed that there was a significant decrease in red blood cells, white blood cells and neutrophils in the exposed workers (Qu et al. 2002).

Benzene can adversely affect both humoral and cellular acquired immunity. Exposure to 9.6–22.3 mg/m^3 (3–7 ppm) benzene along with toluene and xylene for 1–21 years was found to result in increased serum IgM and decreased IgG and IgA immunogobulins (Lange et al. 1973). Levels of leukocytes and lymphocytes can also decrease as a result of long-term benzene exposure. Leukopenia was found in Turkish workers exposed to 48–670 mg/m^3 (15–210 ppm) benzene (Aksoy et al. 1971).

Neurological problems can also arise from chronic benzene exposure. In a group of printing workers in Hong Kong, exposure to benzene (3.5 mg/m^3 [1.1 ppm]) and other organic solvents was shown to cause neurovegetative lability, reduced smell and loss of memory as well as irritation of the mucous membranes (Yu et al. 2004).

5.3.3.2 1,3-Butadiene

1,3-butadiene is an important industrial raw material used in the manufacture of a number of polymers including nitrile rubber and styrene-butadiene latexes. It is also emitted in vehicle exhaust as a product of partial fuel combustion and from tobacco smoke. In ambient air, it is rapidly transformed by reaction with hydroxyl radicals, nitrogen trioxide radicals and ozone. 1,3-butadiene is a known carcinogen (IARC 2008). Levels of 1,3-butadiene in vehicle exhausts have been greatly reduced by the presence of catalytic converters.

Individuals may be exposed to 1,3-butadiene inside vehicles. In a study by Duffy and Nelson (Duffy and Nelson 1997) it was found that during peak hours, interior levels were 12.3 $\mu g/m^3$ (5.5 ppb) in newer cars while older cars without catalytic converters had about twice the level of 1,3-butadiene at 25.8 $\mu g/m^3$ (11.5 ppb).

Mechanisms of Action

Carcinogenicity of 1,3-butadiene is thought to occur when it is metabolised producing DNA alkylating compounds which can then increase cell mutation, chromosomal aberrations and sister chromatid exchanges (Penn and Snyder 2007).

Experimental animal and human liver microsome studies have revealed that 1,3-butadiene is metabolised to 1,2-epoxy-3-butene and subsequently 1,2:3,4-diepoxybutanen by cytochrome P450. The resultant 1,2:3,4-diepoxybutanen is a potent alkylating agent which reacts with DNA in vivo and in vitro causing mutations in somatic and germ cells (IARC 1999).

Carcinogenicity

An American cohort study of workers involved in 1,3-butadiene manufacturing revealed that there was an excess of lymphohaematopoietic cancers based on 42 deaths. In a separate study, workers were found to have an excess of stomach cancers and there were two cases of leukaemia (IARC 1999).

A study of about 15,700 men who had been employed for at least 1 year at a styrene-butadiene rubber plant revealed that 75% of individuals were exposed to 1,3-butadiene. Of these individuals, a higher than expected number died from leukaemia, most of whom had been working in such plants for over 10 years in polymerization, maintenance labour or laboratories where there is a potential for high exposures to 1,3-butadiene (Delzell et al. 1996).

Other Health Effects

Acute exposure to 1,3-butadiene can cause coughing, fatigue, drowsiness, irritation of the eyes, nasal passages and lungs (ATSDR 1992). In addition, neurological effects may occur from 1,3-butadiene inhalation including narcosis and eventual death by respiratory paralysis. Initial symptoms include blurred vision, nausea, paresthesia and mouth, throat and nose dryness with subsequent headache, fatigue, lowered blood pressure, pulse rate and loss of consciousness (ATSDR 1992).

Cardiovascular problems may arise from chronic exposure to 1,3-butadiene, in particular chronic rheumatic and arteriosclerotic heart diseases in middle-aged workers employed in the rubber industry (McMichael et al. 1974). A study by (Matanoski et al. 1990) revealed that exposure to 1,3-butadiene resulted in a 50% increase in standardized mortality ratio for arteriosclerotic heart disease. Mortality data of over 12,000 workers from 1,3-butadiene styrene polymer manufacturing plants across the United States and Canada between 1943–1982 had a standardized mortality ratio of 1.48 for arteriosclerotic heart disease (Matanoski et al. 1990).

5.3.3.3 Formaldehyde

The main source of formaldehyde indoors is from off gassing of building materials such as particle board, plywood resins, carpets, adhesives and resins. It is used in the manufacturing of urea formaldehyde foam insulation (Jones 1999). A study of 96 dwellings in Canada reported that indoor formaldehyde concentrations were in the range of 0.0096–0.09 mg/m^3 (0.0078–0.073 ppm). The higher levels corresponded with low ventilation, painting, varnishing and presence of new wooden or melanin furniture (Gilbert et al. 2006). Formaldehyde, which is a known human carcinogen, is also associated with symptoms such as irritation of eyes, nose and throat, sneezing, coughing, nausea and dyspnoea.

Mechanisms of Action

The exact mechanism of action for formaldehyde is currently unknown. However, formaldehyde is known to combine with unprotonated amino groups in amino acids producing hydroxymethyl amino acid derivatives and a proton which could be related to its germicidal properties (ATSDR 1999a).

Carcinogenicity

The main health concern with formaldehyde exposure is cancer risk. In an investigation by the IARC (IARC 2006), it was shown that there was a statistically significant number of deaths from nasopharyngeal cancer in workers exposed to formaldehyde.

The risk of lung cancer was also found to increase with formaldehyde exposure in a group of formaldehyde workers in ten production plants. It was estimated that they were exposed to about 0.12–2.33 mg/m^3 (0.1–1.9 ppm) formaldehyde. However, no conclusive dose-response was derive from the data (Blair et al. 1986).

Other Health Effects

Controlled experiments of acute exposure to formaldehyde in humans have shown that irritation of the eyes, nose and throat generally occurs at 3.69 mg/m^3 (3 ppm) with an odour threshold of 1.23 mg/m^3 (1 ppm). After the initial irritation, symptoms subside with continued exposure (ATSDR 1999b). In another study, exposure to 0.1 mg/m^3 (0.0813 ppm) of formaldehyde for 30 min in individuals with asthma and dust mite allergies resulted in a significant increase in bronchial responsiveness to the mite (Casset et al. 2006).

Exposure to formaldehyde can also have neurological effects. In a group of 32 men with occupational exposure to the chemical for at least 5 years were found to have a statistically significant decrease in neurological performance in a number of tests with increasing formaldehyde concentration (Bach et al. 1990).

5.3.3.4 Benzo[a]pyrene

Benzo[a]pyrene (BaP) has been often used as a marker for the group of polycyclic aromatic hydrocarbons (PAH) in ambient air (Armstrong et al. 2004). PAH are released from the incomplete combustion of diesel powered vehicles, domestic wood and coal combustion, coal tar, coke and asphalt production, incineration, industrial power generation, domestic cooking and smoking. High concentrations of BaP can occur in busy street canyons and near petrol stations especially during winter (Vardoulakis et al. 2008). BaP can also be released from volcanoes and forest fires (HPA 2008; Agarwal et al. 2009). BaP is a known human carcinogen, which can occur particle-bound and in gas-phase. It is thought that carcinogenic PAH roughly correspond to particle-bound PAH (Levy et al. 2001).

Mechanisms of Action

For most PAHs, their mechanism of action is via covalent binding of PAH metabolites such as epoxide intermediates, dihydrodiols, phenols and quinones to DNA which can subsequently lead to mutation and tumour formation. In the presence of cytochrome P450, the bay region of BaP (between carbon 10 and 11) diol epoxide intermediates is thought to be the main site for carcinogenic activity, therefore the level of P450 in different tissues will influence the effects of BaP (ATSDR 1995).

Carcinogenicity

The genotoxic effects of BaP can be exemplified from a study by Mumford et al. (1993) who have suggested that the high incidence of lung cancer in Xuan Wei, China is related to smoky coal used in homes without chimneys.

Analysis of placental, peripheral and cord white blood cells showed the presence of PAH-DNA adducts which have been suggested as a biomarker to assess PAH exposure.

Other Health Effects

No studies have been identified that report the effects of acute exposure to benzo[a]pyrene in humans. An epidemiological study of 667 employees in a rubber factory has however shown that chronic exposure to BaP can lead to decreased ventilatory function which can be further exacerbated if individuals are also exposed to particulates (Gupta et al. 1993). Such conditions were present in the compounding section of the factory, here, workers showed radiographic abnormalities such as opaque patches, bronchiovascular markings and pleural effusions in the lungs (Gupta et al. 1993).

Immunological effects can occur from BaP exposure. This has been demonstrated in a study which compared a group of coke oven workers exposed to 0.0002–0.50 mg/m^3 BaP with cold-rolling mill workers who were exposed to levels 3–5 orders of magnitude less. It was revealed that coke oven workers had lower serum IgG and IgA than mill workers, also coke oven workers concentrations of IgM decreased and IgE increased (Szczeklik et al. 1994). As there was also exposure to other PAH, it could be argued that the effects observed could be due to exposure to multiple compounds rather than just BaP alone.

5.3.4 Biological Contaminants

Pollen is thought to be the main biological contaminant in outdoor air, others include spores and particles from fungi, bacteria and algae as well as fine debris from other organisms.

5.3.4.1 Pollen

Pollen is produced by a number of plants and is often present in the air with some temporal variation depending on which plants are in season. Different types of pollen may elicit different responses in individuals. Some people may be more allergic to pollen from a particular group of plants than others. Grass pollens are thought to be the most prevalent in causing allergic rhinitis commonly known as hay fever or pollinosis (D'Amato et al. 1998). Pollen counts of 10–50 $grains/m^3$ can lead to hay fever symptoms (Davies and Smith 1973). Other symptoms of exposure include nasal and eye irritation and sneezing (Guillam and Segala 2008). Allergies to pollen occur in 80–90% of asthmatic children and 40–50% of asthmatic adults (Taylor et al. 2007).

Most pollen grains are around 15–50 μm diameter (Burge and Rogers 2000) and as a result rarely penetrate into the bronchiolar airways. However pollen can still cause coughing which is not exclusive to those who are susceptible to allergic rhinitis (D'Amato et al. 2007). Concentrations of pollen fragments in ambient air can be as high as 0.27 mg/m^3 (Miguel et al. 2006).

Pollen grains are dehydrated when they are dispersed in natural pollination. Once they become wet, they undergo metabolic and ultrastructural changes. Allergens in the pollen wall and cytoplasm can be released once they come into contact with mucosal membranes such as in the eyes, nose and mouth causing hay fever in susceptible individuals (D'Amato et al. 2007).

The risk of asthma attacks from pollen can be affected by meteorological conditions. During thunderstorms, asthma epidemics tend to occur as pollen grains rupture due to osmotic shock and release their contents into the atmosphere which includes respirable allergens measuring 0.5–2.5 μm diameter (D'Amato et al. 2007). In June 1994, there was an outbreak of asthma in London which coincided with a heavy thunderstorm. It was also noted that a number of patients who had only suffered from hay fever in the past, had an asthma attack (Venables et al. 1997).

Finally, there is some evidence suggesting that exposure to high levels of pollen such as in the summer months can exacerbate eczema symptoms such as itchiness particularly in those who were already known to be sensitized to pollen (Kramer et al. 2005).

5.3.4.2 Fungi

Outdoors, fungal spores can be found on dead or decaying materials such as fallen trees and compost heaps as well as on other plants and animals as a pathogen. Most of them are in the 2–10 μm size range and as a result can penetrate the lower respiratory system (Burge and Rogers 2000). Ambient levels of fungal spores can be as high as 6,000 spores/m^3 during the autumn months (Li and Kendrick 1995).

Fungal spores from *Aspergillus fumigatus*, *Cladosporium herbarium* and *Alternaria alternata* have had their allergens characterised (Burge and Rogers 2000). Exposure to such spores can cause asthma or hypersensitivity pneumonitis with symptoms of rhinitis, irritation of the eyes and dermatitis (Singh 2005). A study has revealed that fungal spores, not pollen grains, were associated with 8.8% of hospital visits for childhood asthma in Ontario, Canada, with spores from deuteromycetes, basidiomycetes and ascomycetes fungal classes (Dales et al. 2000).

5.3.4.3 Bacteria

Bacteria can become airborne in aerosols such as by the sea, waterfalls or even in water treatment plants. Studies have shown that there is a high prevalence of respiratory problems among those working in wastewater treatment plants (Lundholm and Rylander 1983). Typical levels of airborne bacteria are around

520 × 10^3 EU/m^3 (endotoxin units). Endotoxins produced by bacteria have also been found in aerosols and can range from undetectable levels to as high as 4,000 EU/m^3 while normal levels are in the range of 300–1,000 EU/m^3 (Lee et al. 2006). In cities around the world, bacterial concentrations have been found to be between 175 and 8,500 CFU/m^3 (colony forming units) in Johannesburg, South Africa; 2–4,000 CFU/m^3 in four localities in Sweden; 80–2,813 CFU/m^3 in Marseilles, France; and 71–22,100 CFU/m^3 in Beijing, China (Fang et al. 2007). Such wide variation in concentrations is mostly due to variation in local environmental conditions such as climate and topography.

Soil bacteria such as those in the Actinomycetes group have been suggested to cause hypersensitivity pneumonitis in agricultural environments with irritation of mucosal membranes, bronchitis, rhinitis, asthma and allergic alveolitis (Lacey and Dutkiewicz 1994).

5.3.4.4 Algae

The role of algae as an airborne biological contaminant has been little studied. Algae can be aerosolized allowing them to be transported over vast distances (Sharma et al. 2007). They can be deposited in reservoirs and cause algal blooms resulting in adverse environmental and human health effects particularly if the water is imbibed or comes into contact with skin (Sharma et al. 2007). Like moulds and bacteria, algae are also able to produce toxins (brevetoxin) which can affect human health (Cheng et al. 2005).

Allergic reactions to algal cells is less common than that of pollen, fungi and bacteria (Tormo et al. 2001). Particles 3–30 μm in size in ocean air have been found to cause coughs and nasopharyngeal burning in people who often go to the beach on windy days (Woodcock 1948) while particles <9 μm can be deposited in the nasal, oral and pharyngeal area causing irritation as a result of the particles or presence of algal toxins (Cheng et al. 2005).

During a "red tide" event (large bloom of red algae) in the Gulf of Mexico, brevetoxins in the air have been recorded at 80 ng/m^3 at its highest exposure. Chest tightness, wheezing and shortness of breath were observed in 8% of individuals with low exposure, 11% with moderate exposure and 28% with high exposure. In 33% of the subjects, mild inflammation was recorded in the nasopharyngeal region (Backer et al. 2003).

5.4 Indoor Air Pollutants

People living in developed countries typically spend >90% of their time indoors, with vulnerable individuals (elderly, young children, people with compromised health) spending an even larger proportion of their time indoors. Although indoor air quality levels are highly variable, depending on building materials, indoor combustion sources, use of household products, ventilation rates, etc., there is strong evidence that certain pollutant concentrations (e.g. benzene) in houses may exceed ambient air pollution levels in European cities (Pérez Ballesta et al. 2006; Lai et al. 2007). Indoor air pollution is estimated to cause 36% of all lower respiratory infections and 22% of COPD in the world (WHO 2002).

Sources of indoor air pollution can be broadly classified into the following categories:

1. Emissions from building and furnishing materials, including asbestos from flame retardants and insulation materials, formaldehyde from particleboard and furnishings, and persistent organic pollutants (PCB, PBDE, etc.) from building materials in older houses.
2. Emissions from household consumer products, for example emissions of VOC from DIY products (solvents, adhesives, paints and varnishes), pesticides, insecticides, etc.
3. Fuel and tobacco combustion emissions. These include volatile and semi-volatile organic compounds from fuel combustion, tobacco smoking and cooking; carbon monoxide from incomplete fuel combustion, tobacco smoking, faulty boilers and other gas appliances; respirable particles from fuel combustion and tobacco smoking; NO_X from fuel combustion; and SO_2 from coal combustion.
4. Biological pollutants such as mould, bacteria and viruses, dust mites and animal dander (minute scales from hair, feathers, or skin).
5. Pollutants infiltrating into the house from the soil, such as radon from soil under buildings (and construction materials), methane and other landfill gases from contaminated soil.
6. Infiltration of volatile organic compound from basements, garages, and common hallways.
7. Ambient air pollutants penetrating into the indoor environment, including road traffic-related and industrial pollutants such as NO_2, SO_2 and respirable particles.
8. Ozone can penetrate into the indoor environment from outdoors, but can be also generated indoors from photocopiers and other electronic equipment due to high voltage discharges.

Indoor pollutants can cause acute health effects such as intoxication and death due to short-term exposure to very high concentrations of CO, and chronic health effects such as lung cancer, leukaemia and mesothelioma due to long-term exposure to radon, benzene and asbestos, respectively. Indoor pollutants may cause or aggravate COPD, respiratory infections, cardiac and cardiovascular disease, as well as a range of allergic symptoms, such as atopic dermatitis, rhinitis, conjunctivitis, hay fever, etc. Certain indoor pollutants, such as tobacco smoke and house dust, may aggravate asthma symptoms (Jones 1999).

Apart from health effects mainly related to the home environment, sick building syndrome effects mostly associated with the office environment have been widely reported (Jones 1999). These do not have a known specific cause, and are usually reported as fatigue, muscle pain, headache, nausea, dizziness, dry skin, nasal and chest congestion. However, there has been some research to suggest that sick building syndrome may arise as a result of exposure to mycotoxins produced by *Stachybotrys chartarum* also known as black mould (Assouline-Dayan et al. 2002).

5.5 Indoor Air Pollution in Developing Countries

Combustion by-products represent global human and environmental health challenges that are relevant not only in industrialized nations, but also in developing nations. Unlike in the developed countries where modernization has been accompanied by a shift from biomass fuels to petroleum products and electricity, in most developing countries, nearly 90% of the population still relies on biomass fuels (Bruce et al. 2000). Coal use for cooking and heating is also widespread in countries such as China and South Africa.

The health effects of combustion by-products largely depends on their physical state and chemical composition, which is determined by the type of combustion source, the efficiency of the combustion process, and the fuels that are burned (Avakian et al. 2002). In developing countries, the solid fuels (biomass and coal) are burned in open fires or poorly functioning stoves, generating complex mixtures of products of incomplete combustion (Brook et al. 2004). These substances include PM, CO, nitrous oxides, formaldehyde and PAH (Ezzati and Kammen 2002). Because of the generally poor ventilation conditions under which these fuels are burned, the pollution levels in indoor kitchens using biomass fuels are among the highest ever measured (Dutt et al. 1996). Typical 24-h mean levels of PM_{10} in these environments range from 300 to >3,000 $\mu g/m^3$, and during use of an open fire, levels of up to 20,000 $\mu g/m^3$ have been recorded (Ezzati and Kammen 2001a). It is estimated that 76% of all global particulate matter exposure occurs indoors in the developing world (Fullerton et al. 2008).

The strongest evidence of association between biomass fuels and health has been on respiratory morbidity and mortality (Fullerton et al. 2008). Strong associations have been established between exposure to biomass smoke and respiratory infections in women and children (Bruce et al. 2000; Smith et al. 2000; Ezzati and Kammen 2001a). Based on a meta-analyses of available studies, young children living in households exposed to solid fuels are reported to have two to three times greater risk of developing acute lower respiratory tract infection (ALRI) compared with those living in households using cleaner fuels or suffering less exposure to smoke (Bruce et al. 2000). Acute respiratory infections (ARI) accounts for 20% of all deaths in children under 5 years in developing countries. According to the World Health Report (WHO 2002), indoor air pollution is responsible for 36% of all lower respiratory infections, thus making it one of the top ten risk factors for disease (Ezzati and Kammen 2001b).

Short-term effects of biomass smoke on lung function of children has been shown in a study in Ecuador. In China, lung function tests on children exposed to coal smoke have shown reductions in forced expiratory volume (FEV_1), forced vital capacity and peak flow (Zhang and Smith 2007). Links between biomass smoke and tuberculosis (TB) have been investigated in some studies which have been recently reviewed (Johnson et al. 2007; Schmidt 2008). The existing evidence supports the hypothesis that exposure to biomass smoke increases the risk of TB infection, even though the evidence of causal association is still unavailable. However, given the high levels of PM contained in biomass smoke, and their demonstrated effect of impairment of alveolar macrophages function, the link between biomass smoke exposure and increased risk of TB is biologically plausible (Johnson et al. 2007).

The few studies that have investigated the association between biomass fuels use and asthma have yielded mixed results. In Guatemala, the prevalence of childhood asthma symptoms was higher in households using biomass fuels, with adjusted odds ratios of 2–3.5 for various symptoms. Biomass fuels use was also associated with various asthma symptoms in Nepal (Melsom et al. 2001; Shrestha and Shrestha 2005). Other studies have reported either no association or protective effect (Bruce et al. 2000).

Epidemiological studies in Asia have reported associations between indoor cooking with biomass fuels and cataracts or blindness (Bruce et al. 2000; Fullerton et al. 2008). Evidence from animal studies and studies on tobacco smoke exposure exists to support this hypothesis. In India where three-quarters of all households cook primarily with biomass fuels, a national survey reported an odds ratio of 1.3 for blindness in women in homes using biomass fuels, after correcting for potentially confounding socioeconomic factors. Studies in Guatemala and Zimbabwe have shown that babies born to mothers who cooked using biomass fuels were lighter than those born to mothers who used clean fuels (Boy et al. 2002; Mishra et al. 2004). A survey in India has linked biomass fuels use to anaemia. Prevalence of stunted growth was also higher among users of biomass fuels in comparison to those using cleaner fuels (Mishra and Retherford 2007). Even though association between biomass fuels use and cardiovascular disease has not been explored, a recent study in Guatemala reported an association between exposure to biomass smoke and increased diastolic blood pressure in women (McCracken et al. 2007).

Biomass smoke has also been reported as a possible cause for COPD in non-smoking women living in rural areas (Shrestha and Shrestha 2005; Huttner et al. 2007; Liu et al. 2007), where women exposed to biomass smoke have been found to develop COPD with clinical characteristics, quality of life and increased mortality similar to that of tobacco smokers (Fullerton et al. 2008).

The effect of biomass smoke exposure on cancer has been shown in studies from China, where coal is extensively used for cooking and heating. A review of available studies suggests that domestic coal smoke is a significant risk factor for the development of lung cancer (Mestl et al. 2007; Halonen et al. 2008). Studies from India and Mexico have demonstrated that non-smoking women who have been exposed to biomass smoke for several years have a higher risk of developing adenocarcinoma of the lung (Behera and Balamugesh 2005). The IARC recently termed biomass smoke a probable carcinogen, and coal used as domestic fuel was termed carcinogenic to humans (Straif et al. 2006).

5.6 Multi-pollutant Mixtures

Various sources of air pollution, for example road transport and combustion plants, emit complex mixtures of contaminants which may include carbon-based particles coated with organic compounds such as PAH. In addition, inorganic and organic gases react in the atmosphere in the presence of sunlight to form secondary pollutants such as ozone. As multiple components interact, the chemical composition and consequently the toxicity of the mixture are altered (WHO 2003). Only a small number of these pollutants are regularly monitored in routine air quality monitoring networks to characterise the multi-pollutant mixture and inform related epidemiological studies. As a consequence, the selection of a surrogate for the whole mixtures often depends on monitoring data availability rather than toxicity. For example, NO_2 is commonly used as an indicator of a complex mixture of toxic pollutants emitted from road transport, although NO_2 itself does not represent a significant health hazard in typical ambient air concentrations in developed countries.

The combined exposure of individuals to more than one pollutants may lead to a larger or smaller impact than the additive health effect of the individual pollutants (a process known as *effect modification*). The combined effects of multiple pollutants may be additive (i.e. pollutants act independently), synergistic (i.e. positive interdependence) or antagonistic (i.e. negative interdependence) (WHO 2003). For example, epidemiological studies have established that asbestos exposure and tobacco smoking interact synergistically to increase lung cancer risk. In particular, asbestos enhances the mutagenicity of tobacco carcinogens independently of the initial lung tissue damage responsible for pulmonary fibrosis (Nelson and Kelsey 2002). NO_2 has also been found to modify the effect of particles, with epidemiological studies showing that daily mortality associated with exposure to PM_{10} was higher in cities with higher NO_2 concentrations (Katsouyanni et al. 2001). However, this can be interpreted as an indication that a larger proportion of atmospheric particles originated from road transport in areas with higher NO_2 levels (Katsouyanni 2003).

Effect modification is of public health interest because of its implications for preventing the adverse health effects of air pollution (e.g. certain population subgroups such as smokers may be at particularly high risk), and because it provides further insights into the mechanisms of toxicity of certain pollutants (WHO 2003).

5.7 Health-Based Air Quality Guidelines and Regulations

The Great London Smog (Bell et al. 2004) led directly to the UK Clean Air Act of 1956 which regulated what could be burned in houses and created smoke free zones. Since then, various sets of national and international regulations and guidelines for ambient air quality have been published, including several EU directives (e.g. Directive on Ambient Air Quality and Cleaner Air for Europe 2008) and WHO global and regional air quality guidelines (WHO 2005) (Table 5.1). Although there are regulations

Table 5.1 Recommended guidelines in ambient air various authorising bodies. Averaging times are in parenthesis where available

	Recommended guidelines (mg/m^3)			
Pollutant	WHO	US EPA (NAAQS)	UK (National Air Quality Strategy)	EU
Organic pollutants				
Benzene	–	–	0.005 (annual)	0.005
1,3-Butadiene	–	–	0.00225 (annual)	–
Carbon disulphide	0.1 (24 h)	–	–	–
1,2-Dichloroethane	0.7 (24 h)	–	–	–
Dichloromethane	3 (24 h), 0.45 (weekly)	–	–	–
Formaldehyde	0.1 (30 min)	–	0.1 (30 min)	–
Polycyclic aromatic hydrocarbons (as benzo[a]pyrene)	–	–	2.5×10^{-7}	1.0×10^{-6} (annual)
Styrene	0.07 (30 min)	–	–	–
Tetrachloroethylene	0.25 (long-term)	–	–	–
Toluene	1 (30 min)	–	–	–
Inorganic pollutants				
Arsenic	–	–	–	6.0×10^{-6} (annual)
Cadmium	–	–	–	5.0×10^{-6} (annual)
Hydrogen sulphide	0.007 (30 min)	–	–	–
Lead	0.0005 (annual)	0.0015 (3 months)	0.00025 (annual)	0.0005 (annual)
Mercury	0.001 (annual)	–	–	–
Nickel	–	–	–	20×10^{-6} (annual)
Vanadium	–	–	–	–
Classical pollutants				
Carbon monoxide	10 (8 h) COHb <2.5%[a]	40 (1 h) 10 (8 h)	10 (8 h)	10 (8 h)
Nitrogen dioxide	0.2 (1 h) 0.04 (annual)	0.1 (annual)	0.2 (1 h) 0.04 (annual)	0.2 (1 h) 0.04 (annual)
Ozone	0.10 (8 h)	0.24 (1 h) 0.15 (8 h)	0.10 (8 h)	0.12 (8 h)
Particulate matter	PM_{10} 0.02 (annual) $PM_{2.5}$ 0.01 (annual)	PM_{10} 0.15 (24 h) $PM_{2.5}$ 0.035 (24 h) $PM_{2.5}$ 0.015 (annual)	PM_{10} 0.05 (24 h) PM_{10} 0.04 (annual) $PM_{2.5}$ 0.025 (annual)	PM_{10} 0.05 (24 h) PM_{10} 0.04 (annual) $PM_{2.5}$ 0.025 (annual)
Sulphur dioxide	0.02 (24 h)	0.37 (24 h) 0.08 (annual)	0.35 (1 h) 0.125 (24 h)	0.35 (1 h) 0.125 (24 h)

[a]COHb (carboxyhaemaglobin) levels to be below 2.5%.

Table 5.2 Recommended guidelines for indoor air by various authorising bodies. Averaging times are in parenthesis where available

	Recommended guidelines (mg/m^3)			
Pollutant	WHO	US EPA (NAAQS)	UK (COMEAP)	EU
Benzene	–	–	0.005 (annual)	–
Carbon monoxide	–	–	100 (15 min)	–
			60 (30 min)	
		–	30 (1 h)	
			10 (8 h)	
Environmental tobacco smoke	–	–	–	–
Formaldehyde		–	0.1 (30 min)	
Nitrogen dioxide	–	–	0.3 (1 h)	–
			0.04 (annual)	
Polycyclic aromatic hydrocarbons (as benzo[a]pyrene)	–	–	2.5×10^{-7} (annual)	–

in place for the workplace environment (Occupational Exposure Limit Values Directive 2006), indoor air quality (except radon) is only indirectly regulated through a number of directives on construction materials, energy performance of buildings, gas and heating appliances, product and chemicals safety (Harrison 2002).

The UK Committee on the Medical Effects of Air Pollutants (COMEAP 2004) has recommended guideline values for five pollutants (NO_2, CO, formaldehyde, benzene and BaP) in indoor air. The European project INDEX has also made recommendation for five high priority chemicals which largely overlap with the COMEAP recommendations (Table 5.2). Currently, a WHO working group is developing indoor air quality guidelines focusing on specific chemicals, but also assessing biological agents and indoor pollutants from solid fuel and kerosene combustion (WHO 2006).

To date, no guideline limits have been defined for airborne biological contaminants such as mould by the EPA or WHO as so many species can be present at any time. However, guidelines for modifying conditions to control and prevent microbial growth, such as humidity and condensation, have been developed (WHO 2009).

5.8 Risk Analysis and Health Impact Assessment

Risk analysis and health impact assessment (HIA) are concerned with the determination of the health impacts of policies or interventions which can have an influence on health (although their primary aim may not be health related).

For most inorganic and organic gases and particles (PM_{10}) considered in this chapter, it is usually assumed that the associated health effects are linearly related to pollutant concentrations and that the exposure-response curve passes through zero, which means that there are no thresholds of safe exposure. In quantitative risk analysis, the exposure-response relationships are commonly based on the results of time-series epidemiological studies for inorganic gases (SO_2 and NO_2) and particles (PM_{10}), population cohort studies reflecting chronic exposure to fine particles ($PM_{2.5}$), and occupational cohort studies for exposure to organic gases (benzene and 1,3-butadiene) and PAH.

Health impact assessment focuses on comparing the health impacts/benefits of two or more environmental policy options. For each area unit, the effects of air pollutants on human health are estimated by multiplying the exposure-response coefficient (β) by the change in ambient concentrations reflecting changes in exposure levels (ΔE), the baseline rate for the health outcome considered (M, e.g. deaths per 100,000 population per year) and the total population in this area (P). Thus, the relationship between human health and changes in air pollutant concentrations due to an environmental intervention can be expressed as:

$$\Delta H = \Delta E \times \beta \times P \times M \tag{5.1}$$

As summary of typical exposure-response coefficients derived from time-series studies is given in Table 5.3. COMEAP (2006) has also proposed a single coefficient, represented by $PM_{2.5}$, for the quantification of long-term health effects from exposure to a mixture of air pollutants. It should be noted however that co-variation of pollutants is not usually taken into account. Co-variation means that it is not known which individual pollutant or mixtures of pollutants has caused the recorded health effects or whether some additive or synergistic effects have taken place (WHO 2003).

In cancer risk assessment, the excess cancer cases in a population can be calculated by multiplying the changes in exposure levels (e.g. benzene concentrations) by the unit risk factor (UR) and by the size of the affected population. The unit risk factor represents the excess cancer risk over background (e.g. cancer cases not related to benzene) associated with continuous lifetime exposure to a carcinogen:

$$\Delta C = \Delta E \times UR \times P \tag{5.2}$$

A summary of commonly used cancer risk factors for exposure to air pollution via inhalation are presented in Table 5.4.

Table 5.3 Summary of exposure-response coefficients (COMEAP 1998)

Pollutant (j)	Health Outcome (i)	Exposure-response coefficient (β)
PM_{10}	Deaths brought forward (all causes)	0.00075 per $\mu g/m^3$ (24 h)
	Respiratory hospital admissions	0.00080 per $\mu g/m^3$ (24 h)
SO_2	Deaths brought forward (all causes)	0.00060 per $\mu g/m^3$ (24 h)
	Respiratory hospital admissions	0.00050 per $\mu g/m^3$ (24 h)
NO_2	Respiratory hospital admissions	0.00050 per $\mu g/m^3$ (24 h)

Table 5.4 Summary of inhalation unit risk factors for human carcinogens

Pollutant (*j*)	Health outcome	Inhalation unit risk (*UR*)
Benzene[a]	Leukaemia	6×10^{-6} per μg/m^3 (lifetime)
1,3-butadiene[b]	Leukaemia	3×10^{-5} per μg/m^3 (lifetime)
Benzo[a]pyrene[a]	Lung cancer	9×10^{-2} per μg/m^3 (lifetime)
Formaldehyde[b]	Nasal cancer	1.3×10^{-5} per μg/m^3 (lifetime)

[a]WHO Air Quality Guidelines for Europe – second edition (2000).
[b]US EPA Integrated Risk Information System (IRIS 2002).

5.9 Discussion and Conclusions

Epidemiological evidence has linked inorganic gases (SO_2, NO_2, CO and O_3) as well as inhalable particles (PM_{10}) with acute health effects, such as excess mortality and respiratory morbidity (Katsouyanni 2003). The evidence for chronic effects of inorganic gases is currently weak. By contrast, there is strong evidence that respirable particles ($PM_{2.5}$) have a chronic adverse effect on human health including cardiovascular mortality, and exacerbation of COPD and asthma (Dockery et al. 1989; Pope et al. 1995). Epidemiological studies on general populations and/or occupational groups have provide evidence of carcinogenicity of several volatile and semi-volatile organic compounds, such as benzene, 1,3-butadiene, formaldehyde and BaP.

Despite the relatively well-established epidemiological evidence, the toxicological mechanisms whereby pollutants cause harm to human health are still not well understood, with the exception of CO (which competes with oxygen for the same binding sites on haemoglobin). In most cases, inorganic air pollutants are associated with inflammation in the human respiratory system and airway obstruction, with effects being more noticeable in susceptible individuals with pre-existing respiratory disease. Fine particles may affect human health with a variety of complex mechanisms depending on their size, number, total surface area and chemical composition. The mechanism by which air pollution is playing a part in the rise in asthma rates is still obscure.

Carcinogenesis due to exposure to organic pollutants can be seen as a multistage process including initiation (genotoxic substance interacting with DNA), promotion (e.g. chemicals acting on enzymes) and progression (malignant cells invading other tissues). In addition to environmental factors, genetic predisposition seems to play a role for certain cancers (e.g. lung cancer). There is substantial inter-individual variability in responses to air pollutants even in healthy adults. Although genetic factors certainly play a role, other more subtle factors may be also operating, such as the distribution of ventilation in the different lung compartments and differences in regional deposition of particles within the lungs (Brunekreef and Holgate 2002).

Ambient concentrations of most classical air pollutants have declined in Europe and North America. However, there is still concern about persistently high concentrations of $PM_{10}/PM_{2.5}$ and NO_2 in many urban areas. Atmospheric concentrations of lead bound on particles have also greatly declined in roadside environments in the last 2 decades (mainly due to the ban on leaded petrol in many countries), but there is still some concern about exposure especially of young children to lead through inhalation and ingestion of contaminated soil and road dust.

Although most epidemiological evidence comes from studies carried out in North America and Europe, by far the largest health impact of air pollution is suffered by populations in the developing world, especially those cooking indoors using biomass fuels. Consistent evidence has shown that use of biomass fuels indoors leads to levels of indoor air pollution many times higher than those international ambient air quality standards allow for, exposing household members to a major public health hazard.

Important scientific questions related to the health effects of air pollution are the presence (or lack) of thresholds below which exposure is associated with no further increase in risk, and to what extent air pollution affects life expectancy (Brunekreef and Holgate 2002). Obviously, answering the first question is extremely important for setting air quality objectives, as attention may be shifted from reducing air pollution levels in larger areas to only removing localised pollution hotspots. Regarding the second question, it has been suggested that air pollution is more likely to bring forward the death of elderly people with already compromised health by a few days (harvesting effect). However, renalyses of past air pollution episodes have shown that high mortality rates often persisted beyond the period of the air pollution event, which indicates that there was no substantial harvesting effect (i.e. sharp drop in death rates after the end of the event). In addition, there has been evidence that premature deaths associated with poor air quality do not mainly occur in hospitals, as it would be the case if only people with compromised health were affected.

The challenge nowadays for air pollution epidemiologists and toxicologists is to establish the biological mechanism of health effects at relative low ambient concentrations of air pollutants, as those currently observed in many developed countries. Other key research challenges include the identification of any latency effects (time-lags between exposure and incidence of health effects), establishing appropriate surrogates for mixtures of pollutants in the outdoor and indoor environment, and quantifying the impacts of emerging pollutants such as phthalates, nanoparticles, and several others.

References

Ackermann-Liebrich, U., Leuenberger, P., & Schwartz, J. (1997). Lung function and long term exposure to air pollutants in Switzerland. Study on Air Pollution and Lung Diseases in Adults (SAPALDIA). *American Journal of Respiratory Critical Care Medicine, 155*, 122–129.

Agarwal, T., Khillare, P. S., Shridhar, V., & Ray, S. (2009). Pattern, sources and toxic potential of PAHs in the agricultural soils of Delhi, India. *Journal of Hazardous Materials, 163*, 1033–1039.

Aksoy, M., Dincol, K., Akgun, T., Erdem, S., & Dincol, G. (1971). Haematological effects of chronic benzene poisoning in 217 workers. *British Journal of Industrial Medicine, 28*, 296–302.

Alderman, B. W., Baron, A. E., & Savitz, D. A. (1987). Maternal exposure to neighborhood carbon monoxide and risk of low infant birth weight. *Public Health Reports, 102*, 410–414.

Armstrong, B., Hutchinson, E., Unwin, J., & Fletcher, T. (2004). Lung cancer risk after exposure to polycyclic aromatic hydrocarbons: a review and meta-analysis. *Environmental Health Perspectives, 112*, 970–978.

Assouline-Dayan, Y., Leong, A., Shoenfeld, Y., & Gershwin, M. E. (2002). Studies of sick building syndrome. IV. Mycotoxicosis. *The Journal of Asthma, 39*, 191–201.

ASTDR. (1998). *Sulfur dioxide: health effects*. Atlanta, GA: US Department of Health and Human Services.

Atkinson, R. W., Bremner, S. A., Anderson, H. R., Strachan, D. P., Bland, J. M., & de Leon, A. P. (1999). Short-term associations between emergency hospital admissions for respiratory and cardiovascular disease and outdoor air pollution in London. *Archives of Environmental Health, 54*, 398–411.

ATSDR. (1992). *Toxicological profile for 1,3-butadiene*. Atlanta, GA: US Department of Health and Human Services.

ATSDR. (1995). *Toxicological profile for polycyclic aromatic hydrocarbons*. Atlanta, GA: US Department of Health and Human Services.

ATSDR. (1999a). *Toxicological profile for formaldehyde*. Atlanta, GA: US Department of Health and Human Services.

ATSDR. (1999b). *Toxicological profile for formaldehyde*. Atlanta, GA: US Department of Health and Human Services.

Avakian, M. D., Dellinger, B., Fiedler, H., Gullet, B., Koshland, C., Marklund, S., et al. (2002). The origin, fate, and health effects of combustion by-products: a research framework. *Environmental Health Perspectives, 110*, 1155–1162.

Bach, B., Pedersen, O. F., & Moelhave, L. (1990). Human performance during experimental formaldehyde exposure. *Environment International (USA), 16*(2Medium: X Size), 105–113.

Backer, L. C., Fleming, L. E., Rowan, A., Cheng, Y.-S., Benson, J., Pierce, R. H., et al. (2003). Recreational exposure to aerosolized brevetoxins during Florida red tide events. *Harmful Algae, 2*, 19–28.

Ballester, F., Rodríguez, P., Iñíguez, C., Saez, M., et al. (2006). Air pollution and cardiovascular admissions association in Spain: results within the EMECAS project. *Journal of Epidemiology and Community Health, 60*, 328–336.

Ballester, F., Sáez, M., Pérez-Hoyos, S., & Iñíguez, C. (2002). The EMECAM project: a multicentre study on air pollution and mortality in Spain: combined results for particulates and for sulfur dioxide. *Occupational and Environmental Medicine, 59*, 300–308.

Behera, D., & Balamugesh, T. (2005). Indoor air pollution as a risk factor for lung cancer in women. *Journal of the Association of Physicians of India, 53*, 190–192.

Bell, M. L., Davis, D. L., & Fletcher, T. (2004). A retrospective assessment of mortality from the London smog episode of 1952: The role of influenza and pollution. *Environmental Health Perspectives, 112*, 6–8.

Blair, A., Stewart, P., O'Berg, M., Gaffey, W., Walrath, J., Ward, J., et al. (1986). Mortality among industrial workers exposed to formaldehyde. *Journal of the National Cancer Institute (United States), 6*, 1071–1084.

Bobak, M., & Leon, D. A. (1999). The effect of air pollution on infant mortality appears specific for respiratory causes in the postneonatal period. *Epidemiology, 10*, 666–670.

Boy, E., Bruce, N., & Delgado, H. (2002). Birth weight and exposure to kitchen wood smoke during pregnancy in rural Guatemala. *Environmental Health Perspectives, 110*, 109–114.

Braun-Fahrlander, C., Vuille, J. C., & Sennhauser, F. H. (1997). Respiratory health and long-term exposure to air pollutants in Swiss schoolchildren. *American Journal of Respiratory Critical Care Medicine, 155*, 1042–1049.

Breitner, S., Stölzel, M., Cyrys, J., Pitz, M., Wölke, G., Kreyling, W., et al. (2009). Short-term mortality rates during a decade of improved air quality in Erfurt, Germany. *Environmental Health Perspectives, 117*, 448–454.

Brook, R. D., Franklin, B., Cascio, W., Hong, Y., Howard, G., Lipsett, M., et al. (2004). Air pollution and cardiovascular disease: a statement for healthcare professionals from the expert panel on Population and Prevention Science of the American Heart Association. *Circulation, 109*, 2655–2671.

Bruce, N., Perez-Padilla, R., & Albalak, R. (2000). Indoor air pollution in developing countries: a major environmental and public health challenge. *Bulletin of the World Health Organization, 78*, 1078–1092.

Brunekreef, B. (2007). Health effects of air pollution observed in cohort studies in Europe. *Journal of Exposure Science and Environmental Epidemiology, 17*, S61–S65.

Brunekreef, B., & Holgate, S. T. (2002). Air pollution and health. *Lancet, 360*, 1233–1242.

Burge, H. A., & Rogers, C. A. (2000). Outdoor allergens. *Environmental Health Perspectives, 108*, 653–659.

Burnett, R. T., Stieb, D., Brook, J. R., Cakmak, S., Dales, R., Raizenne, M., et al. (2004). Associations between short-term changes in nitrogen dioxide and mortality in Canadian cities. *Archives of Environmental Health, 59*, 228–236.

Casset, A., Marchand, C., Purohit, A., le Calve, S., Uring-Lambert, B., Donnay, C., et al. (2006). Inhaled formaldehyde exposure: effect on bronchial response to mite allergen in sensitized asthma patients. *Allergy, 61*, 1344–1350.

Cheng, Y. S., Villareal, T. A., Zhou, Y., Gao, J., Pierce, R. H., Wetzel, D., et al. (2005). Characterization of red tide aerosol on the Texas coast. *Harmful Algae, 4*, 87–94.

Chitano, P., Hosselet, J. J., Mapp, C. E., & Fabbri, L. M. (1995). Effect of oxidant air pollutants on the respiratory system: insights from experimental animal research. *European Respiratory Journal, 8*, 1357–1371.

COMEAP. (1998). *The quantification of the effects of air pollution on health in the United Kingdom*. UK: COMEAP

COMEAP. (2004). *Guidance on the effects on health of indoor air pollutants*. UK: COMEAP.

COMEAP. (2006). *Cardiovascular disease and air pollution: a report by the Committee On the Medical Effects of Air Pollutants*. UK: COMEAP.

D'Amato, G., Liccardi, G., & Frenguelli, G. (2007). Thunderstorm-asthma and pollen allergy. *Allergy, 62*, 11–16.

D'Amato, G., Spieksma, F. T., Liccardi, G., Jäger, S., Russo, M., Kontou-Fili, K., et al. (1998). Pollen-related allergy in Europe. *Allergy, 53*, 567–578.

Dales, R. E., Cakmak, S., Burnett, R. T., Judek, S., Coates, F., & Brook, J. R. (2000). Influence of ambient fungal spores on emergency visits for asthma to a regional children's hospital. *American Journal of Respiratory Critical Care Medicine, 162*, 2087–2090.

Davies, R. R., & Smith, L. P. (1973). Forecasting the start and severity of the hay fever season. *Clinical & Experimental Allergy, 3*, 263–267.

Delzell, E., Sathiakumar, N., Hovinga, M., Macaluso, M., Julian, J., Larson, R., et al. (1996). A follow-up study of synthetic rubber workers. *Toxicology, 113*, 182–189.

Department-of-the-Environment. (1996). *The effects of nitrogen dioxide on human health*. Expert Panel on Air Quality Standards – Nitrogen Dioxide. London: The Stationery Office.

Dockery, D.W., Gold, D.R., Rotnitsky, A., Speizer, F.E., Ferris, B.G. Jr. & Ware, J.H. (1989). Effects of age, sex and race on prevalence of respiratory symptoms among children. *American Review of Respiratory Disease, 139*, A23

Dockery, D. W., Pope, C. A., 3rd, Xu, X., Spengler, J. D., Ware, J. H., Fay, M. E., et al. (1993). An association between air pollution and mortality in six U.S. cities. *New England Journal of Medicine, 329*, 1753–1759.

Dockery, D.W., Cunningham, J., Damokosh, A.I., Neas, L.M., Spengler, J.D., Koutrakis, P., Ware, J.H., Raizenne, M. & Speizer, F.E. (1996). Health effects of acid aerosols on North American Children: respiratory symptoms. *Environmental Health Perspectives, 104*, 500–505.

Duffy, B. L., & Nelson, P. F. (1997). Exposure to emissions of 1,3-butadiene and benzene in the cabins of moving motor vehicles and buses in Sydney, Australia. *Atmospheric Environment, 31*, 3877–3885.

Dutt, D., Srinivasa, D. K., Rotti, S. B., Sahai, A., & Konar, D. (1996). Effect of indoor air pollution on the respiratory system of women using different fuels for cooking in an urban slum of Pondicherry. *National Medical Journal of India, 9*, 113–117.

EPA. (2000). *Air quality criteria for carbon monoxide*. Research Triangle Park, NC: EPA.

Ezzati, M., & Kammen, D. (2001a). Indoor air pollution from biomass combustion and acute respiratory infections in Kenya: an exposure-response study. *Lancet, 358*, 619–624.

Ezzati, M., & Kammen, D. M. (2001b). Quantifying the effects of exposure to indoor air pollution from biomass combustion on acute respiratory infections in developing countries. *Environmental Health Perspectives, 109*, 481–489.

Ezzati, M., & Kammen, D. M. (2002). The health impacts of exposure to indoor air pollution from solid fuels in developing countries: knowledge, gaps, and data needs. *Environmental Health Perspectives, 110*, 1057–1068.

Fang, Z., Ouyang, Z., Zheng, H., Wang, X., & Hu, L. (2007). Culturable airborne bacteria in outdoor environments in Beijing, China. *Microbial Ecology, 54*, 487–496.

Forastiere, F., & Faustini, A. (2008). Are we understanding the respiratory effects of traffic related airborne particles? *Thorax, 63*, 574–576.

Fullerton, D. G., Bruce, N., & Gordon, S. B. (2008). Indoor air pollution from biomass fuel smoke is a major health concern in the developing world. *Transactions of the Royal Society of Tropical Medicine and Hygiene, 102*, 843–851.

Galan, I., Tobias, A., Banegas, J. R., & Aranguez, E. (2003). Short-term effects of air pollution on daily asthma emergency room admissions. *European Respiratory Journal, 22*, 802–808.

Gilbert, N. L., Gauvin, D., Guay, M., Héroux, M.-È., Dupuis, G., Legris, M., et al. (2006). Housing characteristics and indoor concentrations of nitrogen dioxide and formaldehyde in Quebec City, Canada. *Environmental Research, 102*, 1–8.

Gonzalez-Flesca, N., Vardoulakis, S. & Cicolella, A. (2002). BTX concentrations near a stage II implemented petrol station. *Environmental Science & Pollution Research, 9*, 169–174.

Greenlee, W. F., Sun, J. D., & Bus, J. S. (1981). A proposed mechanism of benzene toxicity: Formation of reactive intermediates from polyphenol metabolites. *Toxicology and Applied Pharmacology, 59*, 187–195.

Guillam, M., & Segala, C. (2008). Pollen and health effects: synthesis of epidemiological studies revue francaise d allergologie et d immunologie clinique. *Revue française d'allergologie et d'immunologie clinique, 48*, 14–19.

Gupta, P., Banerjee, D. K., Bhargava, S. K., Kaul, R., & Ravi Shankar, V. (1993). Prevalence of impaired lung function in rubber manufacturing factory workers exposed to benzo(a)pyrene and respirable particulate matter. *Indoor and Built Environment, 2*, 26–31.

Harrison, P.T.C., (2002). Indoor air quality guidelines. *Occupational and Environmental Medicine, 59*, 73–74.

Halonen, J. I., Lanki, T., Yli-Tuomi, T., Kulmala, M., Tiittanen, P., & Pekkanen, J. (2008). Urban air pollution, and asthma and COPD hospital emergency room visits. *Thorax, 63*, 635–641.

HEI. (2000). *Reanalysis of the Harvard six cities study and the American cancer society study of particulate air pollution and mortality: a special report of the institute's particle epidemiology reanalysis project*. Boston, MA: HEI.

Hoek, G., Brunekreef, B., Goldbohm, S., Fischer, P., & van den Brandt, P. (2002). The association between mortality and indicators of traffic-related air pollution in a Dutch cohort study. *Lancet, 360*, 1203–1209.

HPA. (2008). *Polycyclic aromatic hydrocarbons (Benzo[a]pyrene)*. HPA Compendium of Chemical Hazards, Health Protection Agency U.K.

Huttner, H., Beyer, M., & Bargon, J. (2007). Charcoal smoke causes bronchial anthracosis and COPD. *Medizinische Klinik, 102*, 59–63.

IARC. (1999). *Re-evaluation of some organic chemicals, hydrazine and hydrogen peroxide* (vol. 77). IARC Monographs on the Evaluation of Carcinogenic Risks to Humans. Lyon, France: IARC.

IARC. (2006). *Formaldehyde, 2-butoxyethanol and 1-tert-butoxypropan-2-ol* (vol. 88). IARC Monographs on the Evaluation of Carcinogenic Risks to Humans. Lyon, France: IARC.
IARC. (2008). *1,3-Butadiene, ethylene oxide and vinyl halides (vinyl fluoride, vinyl chloride and vinyl bromide)* (vol. 97). IARC Monographs on the Evaluation of Carcinogenic Risks to Humans. Lyon, France: IARC.
IRIS. (2002) Integrated risk information system. US: EPA Online [http://www.epa.gov/iris/]
Ishigami, A., Kikuchi, Y., Iwasawa, S., Nishiwaki, Y., Takebayashi, T., Tanaka, S., et al. (2008). Volcanic sulfur dioxide and acute respiratory symptoms on Miyakejima island. *Occupational and Environmental Medicine, 65*, 701–707.
Johnson, E., Langård, S., & Lin, Y. S. (2007). A critique of benzene exposure in the general population. *The Science of the Total Environment, 374*, 183–198.
Jones, A. P. (1999). Indoor air quality and health. *Atmospheric Environment, 33*, 4535–4564.
Karakitsios, S., Delis, V. K., Kassomenos, P. A., & Pilidis, G. A. (2007). Contribution to ambient benzene concentrations in the vicinity of petrol stations: estimation of the associated health risk. *Atmospheric Environment, 41*, 1889–1902.
Katsouyanni, K., Touloumi, G., Spix, C., Schwartz, J., Balducci, F., Medina, S., et al. (1997). Short-term effects of ambient sulphur dioxide and particulate matter on mortality in 12 European cities: results from time series data from the APHEA project. Air pollution and health: a European approach. *BMJ, 314*, 1658–1663.
Katsouyanni, K., Touloumi, G., Samoli, E., Gryparis, A., Tertre, A.L., Monopolis, Y., et al. (2001) Confounding and effect modification in the short-term effects of ambient particles on total mortality: results from 29 European cities within the APHEA2 project. *Epidemiology, 12*, 521–531.
Katsouyanni, K. (2003). Ambient air pollution and health. *British Medical Bulletin, 68*, 143–156.
Kraft, M., Eikmann, T., Kappos, A., Künzli, N., Rapp, R., Schneider, K., et al. (2005). The German view: effects of nitrogen dioxide on human health – derivation of health-related short-term and long-term values. *International Journal of Hygiene and Environmental Health, 208*, 305–318.
Kramer, U., Weidinger, S., Darsow, U., Mohrenschlager, M., Ring, J., & Behrendt, H. (2005). Seasonality in symptom severity influenced by temperature or grass pollen: results of a panel study in children with eczema. *Journal of Investigative Dermatology, 124*, 514–523.
Kunzli, N., Kaiser, R., Medina, S., Studnicka, M., Chanel, O., Filliger, P., Herry, M., Horak, F., Puybonnieux-Texier, V., Quenel, P., Schneider, J., Seethaler, R., Vergnaud, J-C. & Sommer, H. (2000). Public-health impact of outdoor and traffic-related air pollution: a European assessment. *Lancet, 356*, 795–801.
Lacey, J., & Dutkiewicz, J. (1994). Bioaerosols and occupational lung disease. *Journal of Aerosol Science, 25*, 1371–1404.
Laden, F., Neas, L. M., Dockery, D. W., & Schwartz, J. (2000). Association of fine particulate matter from different sources with daily mortality in six US cities. *Environmental Health Perspectives, 108*, 941–947.
Lai, H.K., Jantunen, M.J., Kunzli, N., Kulinskaya, E., Colvile, R. & Niewenhuijsen, M.J. (2007). Determinants of indoor benzene in Europe. *Atmospheric Environment, 41*, 9128–9135.
Lange, A., Smolik, R., Zatoński, W., & Szymańska, J. (1973). Serum immunoglobulin levels in workers exposed to benzene, toluene and xylene. *International Archives of Occupational and Environmental Health, 31*, 37–44.
Lebowitz, M.D. (1991). Populations at risk: addressing health effects due to complex mixtures with a focus on respiratory effects. *Environmental Health Perspectives, 95*, 35–38.
Lee, J., Johnson, J., Reynolds, S., Thorne, P., & O'Shaughnessy, P. (2006). Indoor and outdoor air quality assessment of four wastewater treatment plants. *Journal of Occupational and Environmental Hygiene, 3*, 36–43.
Levy, J. I., Houseman, E. A., Spengler, J. D., Loh, P., & Ryan, L. (2001). Fine particulate matter and polycyclic aromatic hydrocarbon concentration patterns in Roxbury, Massachusetts: a community-based GIS analysis. *Environmental Health Perspectives, 109*, 341–347.
Li, D.-W., & Kendrick, B. (1995). A year-round comparison of fungal spores in indoor and outdoor air. *Mycologia, 87*, 190–195.

Linn, W. S., Gong, H., Jr., Shamoo, D. A., Anderson, K. R., & Avol, El. (1997). Chamber exposures of children to mixed ozone, sulfur dioxide, and sulfuric acid. *Archives of Environmental Health, 52*, 179–187.

Liu, S. M., Zhou, Y. M., Wang, X. P., Wang, D. L., Lu, J. C., Zheng, J. P., et al. (2007). Biomass fuels are the probable risk factor for chronic obstructive pulmonary disease in rural South China. *Thorax, 62*, 889–897.

Lundholm, M., & Rylander, R. (1983). Work related symptoms among sewage workers. *British Journal of Industrial Medicine, 40*, 325–329.

Matanoski, G. M., Santos-Burgoa, C., & Schwartz, L. (1990). Mortality of a Cohort of workers in the Styrene-Butadiene Polymer Manufacturing Industry (1943–1982). *Environmental Health Perspectives, 86*, 107–117.

Maynard, R. (2004). Key airborne pollutants - the impact on health. *Science of the Total Environment, 334–335*, 9–13.

McConnell, R., et al. (2002). Asthma in exercising children exposed to ozone: a cohort study. *Lancet, 359*, 386–391.

McCracken, J. P., Smith, K. R., Diaz, A., Mittleman, M. A., & Schwartz, J. (2007). Chimney stove intervention to reduce long-term wood smoke exposure lowers blood pressure among Guatemalan women. *Environmental Health Perspectives, 115*, 996–1001.

McMichael, A. J., Spirtas, R., & Kupper, L. L. (1974). An epidemiologic study of mortality within a cohort of rubber workers, 1964–1972. *Journal of Occupational Medicine, 16*(7), 458–464.

Medeiros, A. M., Bird, M. G., & Witz, G. (1997). Invited review Potential Biomarkers Of Benzene Exposure. *Journal of Toxicology and Environmental Health, 51*, 519–539.

Melsom, T., Brinch, L., Hessen, J. O., Schei, M. A., Kolstrup, N., Jacobsen, B. K., et al. (2001). Asthma and indoor environment in Nepal. *Thorax, 56*, 477–481.

Mestl, H. E. S., Aunan, K., Seip, H. M., Wang, S., Zhao, Y., & Zhang, D. (2007). Urban and rural exposure to indoor air pollution from domestic biomass and coal burning across China. *Science of The Total Environment, 377*, 12–26.

Miguel, A., Taylor, P., House, J., Glovsky, M., & Flagan, R. (2006). Meteorological influences on respirable fragment release from Chinese elm pollen. *Aerosol Science and Technology, 40*, 690–696.

Miller, K. A., Siscovick, D. S., Sheppard, L., Shepherd, K., Sullivan, J. H., Anderson, G. L., et al. (2007). Long-term exposure to air pollution and incidence of cardiovascular events in women. *New England Journal of Medicine, 356*, 447–458.

Mishra, V., Dai, X., Smith, K. R., & Mika, L. (2004). Maternal exposure to biomass smoke and reduced birth weight in Zimbabwe. *Annals of Epidemiology, 14*, 740–747.

Mishra, V., & Retherford, R. D. (2007). Does biofuel smoke contribute to anaemia and stunting in early childhood? *International Journal of Epidemiology, 36*, 117–129.

Moore, K., Neugebauer, R., Lurmann, F., Hall, J., Brajer, V., Alcorn, S., et al. (2008). Ambient ozone concentrations cause increased hospitalizations for asthma in children: an 18-year study in Southern California. *Environmental Health Perspectives, 116*, 1063–1070.

Morgenstern, H., & Thomas, D. (1993). Principles of study design in environmental epidemiology. *Environmental Health Perspectives, 101*, 23–38.

Mumford, J., Li, X., Chuang, J., Lewtas, J., Hu, F., Shank, T. & Gallagher, J. (1993) Mutagens, polycyclic aromatic hydrocarbon metabolites, DNA adducts and lung cancer from indoor exposure to coal emissions. *Proceedings of the American Association for Cancer Research Annual Meeting, 34*, 152.

Nelson, H.H. & Kelsey, K.T. (2002). The molecular epidemiology of asbestos and tobacco in lung cancer. *Oncogene, 21*, 7284–7288.

Oberdorster, G. (2000). Pulmonary effects of inhaled ultrafine particles. *International Archives of Occupational and Environmental Health, 74*, 1–8.

Pattenden, S., Hoek, G., Braun-Fahrländer, C., Forastiere, F., Kosheleva, A., Neuberger, M., et al. (2006). NO2 and children's respiratory symptoms in the PATY study. *Occupational and Environmental Medicine, 63*, 828–835.

Penn, A., & Snyder, C. A. (2007). 1,3-Butadiene exposure and cardiovascular disease. *Mutation Research/Fundamental and Molecular Mechanisms of Mutagenesis, 621*, 42–49.

Pereira, L. A. A., Loomis, D., Conceição, G. M. S., et al. (1998). Association between air pollution and intrauterine mortality in São Paulo, Brazil. *Environmental Health Perspectives, 106*, 325–329.

Perez Ballesta, P., Field, R.A., Connolly, R., Cao, N., Baeza Caracena, A., De Saeger, E. (2006). Population exposure to benzene: one day cross-sections in six European cities Atmospheric Environment, *40*, 3355–3366.

Peters, A., Doring, A., Wichmann, H. E., & Koenig, W. (1997). Increased plasma viscosity during an air pollution episode: a link to mortality? *Lancet, 349*, 1582–1587.

Peters, J., Hedley, A. J., Wong, C. M., Lam, T. H., Ong, S. G., Liu, J., et al. (1996). Effects of an ambient air pollution intervention and environmental tobacco smoke on children's respiratory health in Hong Kong. *International Journal of Epidemiology, 25*, 821–828.

Pope, C.A., Dockery, D.W., Schwartz, J. (1995). Review of epidemiological evidence of health-effects of particulate air-pollution. *Inhalation Toxicology, 7*, 1–18.

Pope III, C. A., Burnett, R. T., Thun, M. J., Calle, E. E., Krewski, D., Ito, K., et al. (2002). Lung cancer, cardiopulmonary mortality, and long-term exposure to fine particulate air pollution. *JAMA, 287*, 1132–1141.

Qu, Q., Shore, R., Li, G., Jin, X., Chen, L. C., Cohen, B., et al. (2002). Hematological changes among Chinese workers with a broad range of benzene exposures. *American Journal of Industrial Medicine, 42*, 275–285.

Qu, Q., Shore, R., Li, G., Su, L., Jin, X., Melikian, A. A., et al. (2005). Biomarkers of benzene: urinary metabolites in relation to individual genotype and personal exposure. *Chemico-Biological Interactions, 153–154*, 85–95.

Raizenne, M., Neas, L.M., Damokosh, A.I., Dockery, D.W., Spengler, J.D., Koutrakis, P., Ware, J.H. & Speizer, F.E. (1996). Health effects of acid aerosols on North American children: pulmonary function. *Environmental Health Perspectives, 104*, 506–514.

Rinsky, R. A. (1989). Benzene and leukemia: an epidemiologic risk assessment. *Environmental Health Perspectives, 82*, 189–191.

Rinsky, R. A., Hornung, R. W., Silver, S. R., & Tseng, C. Y. (2002). Benzene exposure and hematopoietic mortality: a long-term epidemiologic risk assessment. *American Journal of Industrial Medicine, 42*, 474–480.

Ritz, B., & Yu, F. (1999). The effect of ambient carbon monoxide on low birth weight among children born in southern California between 1989 and 1993. *Environmental Health Perspectives, 107*, 17–25.

Roemer, W., Hoek, G., Brunekreef, B., et al. (1998). Daily variations in air pollution and respiratory health in a multicentre study: the PEACE project. Pollution effects on asthmatic children in Europe. *European Respiratory Journal, 12*, 1354–1361.

Rothman, N., Li, G. L., Dosemeci, M., Bechtold, W. E., Marti, G. E., Wang, Y. Z., et al. (1996a). Hematotoxicity among Chinese workers heavily exposed to benzene. *American Journal of Industrial Medicine, 29*, 236–246.

Rothman, N., Smith, M. T., Hayes, R. B., Li, G. L., Irons, R. D., Dosemeci, M., et al. (1996b). An epidemiological study of early biologic effects in Chinese workers. *Environmental Health Perspectives, 104*, 1365–1370.

Rücker, R., Phipps, R. P., Schneider, A., Frampton, M., Cyrys, J., Oberdörster, G., et al. (2007). Ultrafine particles and platelet activation in patients with coronary heart disease – results from a prospective panel study. *Particle and Fibre Toxicology, 4*, 1. DOI:10.1186/1743-8977-1184-1181.

Salinas, M., & Vega, J. (1995). The effect of outdoor air pollution on mortality risk: an ecological study from Santiago, Chile. *World Health Statics, Q. 48*, 118–125.

Samet, J. M., et al. (2000). Fine particulate air pollution and mortality in 20 U.S. cities, 1987–1994. *New England Journal of Medicine, 343*(24), 1742–1749.

Schelegle, E. S., Eldridge, M. W., Cross, C. E., Walby, W. F., & Adams, W. C. (2001). Differential effects of airway anesthesia on ozone-induced pulmonary responses in human subjects. *American Journal of Respiratory Critical Care Medicine, 163*, 1121–1127.

Schmidt, C. W. (2008). Linking TB *and the environment: an overlooked mitigation strategy. Environmental Health Perspectives, 116*, A 479–485.

Schwartz, J. (1999). Air pollution and hospital admissions for heart disease in eight U.S. counties. *Epidemiology, 10*, 17–22.

Schwartz, J., & Morris, R. (1995). Air-pollution and hospital admissions for cardiovascular-disease in Detroit, Michigan. *American Journal of Epidemiology, 142*, 23–35.

Schwela, D. (2000). Air pollution and health in urban areas. *Reviews on Environmental Health, 15*(1–2), 13–42.

Sharma, N. K., Rai, A. K., Singh, S., & Brown, R. M. (2007). Airborne algae: their present status and relevance. *Journal of Phycology, 43*, 615–627.

Shrestha, I. L., & Shrestha, S. L. (2005). Indoor air pollution from biomass fuels and respiratory health of the exposed population in Nepalese households. *International Journal of Occupational and Environmental Health, 11*, 150–160.

Singh, J. (2005). Toxic moulds and indoor air quality. *Indoor and Built Environment, 14*, 229–234.

Smith, K. R., Samet, J. M., Romieu, I., & Bruce, N. (2000). Indoor air pollution in developing countries and acute lower respiratory infections in children. *Thorax, 55*, 518–532.

Stieb, D. M., Judek, S., & Burnett, R. T. (2002). Meta-analysis of time-series studies of air pollution and mortality: effects of gases and particles and the influence of cause of death, age and season. *Journal of the Air and Waste Management Association, 52*, 470–484.

Straif, K., Baan, R., Grosse, Y., Secretan, B., El Ghissassi, F., & Cogliano, V. (2006). Carcinogenicity of household solid fuel combustion and of high-temperature frying. *Lancet Oncology, 7*, 977–978.

Suh, H., Bahadori, T., Vallarino, J., & Spengler, J. D. (2000). Criteria air pollutants and toxic air pollutants. *Environmental Health Perspectives, 108*, 625–633.

Szczeklik, A., Szczeklik, J., Galuszka, Z., Musial, J., Kolarzyk, E., & Targosz, D. (1994). Humoral immunosuppression in men exposed to polycyclic aromatic hydrocarbons and related carcinogens in polluted environments. *Environmental Health Perspectives, 102*, 302–304.

Taylor, P. E., Jacobson, K. W., House, J. M., & Glovsky, M. M. (2007). Links between pollen, atopy and the asthma epidemic. *International Archives of Allergy and Immunology, 144*, 162–170.

Thurston, G. D., et al. (1997). Summertime haze air pollution and children with asthma. *American Journal of Respiratory and Critical Care Medicine, 155*, 654–660.

Tormo, R., Recio, D., Silva, I., Mu, N., & Oz, A. F. (2001). A quantitative investigation of airborne algae and lichen soredia obtained from pollen traps in south-west Spain. *European Journal of Phycology, 36*, 385–390.

Touloumi, G., Pocock, S. J., Katsouyanni, K., & Trichopoulos, D. (1994). Short-term effects of air pollution on daily mortality in Athens: a time-series analysis. *International Journal of Epidemiology, 23*, 957–967.

Touloumi, G., Samoli, E., & Katsouyanni, K. (1996). Daily mortality and "winter type" air pollution in Athens, Greece —a time series analysis within the APHEA project. In S. St Leger (Ed.), *The APHEA project. Short term effects of air pollution on health: a European approach using epidemiological time series data. Journal of Epidemiology and Community Health, 50*, S47–S51.

Vardoulakis, S., Gonzalez-Flesca, N., Fisher, B.E.A. & Pericleous, K. (2005). Spatial variability of air pollution in the vicinity of a permanent monitoring station in central Paris. *Atmospheric Environment, 39*, 2725–2736.

Vardoulakis, S., Chalabi, Z., Fletcher, T., Grundy, C., & Leonardi, G. S. (2008). Impact and uncertainty of a traffic management intervention: Population exposure to polycyclic aromatic hydrocarbons. *Science of The Total Environment, 394*, 244–251.

Venables, K. M., Allitt, U., Collier, C. G., Emberlin, J., Greig, J. B., Hardaker, P. J., et al. (1997). Thunderstorm-related asthma – the epidemic of 24/25 June 1994. *Clinical & Experimental Allergy, 27*, 725–736.

Verhoeff, A. P., Hoek, G., Schwartz, J., & Van Wijnen, J. H. (1996). Air pollution and daily mortality in Amsterdam. *Epidemiology, 7*, 225–230.

WHO. (1977). *Oxides of nitrogen* (vol. 4). Environmental Health Critera. Geneva: WHO.
WHO. (1993). *Benzene* (vol. 150). Environmental Health Criteria. Geneva: WHO.
WHO. (2002). *The world health report 2002: reducing risks, promoting healthy life*. Geneva: WHO.
WHO. (2003). *Health aspects of air pollution with particulate matter, ozone and nitrogen dioxide*. Report on a WHO Working Group. Bonn, Germany: WHO.
WHO, (2005). *Air quality guidelines. Global update 2005. Particulate matter, ozone, nitrogen dioxide and sulfur dioxide*. Geneva: WHO.
WHO. (2006). *Development of WHO guidelines for indoor air quality*. Report on a Working Group Meeting Bonn, Germany, 23–24 October 2006. Copenhagen, Denmark: World Health Organization.
WHO. (2009). *WHO guidelines for indoor air quality: dampness and mould*. Copenhagen, Denmark: World Health Organization.
Wichmann, H. E., Spix, C., Tuch, T., Wolke, G., Peters, A., Heinrich, J., et al. (2000). Daily mortality and fine and ultrafine particles in Erfurt, Germany part I: role of particle number and particle mass. *Research Repoort (Health Effects Institute), 98*, 5–86 (discussion 87–94).
Wietlisbach, V., Pope, C. A. I., & Ackermann-Liebrich, U. (1996). Air pollution and daily mortality in three Swiss urban areas. *Soz Praeventivmed, 41*, 107–115.
Wong, C. M., et al. (2002). A tale of two cities: effects of air pollution on hospital admissions in Hong Kong and London compared. *Environmental Health Perspectives, 110*, 67–77.
Wong, C-M., Vichit-Vadakan, N., Kan, H., & Qian, Z. (2008). Public Health and Air Pollution in Asia (PAPA): a multicity study of short-term effects of air pollution on mortality. *Environmental Health Perspectives, 116*, 1195–1202.
Woodcock, A. H. (1948). Note concerning human respiratory irritation associated with high concentration of plankton and mass mortality of marine organisms. *Journal of Marine Research, 7*, 56.
Yu, I. T., Lee, N. L., Zhang, X. H., Chen, W. Q., Lam, Y. T., & Wong, T. W. (2004). Occupational exposure to mixtures of organic solvents increases the risk of neurological symptoms among printing workers in Hong Kong. *Journal of Occupational and Environmental Medicine, 46*, 323–330.
Zhang, J. J., & Smith, K. R. (2007). Household air pollution from coal and biomass fuels in China: measurements, health impacts, and interventions. *Environmental Health Perspectives, 115*, 848–855.

Chapter 6
Inhalation Dosimetry Modelling

Christos Housiadas and Mihalis Lazaridis

Abstract The estimation of deposition of inhaled aerosol particles in the respiratory tract during breathing is of importance in environmental health and occupational hygiene assessments. Respiratory deposition and subsequent clearance of the deposited amounts determine the effective doses delivered to the lungs. Realistic inhalation dosimetry modeling provides valuable information on the complete exposure-dose-response relationship, as required in environmental pollutant health effects analyses.

6.1 Introduction

Airborne particulate matter (PM) is a complex mixture of different chemical components. An important aspect in the study of PM is the complexity of the physico-chemical characteristics, their multiple sources, morphology and their dynamics that are correlated with the size of the particles. Particulate matter has very diverse effects ranging from human health to climate forcing (Hinds 1999). In recent years, extensive research effort has been invested in examining the relationship between exposure to ambient particulate matter (PM) (also mixtures of PM and gaseous pollutants) and resulting health effects. Data from epidemiological studies conducted to date demonstrate associations between ambient particulate concentrations and increased morbidity and mortality, while data from toxicological studies have begun to provide potential biological explanations for these observed associations (Pope et al. 1995; Schlesinger 1995; Schlesinger and Cassee 2003; Schwartz 1994; Neuberger et al. 2004; Mauderly and Chow 2008). Specifically, concerning the particulate matter there are recent cohort studies (see for a summary HEI 2001)

C. Housiadas (✉)
"Demokritos" National Centre for Scientific Research,
15310 Agia Paraskevi, Athens, Greece
e-mail: christos@ipta.demokritos.gr

M. Lazaridis
Department of Environmental Engineering, Technical University of Crete,
Polytechneioupolis, 73100 Chania, Greece

M. Lazaridis and I. Colbeck (eds.), *Human Exposure to Pollutants via Dermal Absorption and Inhalation*, Environmental Pollution 17, DOI 10.1007/978-90-481-8663-1_6,

which have followed the successful implementation of the Harvard Six Cities Study (Dockery et al. 1993) and the American Cancer Society (ACS, ACS-CPS II) studies (Pope et al. 1995, 2002) and they show significant effects of particulate matter levels on mortality and morbidity. Long-term average mortality rates ranging from 17% to 26% higher for people living in communities with higher levels of $PM_{2.5}$ were observed (USEPA 2004). Morbidity and mortality and their association with particulate matter exposures occur in fact below the current air quality standards (USEPA 2004; CAFE 2004).

Assessing health risks from exposure to particulate matter (PM) includes consideration of particle-specific mechanisms of chemical disposition, toxicant-target interactions, and tissue response. Including such factors as regional variability in deposition and clearance rates by gender, age and state of health (USEPA 2004; ICRP 1994) makes exposure assessment studies very complex endeavors. Furthermore, the structure and function of the respiratory tract makes it vulnerable to air suspended contaminants and micro organisms which are entering through inhalation. Inhalation is the major route of exposure to PM, irritants, and toxicants for humans.

Respiratory deposition refers to the convective transport processes that determine how much airborne material in the inspired air remains in the respiratory system, and in which part of the system. Such analyses originated for radiological protection purposes, in relation with the assessment of doses incurred to the lungs of workers in nuclear industry. Nowadays they embrace a much broader application range, and are of importance in inhalation toxicology, drug delivery (pharmaceutical aerosols), occupational hygiene, industrial hygiene, or environmental health.

The development of inhalation (dosimetry) models is an important step in understanding exposure – dose – response relationships for ambient PM and can certainly help in assessing human health risks from inhalation of different toxicants in occupational and residential environments. In particular, mechanistic inhalation models can utilize data reflecting intra- and intersubject biological variability by employing statistical distributions for model parameters, enabling prediction of the biological dose distribution in target individuals and specific subpopulations (Kremer and Hancock 2006; Lazaridis et al. 2001; Broday 2004; Balashazy et al. 2007; Balashazy and Hofmann 2000; Finlay and Martin 2008; Yang et al. 2008; Park and Wex 2008; Su et al. 2008).

Numerous mathematical models for predicting particle deposition in the human airways have been developed over the years. In general, available practical (i.e. "routinely operational") models can be classified into three major groups.

The first group follows an approach based on the concept of applying compartmental modeling to the human lung anatomy. Numerous compartmental models have been proposed, differing in the representation of the tracheobronchial tree, the breathing physiology and the resulting air flow, and the expressions used for calculating particle deposition efficiencies (Yeh and Schum 1980; Koblinger and Hofmann 1990; Martonen et al. 1995; Anjilvel and Asgharian 1995; Zhang and Kleinstreuer 2001).

In the second approach, the human lung is modeled as a chamber shaped like a trumpet with variable cross sectional area (Yu and Diu 1983; Nixon and Egan 1987; Darquenne and Paiva 1994). This approach is not intended to provide a realistic morphometric description of the lung but rather to enable efficient implementation of one-dimensional numerical schemes for solving the spatiotemporal concentration field.

The third approach uses a combination of theoretical and empirical expressions to predict particle deposition in the human airways (Rudolf et al. 1986; Freijer et al. 1997). Such is the approach taken by the International Commission on Radiological Protection ICRP66 recommended model (ICRP 1994). Since this model is inherently semiempirical, it has a limited applicability to nonstandard cases that are not included in the database from which the empirical expressions were derived. Indeed, the ICRP66 has a questionable predicting power when aerosol particles cannot be considered inert and stable, that is for particles that do not preserve their composition and size during the breathing cycle. However, despite the above drawbacks ICRP66 is a powerful inhalation dosimetry estimation tool, being simple and reliable and having a user-friendly software package (LUDEP).

Additionally to the above approaches, computational fluid dynamics (CFD) models have been used to model airflow characteristics and detailed deposition patterns in subsections of the human RT (Martonen et al. 2005).

6.2 Physicochemical Properties of Aerosols

Airborne particulate matter (PM) is a complex mixture of different chemical components with its size to range from few nanometers to several 100 μm (Hinds 1999). It is apparent that particulate matter is not a single pollutant and its mass includes a mixture of many pollutants distributed differently at different sizes. The determination of the aerosol size distribution is one of the most important aspects involved both in measuring and modeling aerosol dynamics. Figure 6.1 shows size ranges of aerosols with their definitions.

Particles in the atmosphere have a distribution of sizes and lognormal distributions are used for their description (Hinds 1999). The aerosol size distribution is a crucial parameter that determines the dynamics of aerosols in the atmosphere, their transport, deposition and residence time. Figure 6.2 presents typical atmospheric aerosol distributions by number, surface area and volume.

The entrance of particles inside the human respiratory tract (RT) depends mainly on their size and the air flow characteristics close to the mouth and nose. The entry of particles into the mouth and nose is characterized by the inhalability (I). Inhalability is the intake efficiency at which ambient aerosols enter the mouth and nose. It is defined as the ratio of the particle mass concentration inspired through the nose or the mouth (M_0) for a specific size to the ambient number concentration ($M_{0,amb}$):

$$I = \frac{M_0}{M_{0,amb}} \tag{6.1}$$

Series of experiments have been performed for the determination of the inhalability in low-velocity wind tunnels with the use of a full-size, full-torso mannequin connected to a mechanical breathing machine (Hinds 1999). These experiments have been done using all possible orientations of the mannequin in respect to the

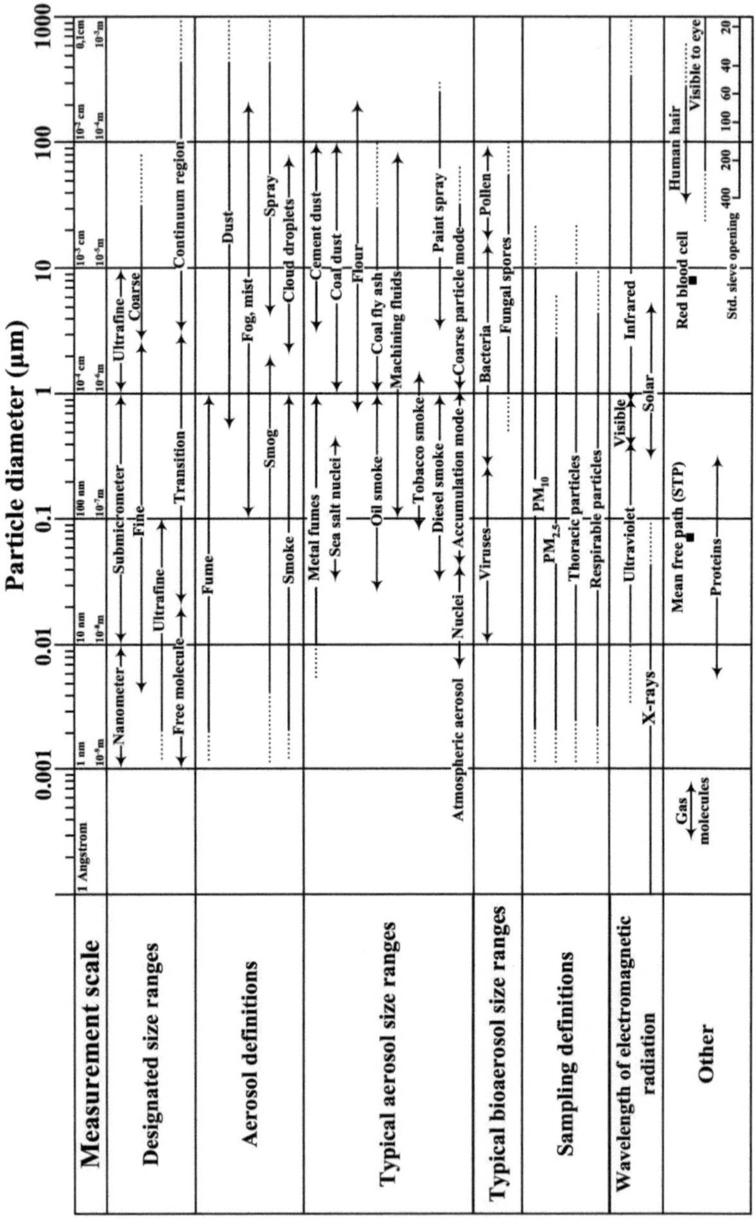

Fig. 6.1 Particle size range for aerosols (Adapted from Hinds 1999)

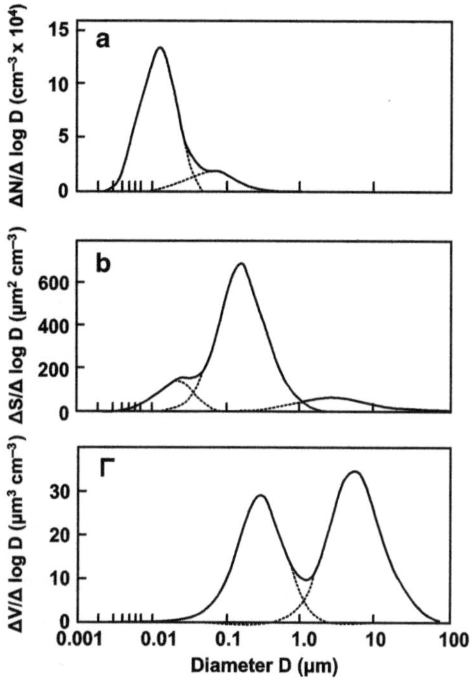

Fig. 6.2 Typical ambient aerosol distributions by (**a**) number (**b**) surface area (Γ) volume (Adapted from Hinds 1999)

wind flow. Orientation averaged data from mouth inhalation are shown in Fig. 6.3. The particles are distinguished at respirable, thoracic and inhalable depending on their ability to entrer specific areas of the human respiratory tract. Specifically, inhalable are the particles that can enter and deposit to the upper respiratory system. Thoracic are the particles which can penetrate the trachea and bronchi. Finally, respirable are the fraction of thoracic that enter the alveoli region. These definitions are performed for practical reasons and its correspondence to the equivalent particle diameter is proposed by the International Standards Organization (ISO 7708) and the American Conference of Governmental Industrial Hygienists (ACGIH 1997). The corresponding curve for PM_{10} (particle mass aerodynamic diameter equal or smaller than 10 μm) is almost identical with the thoracic fraction curve, whereas, the curve for $PM_{2.5}$ (particle mass aerodynamic diameter equal or smaller than 2.5 μm) is found to the left of the respirable fraction (USEPA 2004).

The equation that defines the inhalability for ambient conditions, where the wind velocity is lower than 4 m s^{-1} is given as:

$$I = 0.5\,(1 + \exp\,(-0.06 d_{ae})) \tag{6.2}$$

where d_{ae} is the particle diameter (μm).

At wind velocities higher than 4 m s^{-1} the inhalability can be described by an expression which depends not only on the particle diameter but also from the wind speed U (m s^{-1}) (Vincent et al. 1990):

$$I = 0.5\,(1 + \exp\,(-0.06 d_{ae})) + 10^{-5} U^{2.75} \exp\,(0.055 d_{ae}) \tag{6.3}$$

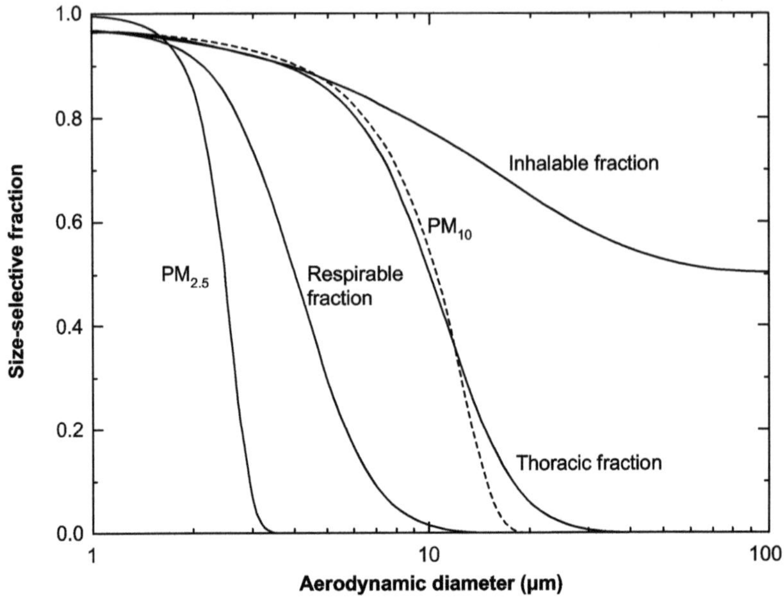

Fig. 6.3 Sampling criteria for respirable, thoracic and inhalable particle fractions (Adapted from ACGIH 1997) together with the PM_{10} and $PM_{2.5}$ curves

The expression for inhalability used in the ICRP66 inhalation dosimetry model is (ICRP 1994):

$$I = 1 - 0.5\left[1 - (7.6 \times 10^{-4} d_{ae}^{2.8} + 1)^{-1}\right] + 10^{-5} u^{2.75} \exp\{0.055 d_{ae}\} \qquad (6.4)$$

The inhalability depends also if the subject inhales from the mouth or nose. Mathematical expression for nose inhalability can be found in the literature even though the data are more scarce (Hinds 1999).

6.3 The Respiratory System

The understanding of the particle deposition in the human respiratory tract (RT) requires the definition of a detailed morphometric model for the respiratory system. The basic aspects of the respiratory system are known so simple inhalation models have been constructed (e.g. ICRP 1994). However, there is limited information available for the detailed geometry of the respiratory tract such as the dimensions of its numerous parts. Besides, there are differences in the RT due to human variability and due to time dependent changes of the respiratory system arising from the breathing cycle.

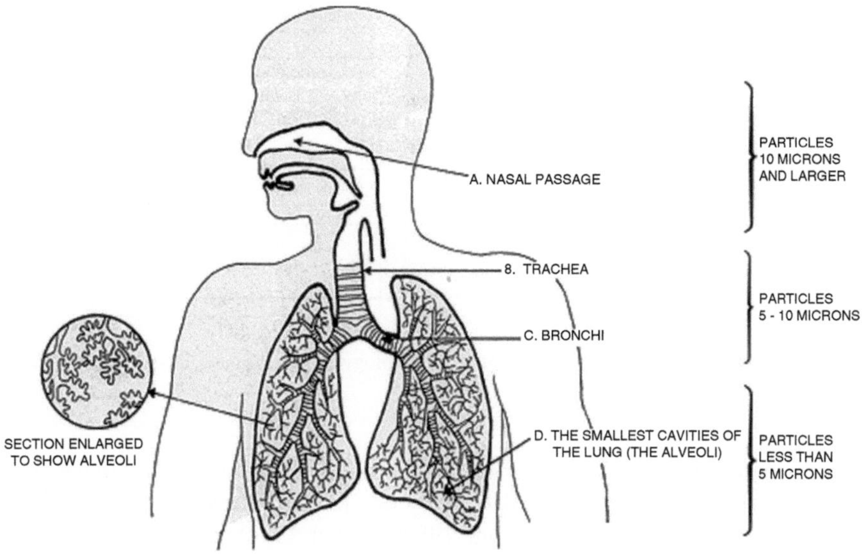

Fig. 6.4 The anatomy of the human respiratory system (Adapted from International Commission on Radiological Protection [ICRP] 1994)

The basic regions of the respiratory system are shown in Fig. 6.4 (ICRP 1994). The human respiratory system can be viewed as three anatomical regions. In each region the deposition characteristics for particles are different as well as their structure. The first region includes the *extrathoracic* airways ET_1 (anterior nose) and ET_2 (posterior nasal passages, larynx, pharynx and mouth). This region serves for the warming and the humidification of the inhaled air. The *extrathoracic* part is known also as *head airways region* or *nasopharyngeal region*.

The second region is the *tracheo-bronchial* region or *lower airways* which includes the bronchial (BB) and the bronchiolar (bb) regions. This region includes the airways which conduct air from the extrathoracic region to the gas exchange regions of the lung. The BB region consists of the trachea and bronchi and the bb region includes the bronchioles and terminal bronchioles. The trachea branches in two airways which are called main bronchi. This branching continues until the terminal bronchioles and the structure of the division is similar to an inverted tree. Both the extrathoracic and tracheo-bronchial airways have also the name of *conducting airways* since they conduct air to the gas exchange regions of the lung.

Finally, the third region is the alveolar-interstitial region (AI), where gas exchange is performed. The AI region is called also *alveolar region*, or *pulmonary region*, or *acinar region*. This region consists of the respiratory bronchioles, the alveolar ducts, the alveoli and the interstitial connective tissue.

There have been several studies to measure the dimensions of the different parts of the respiratory tract which however is a difficult task since the airways have three

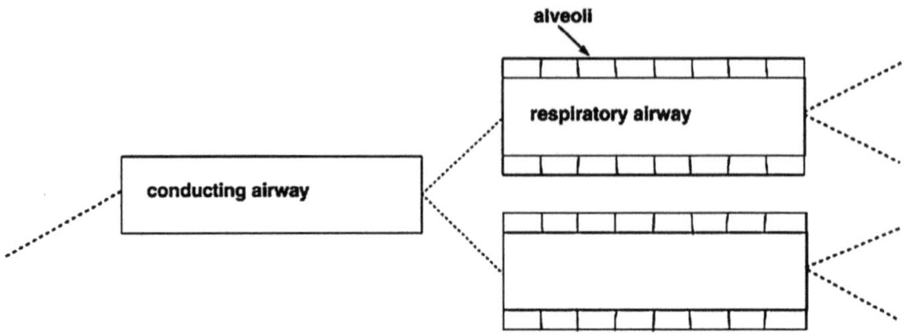

Fig. 6.5 Schematic presentation of the lung morphology as used in lung models with dichotomous branching

dimensional structure. In the scientific literature many researchers have used measurements of casts of normal lungs to define the dimensions of the airways (e.g. Weibel 1963; Yeh and Schum 1980; Weibel 1991). Since many airways resemble cylindrical structure then the definition of the length and the diameter can be used for the definition of them (see Fig. 6.5). It is apparent that the above description of the RT is a simplification of its real structure.

The most extensively used lung model is the symmetric model of Weibel (1963) which is referred as the Weibel A model (Table 6.1). In this model there are 24 airway generations (23 levels of bifurcations), which branch symmetrically (dichotomously), with trachea being generation zero and alveoli generation 23. Sine every generation divides into two branches the number of branches in each generation, z, is $N_z = 2^z$. Table shows the characteristics of the Weibel A model for the structure of the human respiratory tract (ICRP 1994). The generations 0–15 correspond to the conducting airways and the remainder to the gas exchange region. A healthy adult has a number of alveoli between 400 and 1,200 million with a surface for the air exchange close to 80 m² (Hughes 1996).

Since the total cross-sectional area usually increases with penetration distance, the average air velocity in distal airways decreases. Airway gravity angles, which in single pathway lung models are uniform within any generation, vary in accordance with the specific body posture considered (upright position, sleeping position etc.).

There are many simplifications made in the Weibel A model as the symmetric branching and the underprediction of the tracheobronchial airways. Besides, the diameters and lengths of the alveolated airways are too small and also start at high generation number (Finlay 2001). Therefore the Weibel's picture is a simplification of the human bronchial tree, where there is irregular dichotomous branching, and not well defined and invariant over time geometric characteristics of the different bronchial branches. In addition, the tracheobronchial airway surface is lined with ciliated epithelium, but not a constant amount of mucus at the ciliated epithelium exist all the time, making time dependent changes in the geometry of the human airways, which are not considered in the Weibel's model. The main difficulty in the

Table 6.1 Architecture of the human lung according to Weibel A symmetric model for adults with lung volume of 4,800 ml and approximately ¾ increase due to inhalation (Fung 1990; ICRP 1994)

		Generation z	Number, n(z)	Diameter d(z)(cm)	Length l(z) (cm)	Area S(z) (cm^2)	Volume V(z) (cm^3)	Cummulative volume (cm^3)	Flow rate 1 l s^{-1} Speed (cm s^{-1})	Reynolds No.	Residence time (ms)
BB	Trachea	0	1	1.8	12	2.54	30.5	30.5	393	4.350	30
	Main bronchus	1	2	1.22	4.76	2.33	11.25	41.8	427	3.210	11
	Lobar bronchus	2	4	0.83	1.9	2.13	3.97	45.8	462	2.390	4.1
		3	8	0.56	0.76	2	1.52	47.2	507	1.720	
	Segmental bronchus	4	16	0.45	1.27	2.48	3.46	50.7	392	1.110	3.2
	Bronchi with cartilage in wall	5	32	0.35	1.07	3.11	3.3	54	325	690	4.4
		6	64	0.28	0.9	3.96	3.53	57.5	254	434	
		7	128	0.23	0.76	5.1	3.85	61.4	188	277	
		8	256	0.186	0.64	6.95	4.45	65.8	144	164	
bb		9	512	0.154	0.54	9.65	5.17	71	105	99	
		10	1.024	0.13	0.46	13.4	6.31	77.2	73.6	60	
	Terminal bronchus	11	2.048	0.109	0.39	19.6	7.56	84.8	52.3	34	7.4
	Bronchioles with muscle in wall	12	4.096	0.095	0.33	28.8	9.82	94.6	34.4	20	16
		13	8.192	0.082	0.27	44.5	12.45	106	23.1	11	
		14	16.384	0.074	0.23	69.4	16.4	123.4	14.1	6.5	
	Terminal bronchiole	15	32.768	0.066	0.2	113	21.7	145.1	8.92	3.6	31
Al	Respiratory bronchiole	16	65.536	0.06	0.165	180	29.7	174.8	5.40	2.0	60
		17	131.9×10^3	0.054	0.141	300	41.8	216.6	3.33	1.1	
		18	262×10^3	0.05	0.117	534	61.1	277.7	1.94	0.57	
	Alveolar duct	19	524×10^3	0.047	0.099	944	93.2	370.9	1.10	0.31	210
		20	1.05×10^6	0.045	0.083	1.600	139.5	510.4	0.60	0.17	
		21	2.10×10^6	0.043	0.07	3.220	224.3	734.7	0.32	0.08	
		22	4.19×10^6	0.041	0.059	5.880	350	1,084.7	0.18	0.04	
	Alveolar sac	23	8.39×10^6	0.041	0.05	11.800	591	1675	0.09	–	550
	Alveoli, 21 per duct		300×10^6	0.028	0.023		3.200	4.800			

modeling of the RT is the description of a dynamic system as the human respiratory tree, with time dependent changes and variability characteristics between different people. However, the Weibel's model is the most well known model with very extensive applications.

Table 6.1 shows that the majority of the lung volume is contained in the alveolar region. The extrathoracic airways in an adult have a volume of approximately 50 ml, the tracheo-bronchial region a volume of approximately 100 ml, whereas, the remainder of the lung volume (between 2,000 and 4,000 ml during tidal breathing and 6,000 ml when fully inflated for an adult male) is in the alveolar region (Finlay 2001).

The dimensions of the airways are changing due to age, gender, race and weight. The Weibel A model can be viewed as a sequence of filters and the deposition fractions for each region are calculated after accounting for the filtering effect of the airways previously. At each region of the lung the deposition is calculated using empirical functions.

The anatomy and physiology of the RT determine the deposition of particles. Of particular importance are the parameters which determine the rhythm and process of the breathing cycle (physiology). The breathing mode (nose breathers, nasal augmenters) intensity depends upon activity. Inhalability depends also on the breathing mode. The lung physiology is based on several parameters such as (see also Fig. 6.6):

- The tidal volume V_T, which is the air volume inhaled and exhaled during a periodic breathing (500–3,800 ml depending on exercise level)
- The breathing frequency *f*, which is defined as the number of tidal breaths per minute (12–18 min^{-1} for normal activity)
- The total lung capacity (*TLC*), which is the total volume of the airspaces in the lung when is totally inflated (maximum breath) (about 6,000–7,000 ml)
- The functional residual capacity *FRC*, which is the lung volume at the end of normal exhalation or at the start of normal inhalation (close to 3,000 ml for adults)
- The residual volume (*RV*), which is the remaining volume of the airspaces when someone exhales the maximum volume of air as one can
- The vital capacity (*VC*), which is maximum air volume that someone can inhale (close to 4,000 ml for adults)
- The forced expiratory volume in 1 s (FEV_1), which is the maximum volume of air that someone can exhale during 1 s when there is the maximum volume in the airspaces (*TLC*)

The age plays an important role for the dimensions of the different volume parameters in the RT. For an elderly person the residual volume and the functional residual capacity are increased, whereas, the vital capacity, the inspiratory reverse volume and the expiratory reserve volume are decreased. In addition, the dead volumes at each part of the RT are changed since the alveolar region losses its elasticity and the entrained air is not divided uniformly at the different airways. Therefore a

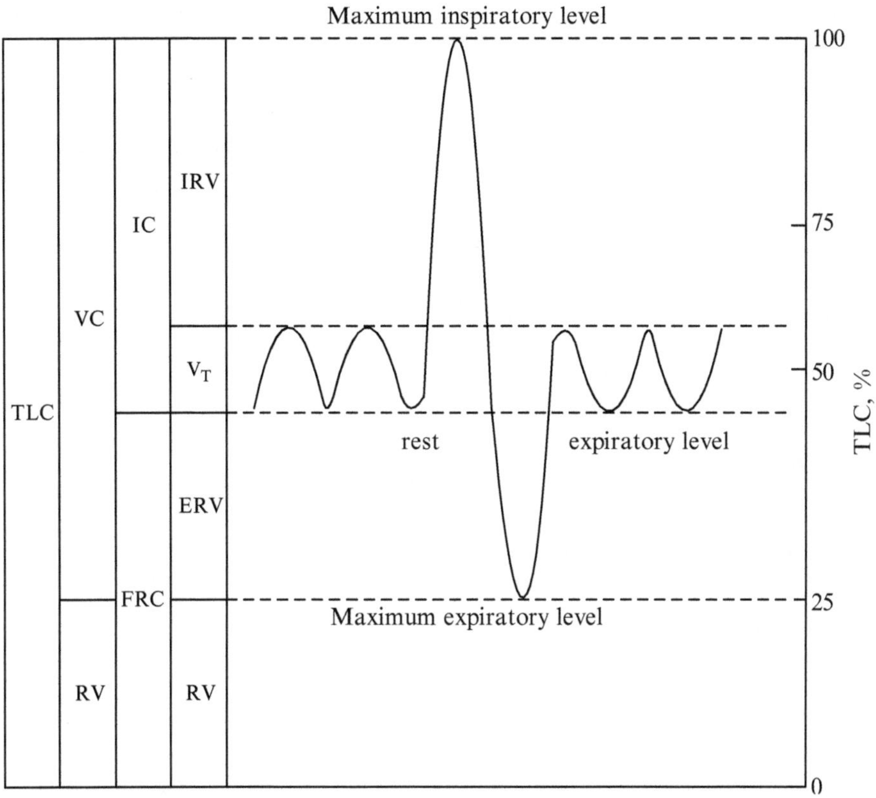

Fig. 6.6 Diagram of the different air volume parameters (respiratory volumes) used in the RP for the description of the breathing cycle. TLC = total lung capacity, VC = vital capacity, RV = residual volume, FRC = functional residual capacity, IRV = inspiratory reverse volume, ERV = expiratory reverse volume, V_T = tidal volume, IC = inspiratory capacity (Adapted from ICRP 1994)

number of alveoli do not receive continuously air and as a result the air volume that do not involved actively in the breathing cycle is increased.

The physiology parameters which are used for the description of the breathing cycle are dependent on the race and the age of the human subject. For example an adult person inhales more air than a child, whereas, the breathing frequency is decreased. Table 6.2 presents representative values for physiology parameters of Caucasian persons under different activities. In addition, the main anatomical parameters which are used for the calculations of particle deposition in the RT are given for male, female and a child.

Table 6.2 Representative values of anatomy and physiology parameters which are used for the calculation of particle deposition in the RT (ICRP 1994)

	Exposed person		
Parameter	Male	Female	Child (3 months)
FRC: Functional residual capacity (ml)	3,301	2,681	148
V_D(ET): Anatomical dead space of the Extrathoracic region (ml)	50	40	2.6
V_D(BB): Anatomical dead space of the trachea and bronchi (ml)	49	40	4.5
V_D(bb): Anatomical dead space of bronchioli (ml)	47	44	6.8
d_0 :Diameter of Trachea (cm)	1.65	1.53	0.616
d_9 :Diameter of the first bronchioles (cm)	0.165	0.159	0.099
d_{16} : Diameter of the terminal bronchiole (cm)	0.051	0.048	0.020
Mean values during sleeping			
B : Volumetric flow rate during inhalation (m^3/h)	0.45	0.32	0.99
V_T : Air volume during inhalation (ml)	625	444	39
V : Volumetric flow rate (ml/sec)	250	178	50
f: Breathing frequency (breaths/min)	12	12	38
Mean values at rest			
B : Volumetric flow rate during inhalation (m^3/h)	0.54	0.39	–
V_T : Air volume during inhalation (ml)	750	464	–
V : Volumetric flow rate (ml/s)	300	217	–
f: Breathing frequency (breaths/min)	12	14	–
Mean values for light exercise			
B : Volumetric flow rate during inhalation (m^3/h)	1.5	1.25	0.19
V_T : Air volume during inhalation (ml)	1,250	992	66
V : Volumetric flow rate (ml/s	833	694	106
f: Breathing frequency (breaths /min)	20	21	48
Mean values for heavy exercise			
B : Volumetric flow rate during inhalation (m^3/h)	3	2.7	–
V_T : Air volume during inhalation (ml)	1,920	1,364	–
V : Volumetric flow rate (ml/s)	1,670	1,500	–
f: Breathing frequency (breaths/min)	26	33	–

6.4 Deposition and Clearance Mechanisms

6.4.1 Deposition of Particles in the Human Respiratory Tract

The deposition characteristics of particles in the RT depend from their size, their physico-chemical properties and the physiology of the person. The main deposition mechanisms in the RT are inertial impaction, gravitational settling and Brownian diffusion (see Fig. 6.7). The deposition mechanisms of interception and electrostatic deposition are important in specific cases (Hinds 1999; Finlay 2001). The total deposition is calculated as a superposition of independent deposition efficiencies of the different mechanisms.

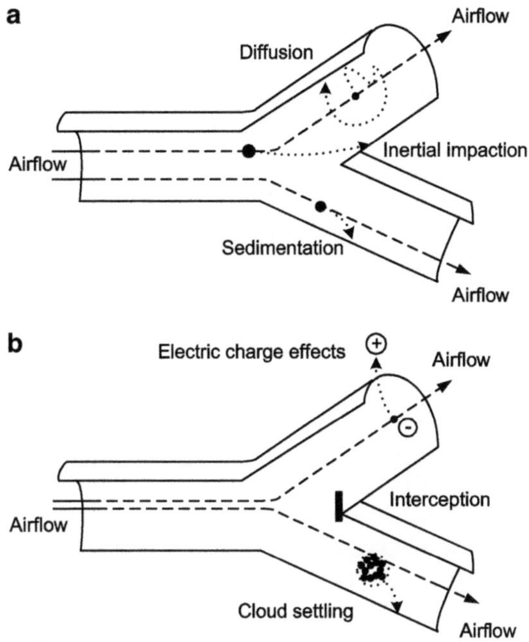

Fig. 6.7 Schematic description of deposition mechanisms of particles in the RT (Adapted from Isaacs et al. 2005)

The dose of particles (for a usual bimodal size distribution consisting of fine and coarse fraction) in each part of the RT can be calculated using the following expression:

$$H = n_0 c_A B\left(a\sum_i n_{fine,i} + (1-a)\sum_i n_{coarse,i}\right) \tag{6.5}$$

where, H is the dose-rate (μg/h), n_0 is the inhalability ratio, c_A is the aerosol concentration in air, B is the ventilation rate, α is the fine mode mass fraction, $n_{\text{fine},i}$ is the retention in region i of lungs for fine particles and $n_{\text{coarse},i}$ is the retention in region i of lungs for coarse particles (ICRP 1994).

The *inertial impaction* mechanism is a result of the inadequacy of particles to follow the streamlines at places that air changes abruptly direction inside the RT. As a result the particles continue their original direction for a short distance due to their inertia. This mechanism is important for particles having an aerodynamic diameter larger than 0.5 μm mainly in the upper RT (ICRP 2002). The probability of particle deposition due to impaction is proportional to the airflow velocity, the inhalation frequency and the particle size and density (Schlesinger 1988). Inertial impaction is important in areas of sharp curvatures in the streamlines, such as in the nasal turbinates and airway bifurcations. Inertial deposition is proportional to the Stokes number $\left(St = \frac{\tau \overline{u}}{l} \right)$ of the particles that enter in the RT (τ is the particle relaxation time and the ratio $l / \overline{u}$ is the average residence time of the particle in the airway, where l the length of the airway and $\overline{u}$ is the airflow velocity). The highest probability of particles to deposit by impaction is in the bronchial region (Hinds 1999).

Gravitational settling refers to the deposition of particles in the RT due to gravity and together with the mechanism *inertial impaction* are the most important mechanisms for deposition of particles with aerodynamic diameter greater than 0.5 μm. The mechanism of *gravitational settling* is important for deposition in small airways and the alveolar region (Schlesinger 1988). The probability of particle deposition due to gravity is proportional to the residence time in the airways, their size and density and inversely proportional to the breathing frequency. Gravitational settling and inertial impaction are competing deposition mechanisms in the RT (Hinds 1999). The inertial impaction is most important for increased flow rates. The relative contributions of these two mechanisms to the particle deposition at the bifurcations of the first generations of the alveolar region are about the same. However, their difference is at the deposition location with the major fraction of deposited particles due to impaction at the airway bifurcations, whereas, the particles deposited due to gravitation are located along the whole airway (Balashazy et al. 1999).

Particles smaller than 0.2 μm have a Brownian motion and may deposit at the airway walls especially in smaller airways (alveolated airways). *Diffusional deposition* is main deposition mechanism for particles in the bronchioli and the alveolar region where the flow velocity is low and the residence times long. In addition, the deposition of ultrafine particles (smaller than 0.01 μm) due to diffusion is significant at the head airways due to high diffusion coefficients (ICRP 2002).

Electrically charged particles attract to airway walls in the RT due to electrostatic image charge induced in the airway surfaces. The *electrostatic attraction* is inverserly proportional to the size of the particles and the velocity air flow. However, the concentration of electrical charged particles in the atmosphere is not high since the particles are neutralized from the atmospheric ions. Therefore the importance of the mechanism of electrostatic attraction is relative small compared to the other deposition mechanisms. Experimental data have shown that the deposition mechanism of electrostatic attraction is a main deposition mechanism for ultrafine particles in the lower tracheobronchial area (USEPA 2004).

Another deposition mechanism is the particle *interception* at which particles are having a contact with the airway surfaces due their size. The interception is dependent on the particle size and its morphology. This mechanism is important for fibrous particles of large aspect ratio, and for highly elongated fibers (ICRP 1994).

There are several dosimetry models which were developed to predict the total and regional deposition of particles in the RT. The most used models are those of the International Commission on Radiological Protection (ICRP 1994) and the

National Council on Radiation Protection and Measurement (NCRP 1994). Deposition differences predicted by the two models are small compared to the subject variability differences and are both recommended for particle deposition calculations.

The ICRP model has been used here to study the particle deposition at different parts of the RT versus their diameter for a man, woman and a child (3 months old) (see Figs. 6.8–6.10). The results show that at the upper respiratory tract (ET1 and ET2 regions) the deposition of particles with a diameter smaller than 0.2 μm is higher for men compared to women due to higher volumetric flow rates For larger particles the deposition percentage for males and females is almost equal.

In the BB part of the tracheobronchial region (bronchi) coarse particles have the same deposition characteristics for males and females. However, particles with a diameter smaller than 0.002 μm deposit with higher probability in the female's RT, whereas, particles with diameter in the range between 0.002 and 0.2 μm deposit with higher probability in the men's RT. In the bb part (bronchioli) and in the alveolar region there is the same behavior as the BB region. Particles with diameters smaller than 0.006 and 0.021 μm deposit more in the RT of women for the regions

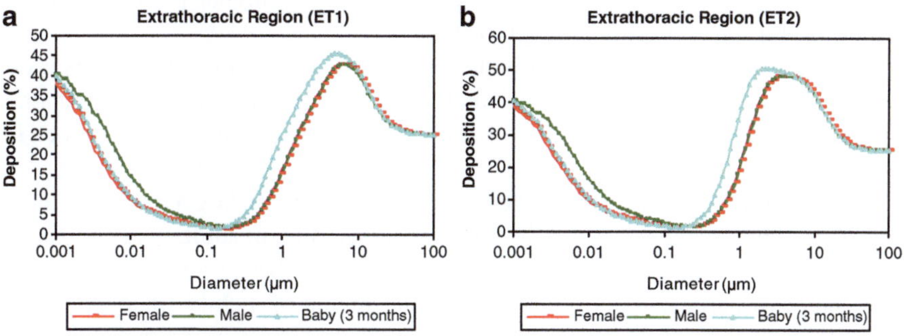

Fig. 6.8 Mass particle percentage which deposits in the (**a**) ET1 and (**b**) ET2 thoracic regions of the RT for a man, woman and a child (3 months) in relation to the particle diameter

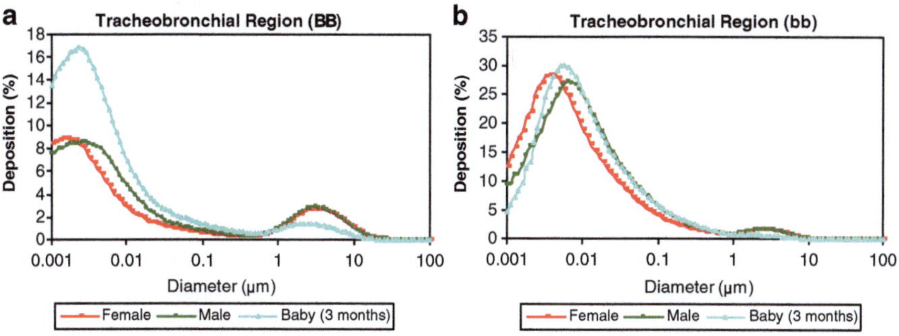

Fig. 6.9 Mass particle percentage which deposits in the (**a**) BB and (**b**) bb tracheobronchial regions of the RT for a man, woman and a child (3 months) in relation to the particle diameter

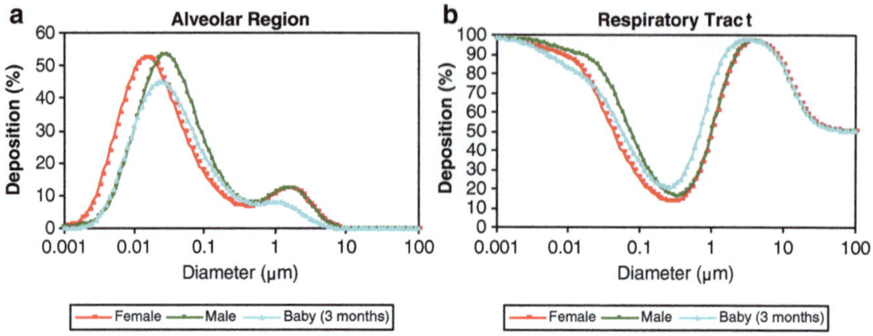

Fig. 6.10 Mass particle percentage which deposits in the (**a**) alveolar region of the RT and (**b**) total deposition of particles in the RT for a man, woman and a child (3 months) in relation to the particle diameter

bb and AL (alveolar) respectively. Particles with diameter in the ranges between 0.006 and 0.85 μm and 0.021 and 0.7 μm deposit with higher probability in the men's RT for the regions bb and AL respectively.

Smaller particles deposit easier in the lower parts of the female's RT. This is due to anatomical differences between women and men. The total deposition fraction is larger for men for the small particles and almost equal between men and women for the coarse fraction.

A 3 months child has higher deposition percentage of larger particles in the upper respiratory and extrathoracic regions compared to an adult, whereas, for the smaller particles the deposition profile is the same as that of an adult. A smaller percentage of particles deposits in the alveolar region of a child since the dimensions of this region are smaller than that of an adult. However, the deposition particle mass per surface area in the alveolar region of a child is large (USEPA 2004). Finally, the total particle deposition percentage of a child compared to an adult is lower for small particles and higher for larger particles. In addition, due to child activities during a day the volumetric flow rate is high and the particle dose in the RT is also higher compared to the average activity conditions applied to adults.

Elderly and individuals with chronic obstructive pulmonary disease (COPD) breath with higher frequency, not always in a symmetric way with respect to inhalation/exhalation, and have higher deposition percentage in the tracheobronchial region compared to healthy individuals (USEPA 2004). In addition, individuals with COPD have also higher percentage of particle deposition in the alveolar region compared to healthy people. However, the volumetric flow rate for individuals with COPD is not evenly distributed due to obstruction of parts of the RT. Therefore, whereas some parts of the alveolar region are not reachable some other parts have locally intense deposition. The semi-empirical ICRP model cannot be used for COPD breathing conditions. In such circumstances the use of mechanistic respiratory deposition models seems in order.

6.4.2 Particle Clearance in the Human Respiratory Tract

The deposited particles in the RT undergo further clearance (Schlesinger 1988; Gradon et al. 1996). The clearance depends on the particle solubility, its chemical composition and size as well as on physiological parameters (metabolic rate, health condition, age, gender). The calculation of clearance is a necessary step in dosimetric assessments because the internally delivered biological dose to the lung tissue is effectively the difference between the deposited amount and the cleared amount.

Inhaled pollutants deposit in the airways by various deposition mechanisms, and subsequently interact with the epithelium depending on their reactivity and solubility. The amount and site of PM deposition and of the clearance rate are affected by (Smith et al. 2008):

- Particle geometric and physicochemical characteristics (size, shape, charge, density, chemical composition)
- The morphology and inflammatory state of the respiratory tract
- Physiological and respiration parameters (breathing pattern, ventilation rate, exercise level, metabolic rate, health condition, body posture, age, gender)
- Environmental conditions (temperature, humidity, ambient concentration)

The particle clearance from the RT is a natural defence mechanism of the human body. The clearance translocates the particles to the gastrointestinal (GI) region, the regional lymph nodes (LN) via lymphatic channels and into the blood by the absorption mechanism (ICRP 1994). The clearance rate by each route depends on the region in which particles were deposited, the physicochemical properties of the particles (e.g. solubility, particle size, chemical composition) and their mass/number concentration.

The clearance mechanisms of deposited particles in the RT may be performed by two mechanisms which act competitive (particle transport and absorption into blood). These are also the main mechanisms used in the ICRP model to study particle clearance:

1. *Particle transport*: Particles are transported to the GI tract and the lymph nodes and from one part of the RT to another. The transport is performed with the mucus layer or uptake by macrophages. At the upper RT additional mechanisms such as coughing contribute also to particle removal.
2. *Absorption into blood*: The movement of materials from the deposited particles into blood occurs with the dissociation of particles and the absorption into blood. This mechanism refers also to the movement of ultrafine particles directly into the blood (ICRP 1994). This direct route has been occasionally reported in animal toxicokinetic experimental studies performed with engineered nanoparticles (Oberdörster et al. 2002).

There are different clearance mechanisms which dominate at each part of the RT. In particular:

- *Upper Respiratory Tract (ET1 and ET2 regions)*
  In the upper respiratory tract and especially in the posterior nasal passages insoluble particles are cleared by mucus flow (USEPA 2004). The mucus is transported to pharynx and swallowed. Otherwise, particles can be also removed by sneezing. In the anterior part of the respiratory tract particles are mainly removed by extrinsic means (nose blowing) (ICRP 1994). The mucus transport velocity in the anterior part is small (close to 2 mm/h). The time which is needed for particles to transported from the anterior to the posterior part is close to 10–20 min (USEPA 2004). Soluble particles in the nasal passages are absorbed to the blood through the epithelium. Also, in animal studies it has been observed that very small particles of the ultrafine range (i.e. nanoparticles) deposited in the nasal region may migrate to the central nervous system (brain) through the olfactory lobe (Oberdörster et al. 2004). There is uncertainty as to whether this mechanism occurs also in human.
- *Tracheobrobcial region (BB and bb regions)*
  This region is covered by mucus and mucociliary clearance to the GI and pharynx is the main mechanism for particle transport. In addition, insoluble particles can be also removed by macrophage and the soluble particles can be transported into blood. In this region the mucus transport is reduced from the trachea region to the regions deeper in the RT. The clearance from this region is quick (close to 24 h) but some insoluble particles may stay for time periods larger than 24 h (USEPA 2004).
- *Alveolar – interstitial region (Al region)*
  In the alveolar region particles are removed with (1) absorption to blood and (2) phagocytosis from the macrophage and sequential transport to the mucus layer and transport to the GI region or transport to the blood through the lymph system and the interstitium (see Fig. 6.11).

Lung retention experimental data for insoluble particles are scarce but data show a high retention time with some material to remain in the lungs for decades (ICRP 1994). The clearance mechanism can be described as a multiphase process with each phase to correspond to a separate clearance mechanism with high retention times (Tran et al. 1999a,b). The ICRP model (ICRP 1994) describes the retention in the alveolar region as a two-component exponential function, with about 30% of particles to have a half-time of 30 days and the remaining a half-time of several 100 days.

The clearance through the mucus layer is dependent on the particle size and deposition area. Macrophages are effective for the phagocytosis of particles except of particles of asbestos and silica. These particles produce an inflammation to the epithelium cells and the macrophage which have an effect on the phagocytosis rate (Donaldson et al. 2000). Increased dose in the alveolar region (particle mass/number) have also an effect on the retention time (USEPA 2004). In general, the clearance from the alveolar region follows an exponential decay and can be represented by three separate rates, the fast (Al_1 with a rate close to 0.02 day^{-1}) with the transport of macrophage to the mucus layer, the medium (Al_2 with a rate close to 0.001 day^{-1}) with the transport of macrophage through the

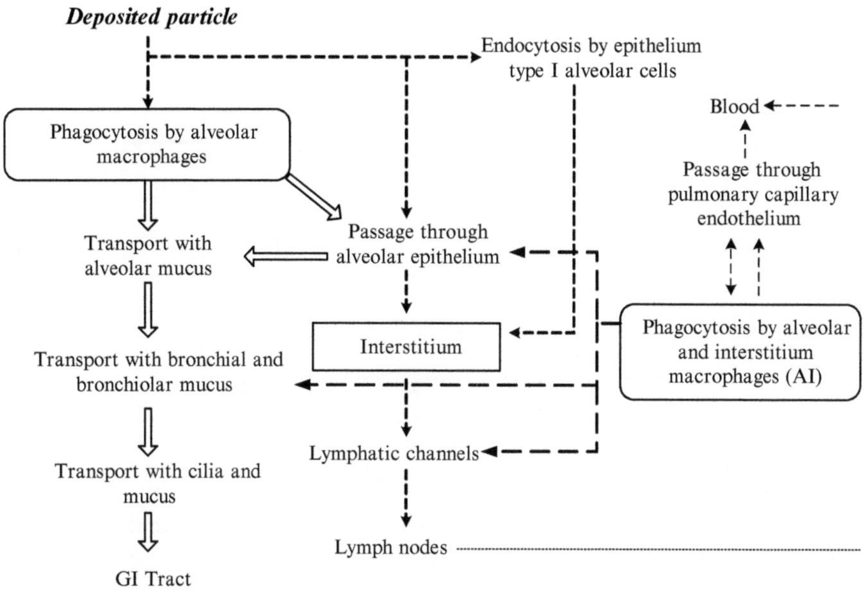

Fig. 6.11 Mechanisms for particle clearance from the alveolar region for insoluble particles (Adapted from Schlesinger 1988; USEPA 2004)

interstitium and the slow one (Al_3 with a rate close to 0.0001 day^{-1}) which is performed with the particle breakup and further absorption and transport (ICRP 1994; Schlesinger 1995).

Finally, the available dose due to particle exposure and inhalation can be calculated as the product of the mean concentration of particles in air, the mean breathing frequency and the exposure time. Therefore, the delivered dose rate (D_{app} – μg h^{-1}) from the inhalation of polydisperse particles is given by:

$$D_{app} = \sum_{i=1}^{n} B(a_i IF_{0,i} C_{A,i}) n_{i,j} \tag{6.6}$$

where, $IF_{0,i}$ is the particle inhalability, $C_{A,i}$ is the particle concentration in air (μg m^{-3}), B is the breathing frequency (m^3 h^{-1}), α_i is mass fraction of particles and $n_{i,j}$ is the retention of particles with size i to the region j of the RT. The term i refers to the particle mean diameter.

The recommended approach for the clearance mechanism in the RT includes calculations of particle clearance in the lung following deposition for soluble, readily absorbed into blood, and relatively insoluble and nontoxic solid particles (ICRP 1994). Particles deposited in the respiratory tract are cleared by three main routes: by absorption into blood, to the gastrointestinal tract via the pharynx and to lymph nodes via lymphatic channels. The material travels in the respiratory tract following the course in the Fig. 6.12.

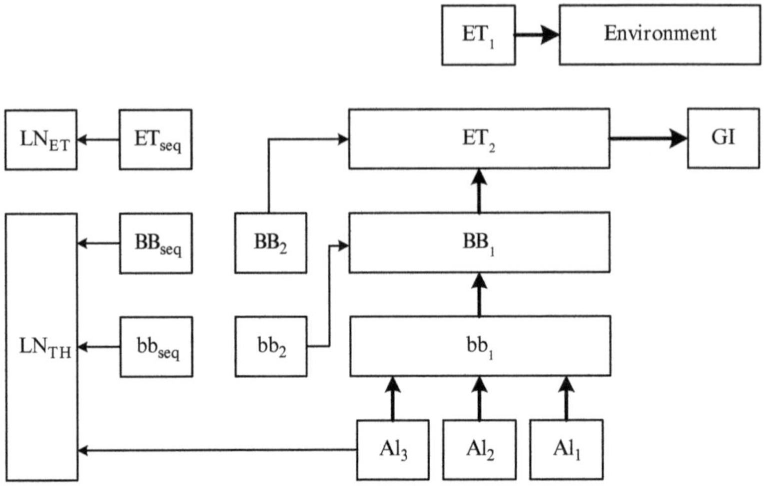

Fig. 6.12 Model of particle movement in the different regions of the RT, where each region consists of specific parts which describe the time dependent transport rate (Adapted from ICRP 1994). Details on the different regions are given in Tables 6.3 and 6.6

Each region of the respiratory tract is divided in a combination of compartments cleared with a different constant rate so that the overall clearance approximates the time-dependent clearance behaviour of the region. Clearance from ET_1 to Environment is not considered in the model because the material deposited there is removed by extrinsic means (nose blowing, wiping); a reference value for residence time is 1 day^{-1} as proposed in the ICRP model. Particle residence times and clearance rates for mechanical transport are given for each compartment in the ICRP Publication 66 and are presented in the Table 6.3.

In the model there are not included calculations for the transport of particles from the ET1 region to the Environment since this rate is dependent from parameters which cannot be predicted such as the frequency and duration of coughing (Chang et al. 1991).

The fraction of the initial deposit assigned to each compartment is given in the Table 6.4 where $f_s = 0.5$, for $d_{ae} \leq 2.5\sqrt{\rho/\chi}$ μm, and $f_s = 0.5\exp\left(-0.63\left(d_{ae}\sqrt{\chi/\rho} - 2.5\right)\right)$, for $d_{ae} > 2.5\sqrt{\rho/\chi}$ μm.

Absorption into blood is assumed to occur at the same rate in the regions considered. Dissolution of particles and uptake into blood following deposition can occur with variable rates dependent on the aerosol physical and chemical characteristics. Though, absorption behavior of particles has been classified in three main categories; fast, moderate and slow for particles that take days, weeks or years to dissociate and enter the veins. Particles dissolution rates used in the model are according to the ICRP classification for fast and slow cleared particles and are presented in the Table 6.5.

The model approach, as mentioned before, does not estimate clearance for particles exhibiting moderate absorption behavior. In addition retention for slow cleared

Table 6.3 Rates of clearance transport of particles together with their half time in the different regions of RT

Clearance rates				
Pathway	From	To	Rate (day^{-1})	Half-time
$m_{1,4}$	Al_1	bb_1	0.02	35 days
$m_{2,4}$	Al_2	bb_1	0.001	700 days
$m_{3,4}$	Al_3	bb_1	0.0001	7,000 days
$m_{3,10}$	Al_3	LN_{TH}	0.00002	–
$m_{4,7}$	bb_1	BB_1	2	8 h
$m_{5,7}$	bb_2	BB_1	0.03	23 days
$m_{6,10}$	bb_{seq}	LN_{TH}	0.01	70 days
$m_{7,11}$	BB_1	ET_2	10	100 min
$m_{8,11}$	BB_2	ET_2	0.03	23 days
$m_{9,10}$	BB_{seq}	LN_{TH}	0.01	70 days
$m_{11,14}$	ET_2	GI tract	100	10 min
$m_{12,13}$	ET_{seq}	LN_{ET}	0.001	700 days

Table 6.4 Deposition fraction of particles at different compartments of the RT (ICRP 1994)

Deposition site	Compartment	Fraction of deposit in region assigned to compartment
ET_2	ET_2	0.9995
	ET_{seq}	0.0005
BB	BB_1	0.993-f_s
	BB_2	f_s
	BB_{seq}	0.007
bb	bb_1	0.993-f_s
	bb_2	f_s
	bb_{seq}	0.007
Al	Al_1	0.3
	Al_2	0.6
	Al_3	0.1

particles is estimated with the assumption that all particles dissolute slowly and that the fraction of material readily absorbed is zero.

The particle absorption into blood can be differentiated in two separate stages: (1) the breakup into substances which can be absorbed into blood and (2) the direct absorption into blood.

Retention in each region is given by the mass balance equation $\frac{dR_i(t)}{dt} = -\lambda_i(t)R_i(t)$ where $\lambda_i(t)$ is the instantaneous clearance rate of the deposit in compartment i and $R_i(t)$ the retained mass after time t. The rate is the sum of mechanical movement and absorption into blood rates $\lambda_i(t) = m_i(t) + s_i(t)$. Transport equations for particles are solved numerically per minute[1] (matrix method) to determine retention.

[1]Calculations with another timestep have been also tested by the authors. Solving the system for every breath gives almost the same results for retention but expends more time and computer resources. Results variation for timesteps greater than 1 min (checked for 5 and 10 min) is relatively large.

Table 6.5 Absorption rates used in the ICRP model for fast and slow clearance (ICRP 1994)

ICRP publication 30 classification	D (days)	W (weeks)	Y (years)
Type of absorption behavior:	*F (fast)*	*M (moderate)*	*S (slow)*
Fraction dissolved rapidly, f_r	1	0.1	0.001
Approximate dissolution rate:			
Rapid (day^{-1}), s_r	100	100	100
Slow (day^{-1}), s_s	–	0.005	0.0001
Model parameters			
Initial dissolution rate (day^{-1}), s_p	100	10	0.1
Transformation rate (day^{-1}), s_{pt}	0	90	100
Final dissolution rate (day^{-1}), s_t	–	0.005	0.0001
Fraction of material bound, f_b	0	0	0
Uptake rate of bound material (day^{-1}), s_b	–	–	–

Table 6.6 The relationships between target tissues and clearance components. Retention results are assigned to the source components

Target	Source component	Equivalent clearance components
ET_1	Surface deposit	14
ET_2	Surface fluid (ET_2_sur)	$11 + 11_T$
	Particles in airway wall (ET_seq)	$12 + 12_T$
	Bound material	ET_b
LN_{ET}	All transformations	$13 + 13_T + LN_{ETb}$
BB	Mucous gel (BB_gel)	$7 + 7_T$
	Mucous sol (BB_sol)	$8 + 8_T$
	Particles in airway wall (BB_seq)	$9 + 9_T$
	Bound material	BB_b
	Alveolar interstitium	$1 + 2 + 3 + 1_T + 2_T + 3_T + AI_b$
bb	Mucous gel (bb_seq)	$4 + 4_T$
	Mucous sol (bb_sol)	$5 + 5_T$
	Particles in airway wall (bb_seq)	$6 + 6_T$
	Bound material	bb_b
	Alveolar interstitium	$1 + 2 + 3 + 1_T + 2_T + 3_T + AI_b$
AI	All transformations (in AI and LN_{TH} as Total_AI)	$1 + 2 + 3 + 1_T + 2_T + 3_T + AI_b + 10 + 10_T + LN_{THb}$
LN_{TH}	All transformations	$10 + 10_T + LN_{THb}$

The relationship between target tissues and clearance components is given in the Table 6.6. Retention results are assigned to the source components.

The results stand for retention and are available in two types: the retained mass in the tracheobronchial, alveolar-interstitial regions, lymph nodes and mass to the gastrointestinal system during exposure and post-exposure times, and the relative retention in the above regions for exposure and post-exposure times.

6.5 Particle Deposition Measurements

The respiratory tract deposition has been studied in the literature using various aerosolized dusts in the human and animal lungs as well as at lung airway replicas. Recent measurements on particle deposition in the human t airways has been performed by Zhang and Finlay (2005), Zhou and Cheng (2005), Londahl et al. (2007), Morawska et al. (2005). An excellent overview of past measurement aerosol deposition studies is presented by USEPA (2004). Comparison of two lung deposition models using aerosol dynamics (Lazaridis et al. 2001; Mitsakou et al. 2005) with available experimental data is presented in the current section.

Chan and Lippmann (1980) used monodispersed Fe_2O_3 particles with mass median diameter larger than 2 μm and density of 2.56 g/cm^3 for in vivo deposition determination in 26 healthy nonsmokers and for in vitro measurements in hollow casts. The tidal volume was approximately 1,000 cm^3. A mechanistic dosimetry model was used to simulate the deposition measurements from Chan and Lippmann (1980) (Lazaridis et al. 2001). In this dosimetry model the Aerosol General Dynamic Equation is solved numerically during inhalation using a discrete-nodal point method for describing the particle size distribution. This model incorporates explicitly the mechanisms of nucleation, condensation, coagulation, convection and deposition of gases and particles, as well as a module for considering gas phase reactions.

To compare model predictions with experimental data it was assumed that the tidal volume was inspired in 1 s, corresponding to an air flow rate of 1,000 cm^3/s. This is relatively high air flow, corresponding to nonsedentary breathing conditions under which respired particles undergo enhanced inertial deposition. The model predicts the evolution of the size distribution and composition of inhaled particles and their deposition characteristics for each generation of the human airways. The model has modular structure and the user has the flexibility to include or exclude specific physical processes in a particular simulation. Predictions of the dosimetry model compared with the experimental data for tracheobronchial deposition of Chan and Lippmann (1980) are presented in Fig. 6.13. Model predictions were obtained for three typical bifurcation angles θ, the only parameter not rigidly fixed in the Weibel's lung model. Typical deposition conditions correspond to bifurcation angles of 45° (typical deposition conditions), 65° (maximum deposition conditions), and 25° (minimum deposition conditions). Figure 6.13 shows that particle deposition data in the tracheobronchial region seems bound by the theoretical results for the most and least favorable deposition conditions. Varying the bifurcation angles affects only inertial deposition, which takes place mainly in the proximal airways. Therefore, alveolar deposition is less affected by these morphological variations (Fig. 6.14). Overall, the natural variability of the lung morphology, which is transformed into model uncertainty with respect to the bifurcation angles, significantly affects the particle deposition profile along the tracheobronchial airways. Indeed, deposition of large particles in the upper airways can double under adverse conditions (i.e. larger surfaces for impaction in the carinal ridges). In contrast, deposition

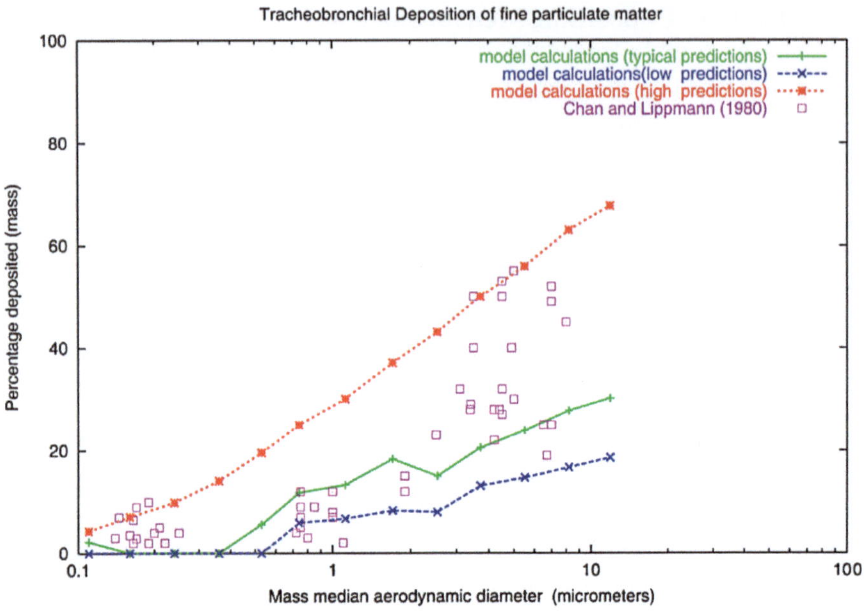

Fig. 6.13 Comparison of model predictions with the experimental data for tracheobronchial deposition from Chan and Lippmann (1980)

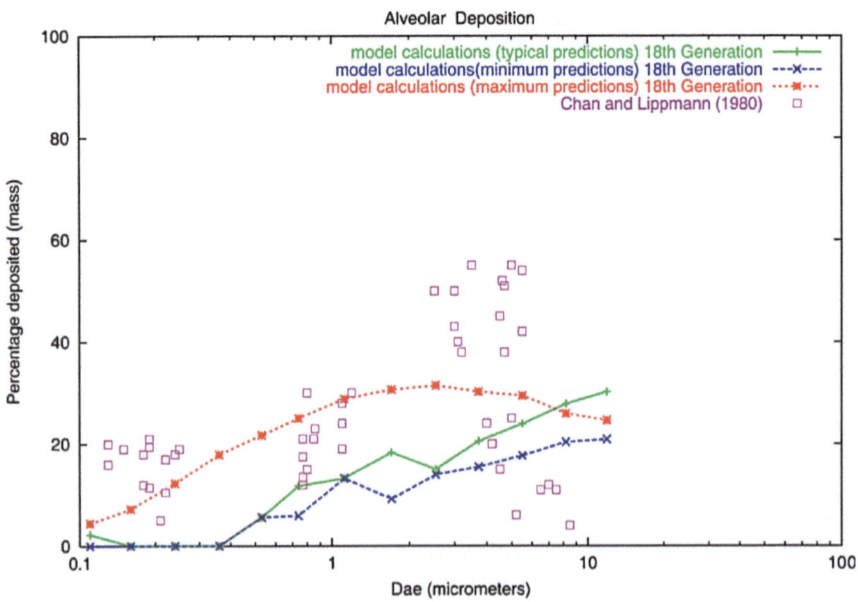

Fig. 6.14 Comparison of model predictions with the experimental data for alveolar deposition from Chan and Lippmann (1980)

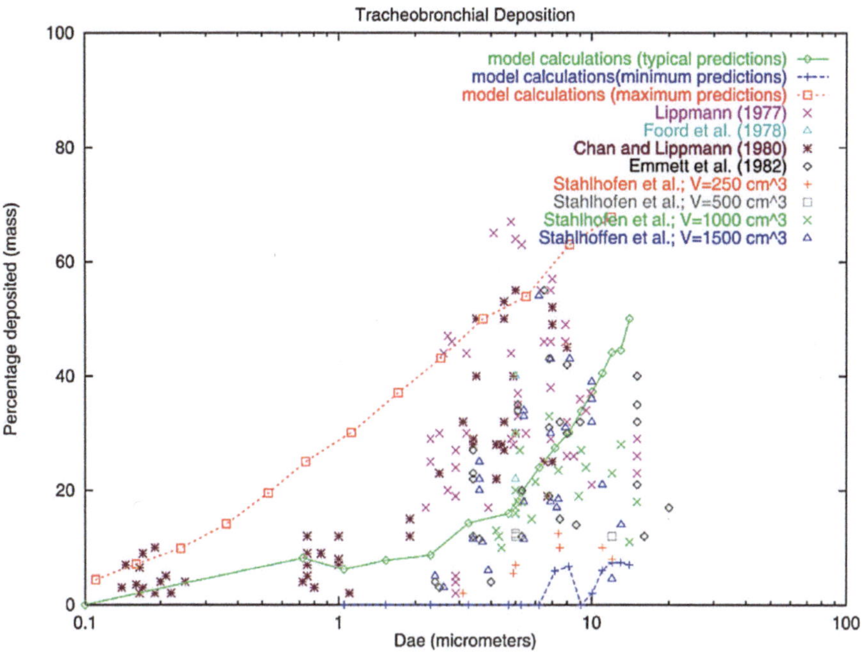

Fig. 6.15 Predicted tracheobronchial deposition vs experimental data (Adapted from Lazaridis et al. 2001)

of small particles, which are less affected by inertial impaction, is less affected by the variation in branching angles.

Figures 6.15 and 6.16 show model predictions of deposition in the tracheobronchial and the alveolar regions respectively, versus data from different studies (Chan and Lippmann 1980; Emmett et al. 1982; Stahlhofen et al. 1986; Foord et al. 1978; Lippmann 1977). The experimental data were obtained by tracking radioactive labelled poorly soluble particles of diameter larger than 0.1 μm. For example, Emmett et al. (1982) measured total and regional deposition of monodisperse aerosols in the RT of 12 healthy subjects that breathed through the mouth in an upright position. Radioactively labeled polystyrene particles with aerodynamic diameters ranging between 3.5 and 10.0 μm and density of 1,060 kg/m^3 were used. The data show that total deposition rate increased with increasing particle size.

In vivo measurements were based on the amount of radioactivity retained in the lung as a function of time, where the "fast-cleared" and "slow-cleared" deposition fractions correspond to deposition in the ciliated tracheobronchial and alveolar regions, respectively (USEPA 2004). Model parameters include air flow of 1,000 cm^3 s^{-1}, log-normally distributed inert aerosol with an initial 1 μm count median diameter, and a 1.7 geometric standard deviation. Although experimental data are quite scattered, owing both to nonconsistency in methodology (Stahlhofen et al. 1986) and to inter-subject variability (USEPA 2004), the theoretical predictions are in fair agreement with the general deposition pattern in the tracheobronchial region.

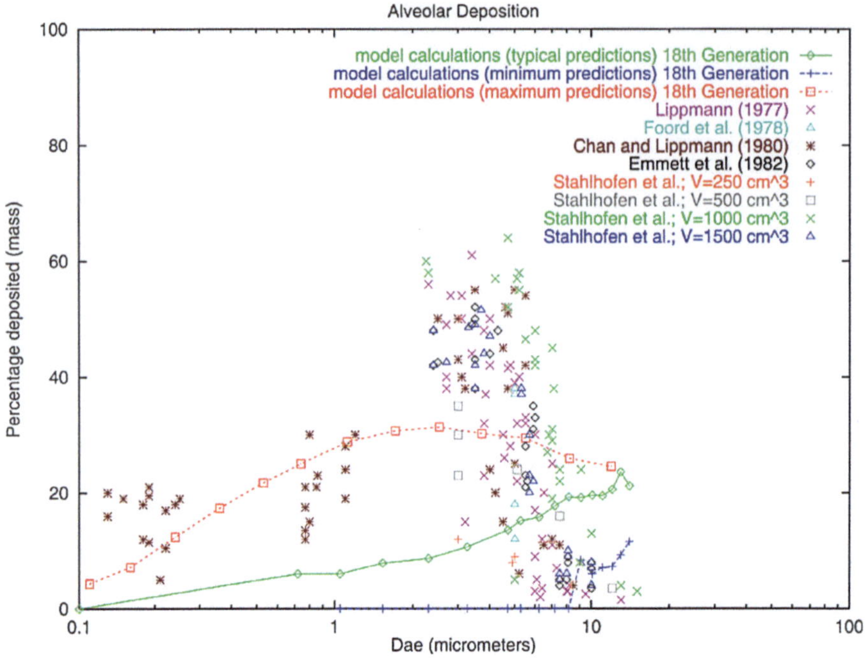

Fig. 6.16 Predicted alveolar deposition vs experimental data (Adapted from Lazaridis et al. 2001)

Predictions of the dosimetry model versus available experimental data for the tracheobronchial and alveolar regions of Long-Evans rats are shown in Figs. 6.17 and 6.18. The model calculations were performed using the recommended lung model for Long-Evans rats (Raabe et al. 1975; Phalen et al. 1974) without changing the bifurcation angles (typical predictions). The model calculations show a reasonable agreement with the experimental data but additional experimental data under controlled conditions are needed for a comprehensive model comparison.

Mitsakou et al. (2005) validated also the developed model by comparing with available experimental and numerical results.

Mitsakou et al. (2007) developed a model for the aerosol deposition in the mouth and throat regions of a human. The model has been based on a mechanistic description of deposition without any tuning to extrathoracic deposition measurement data. Figure show the comparison between the predictions of the model and many measurements, as obtained in different in vivo experimental studies. The simulation results have been obtained assuming aerosol particles of unit density ($\rho_p = 1$ g cm^{-3}), having diameters between 1 and 17 μm. Steady inhalation flow-rates are used, corresponding to ventilatory conditions typical to those encountered in the experiments (Fig. 6.19a: 1,000 cm^3 tidal volume, 15 min^{-1} breathing frequency; Fig. 6.19b: 1,500 cm^3 tidal volume, 10 min^{-1} breathing frequency). As expected, particle deposition increases sharply with increasing values of the impaction

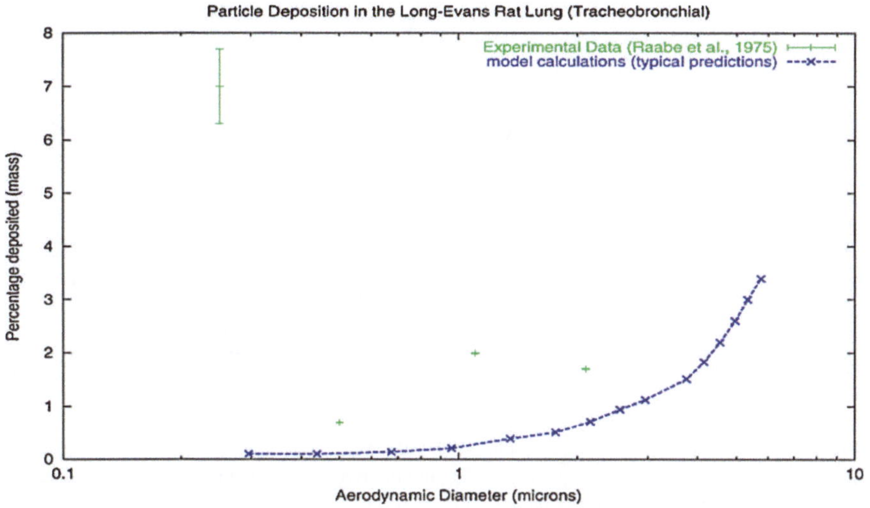

Fig. 6.17 Predicted tracheobronchial deposition versus experimental data in the Long-Evans rat lung

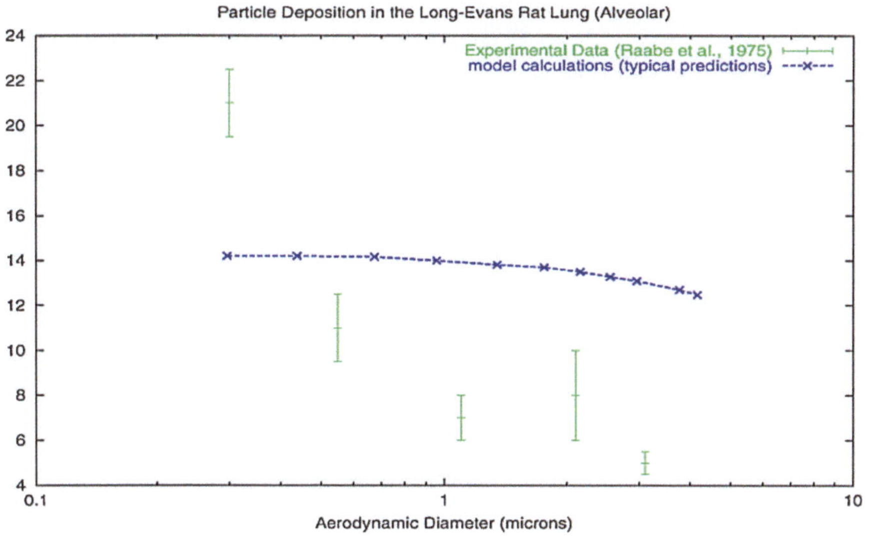

Fig. 6.18 Predicted alveolar deposition versus experimental data in the Long-Evans rat lung

parameter, which is a consequence of the dominance of inertial impaction as deposition mechanism. Also, as can be observed, there is a large scatter in the data between the various experimental works. Possible reasons for this variability are the different experimental techniques used, as well as the high degree of variability among subjects.

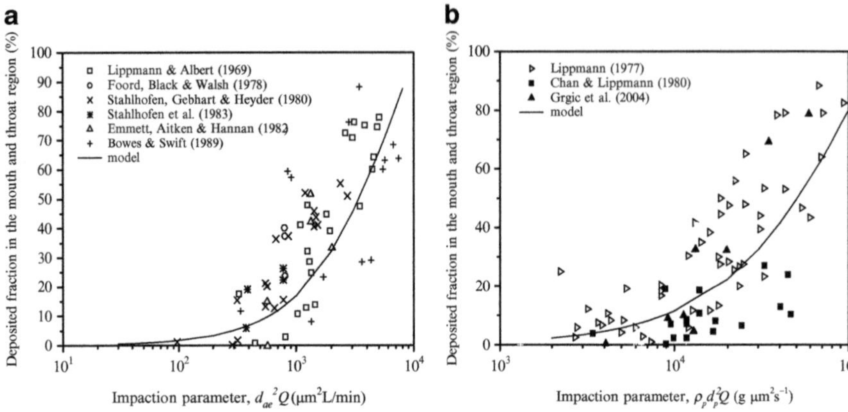

Fig. 6.19 Deposited fractions in the mouth and throat region as function of the impaction parameter, expressed as d_{ae}^2Q (**a**) or $\rho_p d_p^2 Q$ (**b**), from many in vivo measurement studies, and comparison with the predictions of the present model

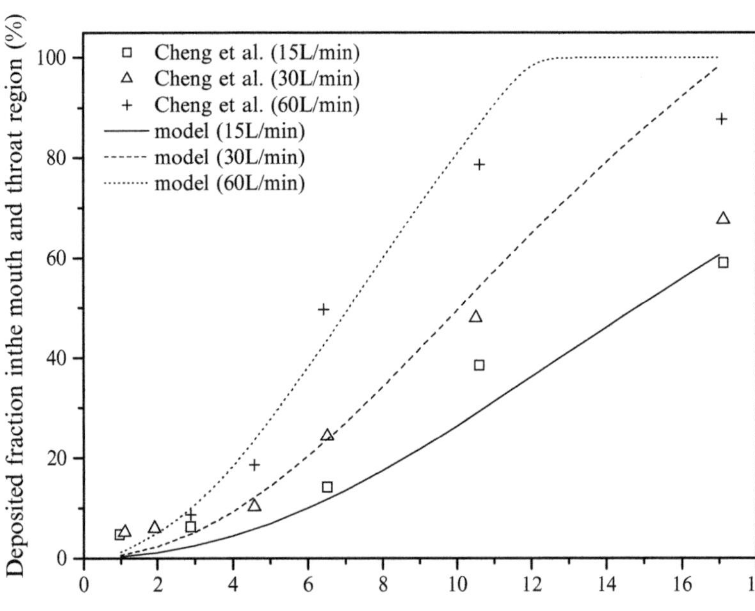

Fig. 6.20 Mouth-throat deposition as a function of particle size for various flow rates, as experimentally determined by Cheng et al. (1999), and as predicted with the present model

In order to assess the parametric effect of inspiratory flow rate, we compare the model predictions with the measurements of Cheng et al. (1999), in which deposition is measured at various flow rate conditions. In these tests, deposition of monodisperse particles with sizes between 1 and 17 μm and for steady flow rates in the range 15–60 L min^{-1} has been determined as a function of particle aerodynamic diameter, using a cast of human oral airways which mimic the realistic geometry. The simulations are performed using the same experimental conditions. The deposited fraction in the mouth and throat region, as experimentally determined and predicted by the model, is shown in Fig. 6.20. The measurements show that particle deposition increases with increasing particle diameter and flow rate, as result of the strong inertial impaction characterizing particles larger than 1 μm. This trend is well reproduced by the model. Model predictions agree closely with the experimental values, especially at the flow rate $Q = 15$ L min^{-1}.

6.6 Computational Methods

6.6.1 Classification of Models

In respiratory deposition modelling one may distinguish between two major categories of models, namely, empirical and mechanistic models.

The empirical approach uses a combination of theoretical and empirical expressions to predict particle deposition in the respiratory tract (RT). The human RT is considered as a series of anatomical compartments through which the aerosol passes during a breath (Chan and Lippmann 1980; Rudolf et al. 1986; Chang et al. 1991; Phalen et al. 1991; ICRP 1994). Each compartment is seen as a filter, hence, respiratory deposition is seen as a stationary filtering process. The basic characteristic of the empirical compartmental models is that no consideration is given to fluid and particle dynamics. Algebraic expressions are used to describe the filtering efficiencies of the compartments, which derive from, and therefore fit on, experimental data. Such models are attractive because they are simple, robust and computationally expedient. In particular, the model recommended by the International Commission on Radiological Protection (ICRP 1994) finds wide application, and is considered as a standard model for routine inhalation dosimetry assessments. Clearly, however, the application range of such models is strictly limited to the specific morphology, physiology and lung conditions for which the model parameters were adjusted, and therefore it is not possible to investigate different conditions and exposure scenarios. Also, empirical models have a questionable predicting power when the particles are not inert, that is, for particles that do not preserve their size during breathing due to aerosol dynamics effects like, for example, hygroscopic growth.

Mechanistic models calculate respiratory deposition on a basis of a more realistic description of lung structure and physiology, and as a result of physical processes. The essential benefit of mechanistic modelling is the ability to address a variety of conditions. For example, it is possible to investigate breathing pattern variations (important in the study of subjects with chronic obstructive pulmonary diseases – COPDs), or effects connected with the respiratory tract morphology (important in extrapolating from laboratory animals toxicological studies to humans). Mechanistic models can be classified in a number of ways, employing criteria with respect to the frame of reference used, the lung morphometry used, the physical processes accommodated etc. In the present review we categorize them into two broad classes, the whole-RT 1D (one-dimensional) models and the CFD-based models. The above distinction is based on the level of complexity in modeling the geometry and airflow in the respiratory system. It may serves as the basic categorization criterion because it permits to bring out the most salient features of each category, illustrating the basic advantages and drawbacks of each category.

The respiratory system consists of organs of complicated anatomy, morphology and physiology. An accurate description of the convective aerosol transport requires complicated modeling of geometry and flow, involving computational fluid dynamics (CFD). However, a CFD-based computational approach cannot be implemented for the whole respiratory system because of its complexity and computing resources requirements. For treating the respiratory system as a whole, non-CFD-based approaches need to be implemented, requiring a simplified description of flow and geometry. Several such mechanistic models exist, which calculate deposition along the whole respiratory tract assuming flow and geometry idealizations (Taulbee and Yu 1975; Egan and Nixon 1985; Anjilvel and Asgharian 1995; Balashazy and Hofmann 1993a, b). These models can be called whole-RT 1D mechanistic models because their basic characteristic is to solve the convective aerosol transport along an one dimensional flow path that reflects the distal distance (i.e. the depth) into the respiratory system. These models rely on idealized, Lagrangian or Eulerian, one-dimensional pipe flow analyses by ignoring the swirling secondary air flow patterns present in the respiratory tract (Finlay 2001). In the next level of complexity, that is in the CFD-based mechanistic models, the velocity field is calculated in detail inside the respiratory tract and, consequently, deposition patterns are determined in a refined way. This is done by employing, usually three-dimensional, computational fluid dynamics methodologies (Finlay et al. 1996; Oldham et al. 2000; Zhang et al. 2002a,b). This approach is computationally intensive and requires precise description of the human lung geometry, which implies an extremely bulk data set. As mentioned before, due to these difficulties CFD-based models cannot deal with the whole respiratory system but, instead, with selected elements, in isolation from the rest of the system (e.g. a bifurcating airway, or a single alveolated duct). However, these models permitted to reveal fundamental features in respiratory deposition mechanics, as for example the presence of deposition hot spots in bifurcating airways. In general, the CFD-based models are very useful in investigating the microphysical details of respiratory deposition. On the other hand, the whole-RT 1D models are very useful in investigating the exposure-dose-response relationships, either at individual level or population-aggregate (collective) level. The purely mechanistic basis of these models permits to accommodate parametrically the physiological and morphological attributes. Therefore these models can utilize data for specific types of individuals permitting to differentiate the inhalation doses between, for instance, susceptible individuals, subpopulations of specific age or health condition etc. This is a significant improvement in terms of resolution capacity

Table 6.7 Basic categories of RT deposition models

Model category	Advantages	Limitations
Semi-empirical compartmental (ICRP model)	• Simple • Validated • Operational	• Empirical • Based on stationary filtration theory • Applies only to symmetric, 2-stage breathing in humans • Inert aerosols (no aerosol dynamics)
Whole-RT 1D mechanistic	• Good physical basis • Complete distribution of deposition • Not very demanding computationally • Parametric with respect to physiology & morphology	• Cannot simulate microphysical details • Difficult to account for lung radial non-homogeneities • Cannot detect hot spots
CFD-based mechanistic	• Strong physical basis • Detailed description of deposition (hot spots)	• Complicated • Require large computing resources • Apply to single elements of the RT • Difficult to use for operational purposes

in comparison to the capacity of the empirical ICRP model, which permits only a gross differentiation between male and female or adult and children. For this reason whole-RT 1D models are usually employed in the large source-to-dose computational systems, as for example, in the urban-exposure version of the AirQuis air quality management system (Coulson et al. 2005). The ability of whole-RT 1D models to reflect parametrically the intra- and inter-subject biological variability makes them suitable in predicting the doses at population level, by employing statistical distributions for model parameters. As an example, Georgopoulos et al. (2005) used whole-RT 1D modelling to calculate the inhalation dosimetry step in an integrated study of population exposure to PM in the city of Philadelphia.

Table 6.7 presents the three categories of models according to the classification above, and summarizes the more significant advantages and limitations of each category.

6.6.2 Empirical Compartmental Modeling: The ICRP Model

Several empirical models have been presented in the literature (Chan and Lippmann 1980; Rudolf et al. 1986, 1994; Davies 1982; Yu et al. 1992; Chang et al. 1991; Phalen et al. 1991; ICRP 1994). The thrust to empirical compartmental modeling originated for radiological protection purposes, in relation with the assessment of doses incurred to the lungs of workers in nuclear industry. A significant step was made by the Task Group of the International Commission on Radiological Protection (ICRP), which in 1966 presented a complete compartmental model suitable for estimating lung deposition and clearance of radioactive aerosols. An improved version was proposed in 1994, known as the ICRP66 standard (ICRP 1994). The US National Council on Radiation Protection and Measurements (NCRP) proposed soon after a similar model (NCRP 1994). Below, we present in more detail the model proposed by ICRP. This methodology is recommended to be used one for quick, practical calculations of respiratory deposition.

The ICRP model permits to determine the so-called *regional deposition*, that is deposition in regions ET1, ET2, BB, bb, and AI. The model considers the RT as a filtering system of five filters in series, each corresponding to one region, followed by the same filters in reverse order to simulate the expiration phase. The filtering efficiencies are semiempirical algebraic expressions derived from experiments, theory, and fittings to numerical results from more advanced models. To implement the model only algebraic calculations are required, quite intensive though, to be performed by hand. The interested reader may find all necessary details to set up a programme for the calculation of regional deposition in Chapter 5 of (ICRP 1994). There are also available relatively inexpensive commercial software packages (LUDEP) suitable for RT deposition calculations, having a user-friendly interface and default input parameters. Table 6.8 provides detailed regional deposition data for an adult, a 5-year child and a 10-year child, as obtained with the recommended ICRP model. The data provided are deposited fractions (%) of the inhaled particulate matter (effect of inhalability included), and refer to an average for male and female results. In each row (particle diameter), the upper, middle and lower values correspond to the "heavy exercise", "light exercise" and "sleep" activity levels respectively. Total deposition is obtained by adding together the five regions. Detailed data are given for particles with $d_a < 10$ μm. Larger particles, if they achieve to enter the nose or mouth, are all collected by the head airways of the extrathoracic regions (regions ET1 + ET2). Hence, for $d_a > 10$ μm the total deposited fraction in the respiratory tract is practically equal to the inhalable fraction, see below Eqs. 6.7–6.10.

Figure 6.21 shows graphically the total deposition and the depositions in the extrathoracic region (regions ET1 + ET2), the tracheobronchial region (BB + bb), and the alveolar region (AI) as function of particle size, as inferred from the ICRP model. Monodisperse spheres of standard density (1,000 kg m^{-3}) are considered. The data refer to an average for males and females at the "light exercise" level. The trend is similar to what is typically observed in aerosol filtering: large particles, as well as small particles deposit very efficiently, and in-between, there is a size range of decreased retention. Roughly, particles with $d_a = 0.3$ μm are deposited in the respiratory tract by less than, say 15% (i.e. more than 85% of the inhaled particulate matter is exhaled). Particles larger than 1 μm and ultrafine particles smaller than 0.003 μm deposit practically in the extrathoracic region. The particles that deposit preferentially in the alveolar region, which is the most sensitive with respect to health effects, are in the range $0.007 < d_a < 0.3$ μm.

Table 6.8 Regional deposition data calculated with the ICRP model (From Drossinos and Housiadas 2006)

	Adult					Child 5 years					Child 10 years				
d_a (μm)	ET1	ET2	BB	bb	AI	ET1	ET2	BB	bb	AI	ET1	ET2	BB	bb	AI
0.001	19.85	45.88	9.99	20.12	2.12						20.00	46.32	10.67	18.69	1.90
	39.76	40.31	8.02	10.11	0.35	40.70	41.05	9.44	6.99	0.11	39.78	40.32	8.19	9.73	0.39
	42.23	42.25	9.77	3.96	0.01	42.33	42.33	10.47	3.26	0.00	42.25	42.26	9.99	3.74	0.01
0.002	14.79	32.68	8.10	28.89	11.66						14.96	33.09	8.80	27.67	10.83
	29.67	31.78	8.58	22.22	4.32	30.75	32.75	11.01	18.87	2.27	29.68	31.79	8.80	21.32	4.59
	32.62	34.40	13.05	14.80	0.44	32.73	34.51	14.48	13.45	0.39	32.63	34.42	13.45	14.36	0.48
0.005	2.67	16.38	3.98	23.81	40.98						7.80	16.68	4.39	23.45	38.88
	15.39	17.36	5.23	26.46	27.66	16.16	18.21	7.21	27.01	20.51	15.41	17.38	5.39	25.09	27.91
	17.54	19.69	9.55	30.56	10.67	17.63	19.79	10.89	29.33	10.28	17.55	19.71	9.89	29.57	11.15
0.01	4.35	9.14	2.05	15.26	57.30						4.54	9.62	2.36	15.80	51.53
	8.58	9.68	2.83	18.77	48.14	9.27	10.55	4.14	21.53	36.33	8.74	9.91	3.00	18.29	45.52
	9.80	11.07	5.42	26.55	28.87	9.95	11.27	6.32	26.46	26.65	9.85	11.14	5.65	25.74	28.87
0.02	2.72	5.64	1.19	9.81	51.29						2.92	6.10	1.40	10.42	42.27
	5.17	5.75	1.64	12.43	50.00	5.86	6.58	2.53	15.40	34.16	5.47	6.12	1.80	12.52	42.74
	5.62	6.27	3.03	18.67	38.92	5.99	6.73	3.75	19.86	31.56	5.81	6.50	3.27	18.63	35.96
0.05	1.51	3.02	0.63	5.02	27.40						1.59	3.19	0.73	5.19	21.01
	2.91	3.07	0.91	6.71	29.62	3.24	3.45	1.37	8.13	18.31	3.04	3.22	0.97	6.58	23.19
	3.15	3.36	1.71	10.84	27.61	3.38	3.62	2.11	11.33	19.58	3.28	3.50	1.84	10.71	23.52
0.1	0.92	1.78	0.40	2.79	14.65						0.96	1.85	0.45	2.83	10.97
	1.81	1.81	0.58	3.84	16.38	1.98	1.99	0.85	4.53	9.78	1.87	1.86	0.61	3.68	12.44
	2.00	2.04	1.12	6.47	16.20	2.10	2.14	1.35	6.55	10.95	2.05	2.09	1.19	6.27	13.40
0.2	0.74	1.14	0.26	1.56	7.93						0.85	1.23	0.29	1.56	5.87
	1.44	1.23	0.37	2.17	9.01	1.77	1.48	0.53	2.52	5.27	1.76	1.45	0.38	2.04	6.72
	1.28	1.22	0.72	3.76	9.19	1.39	1.30	0.86	3.73	6.06	1.32	1.25	0.76	3.59	7.49
0.5	2.88	3.02	0.62	0.76	5.88						3.57	3.86	0.70	0.75	4.24
	5.51	5.18	0.32	1.05	6.62	7.50	7.54	0.38	1.18	3.70	7.73	7.83	0.38	0.95	4.73
	1.75	1.35	0.40	2.00	7.20	3.45	2.94	0.47	1.86	4.56	2.44	1.96	0.42	1.83	5.75

(continued)

Table 6.8 (continued)

	Adult					Child 5 years					Child 10 years				
d_a (μm)	ET1	ET2	BB	bb	AI	ET1	ET2	BB	bb	AI	ET1	ET2	BB	bb	AI
1.0	8.18	10.99	2.64	0.78	10.01						10.01	13.61	2.91	0.74	7.08
	15.86	19.79	1.00	0.87	10.32	19.91	26.77	–0.97	0.81	5.44	20.21	27.34	1.19	0.71	6.85
	5.65	5.41	0.35	2.25	12.80	10.99	12.28	0.50	1.48	7.75	7.98	8.24	0.41	1.66	10.09
2.0	14.56	25.18	8.89	2.74	15.20						15.81	28.49	9.47	2.78	10.53
	28.82	42.53	2.42	1.60	11.43	32.55	48.95	2.06	1.26	5.39	32.70	48.61	2.46	1.25	6.50
	14.83	18.43	0.94	4.41	20.91	23.80	34.05	1.29	2.48	10.96	19.29	25.85	1.13	3.12	15.63
5.0	21.04	43.68	19.54	4.40	5.74						21.63	47.53	18.00	4.17	3.32
	41.89	48.10	2.42	1.07	2.60	43.56	48.84	1.77	0.80	1.18	43.72	48.66	1.99	0.77	1.19
	31.29	40.26	2.20	4.57	11.73	38.46	47.20	1.88	1.99	4.40	35.19	44.54	2.09	2.94	7.54
10.0	20.19	55.48	7.81	0.08	0.01						20.35	57.19	6.01	0.05	0.00
	40.32	42.17	1.07	0.12	0.06	40.76	42.16	0.72	0.07	0.02	40.80	42.15	0.74	0.05	0.01
	36.71	41.34	1.85	1.55	1.31	39.29	42.10	1.18	0.49	0.37	38.20	41.87	1.50	0.87	0.74

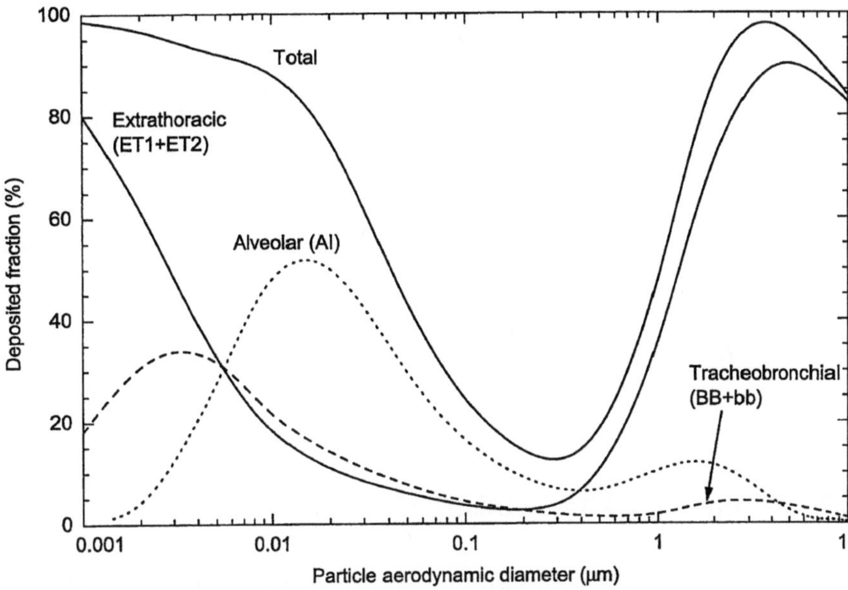

Fig. 6.21 Deposition fractions in different regions of the respiratory tract for an adult undergoing "light exercise" as calculated with the ICRP model (ICRP 1994)

For quick estimates, simplified equations fitted to the ICRP model can be used, as proposed by Hinds (1999). Monodisperse spheres of standard density are considered. The data refer to an average for males and females at an average exercise level. These simplified fits are given below for, respectively, the head airways (regions ET1 + ET2), the tracheobronchial region (BB + bb), the alveolar region (AI), and total deposition. Particle diameter, d_p, must be in μm.

$$DF_{HA} = IF\left(\frac{1}{1+\exp(6.84+1.183\ \ln d_p)} + \frac{1}{1+\exp(0.924-1.885\ \ln d_p)}\right) \tag{6.7}$$

$$DF_{TB} = \left(\frac{0.00352}{d_p}\right)\left[\begin{array}{l}\exp\left(-0.234\ (\ln d_p + 3.40)^2\right) \\ +\ 63.9\ \exp\left(-0.819\ (\ln d_p - 1.61)^2\right)\end{array}\right] \tag{6.8}$$

$$DF_{AL} = \left(\frac{0.0155}{d_p}\right)\left[\begin{array}{l}\exp\left(-0.416\ (\ln d_p + 2.84)^2\right) \\ +\ 19.11\ \exp\left(-0.482\ (\ln d_p - 1.362)^2\right)\end{array}\right] \tag{6.9}$$

$$DF = IF\left(0.0587 + \frac{0.911}{1+\exp\ (4.77+1.485\ ln\ d_p)} + \frac{0.943}{1+\exp\ (0.508-2.58\ ln\ d_p)}\right) \tag{6.10}$$

In the above expressions *IF* is the inhalable fraction, and is given be Eq. (xx) below. Not all particles present in the inspired air enter through the nose or mouth during inhalation. The situation is similar to what is normally encountered in aerosol sampling, where, losses exist at the inlet of the sampler because of distortion of the streamlines and particle inertia. The inhalable fraction, *IF*, is the ratio between particle concentration in the inhaled air and particle concentration in the ambient air. The ICRP model gives the following expression for the inhalable fraction, as function of particle aerodynamic diameter d_a (expressed in μm) and wind speed U (in m s^{-1})

$$IF = 1 - 0.5\left(1 - \frac{1}{7.6\times10^{-4} d_a^2 + 1}\right) + 10^{-5} U^{2.75} \exp(0.055 d_a) \qquad (6.11)$$

6.7 Whole-RT 1D Mechanistic Modeling

Numerous whole-RT 1D mechanistic models have been presented in the literature, dating as back as 1935 (Findeisen 1935). Obviously, with so many models available, and variants in-between them, a strict categorization is not possible. We will therefore distinguish by presenting the relevant computational approaches employed by the models (sometimes in the same model a combination of these approaches is implemented), rather than presenting clear-cut model sub-classes. In such terms, one may distinguish between Eulerian and Lagrangian modeling, depending on the frame considered in describing the fluid and particle motion (stationary or moving with the flow, respectively). Also, one may distinguish between deterministic and stochastic modeling, depending on the description of the flow pathway in the lung. Deterministic are the approach that employs a single predefined path representing average lung conditions, whereas stochastic models implement Monte Carlo techniques to account for lung asymmetry and path variations. One may also distinguish between models accommodating particle dynamical effects (i.e. assuming a changing size distribution during breathing) and models ignoring such effects by considering an inert aerosol. Below, we describe the above modeling approaches.

6.7.1 Lagrangian Modelling

In Lagrangian modeling the fluid (air) and particle motion is described by using a moving frame of reference, which follows the inhaled "aerosol bolus" in its way through the respiratory tract. The Lagrangian approach is conceptually more simple in calculating lung deposition. In general, the majority of the respiratory deposition models presented in the literature are Lagrangian. Examples are the models of Anjilvel and Asgharian (1995), Finlay and Stapleton (1995), Darquenne and Paiva (1994) and Koblinger and Hofmann (1990) among others.

The basic idea behind Lagrangian modeling is to track deposition as particles propagate along the RT pathway, by considering a chain of well-mixed airway chambers, each belonging to a different generation. These chambers are usually considered as tube segments and their geometric inputs are extracted directly from morphometric datasets of airway dimensions. For each airway chamber (generation), the overall deposition efficiency is calculated. The deposition efficiency P can be defined as the fraction of the inflowing particles that remains in the system. It can be written as

$$P = 1 - \frac{C_{\text{out}}}{C_0} \tag{6.12}$$

where, C_{out} and C_0 are the particle concentrations (either mass or number) at the outlet and inlet of the airway, respectively. The deposition efficiency can be interpreted as the probability of each individual particle within a uniform concentration particle-cloud of non-interacting particles to deposit. Under the common assumption that the different deposition mechanisms act independently of each other, the overall deposition efficiency in the *ith* airway can be calculated as

$$P_i^* = i - \prod_k (1 - P_{i,k}) \tag{6.13}$$

where $P_{i,k}$ correspond to the independent efficiencies pertinent to the kth deposition mechanism (e.g. Brownian diffusion, gravitational settling etc.). A more rigorous expression than Eq. 6.13 to sum the individual efficiencies is the so-called L_p-norm, which is based on probability theory principles:

$$P_i^* = \left(\sum_k P_{i,k}^a \right)^{1/a} \tag{6.14}$$

Various values for the exponent a have been proposed in the literature. For example, Asgharian and Anjivel (1994) use either $a = 3$ or $a = 1.4$, ICRP (1994) suggests $a = 2$, while other models employ $a = 1$ (see, in Balashazy et al. 1990). Finlay (2001) performed a sensitivity analysis on the value of the exponent a. He showed that the differences are largest only for particles having diameter near 1 μm. Note that at this size range most of deposition is in the extrathoracic part of the RT, while lung deposition is minimum (see Fig. 6.21). Outside this range the differences resulting from using different exponent a are within the uncertainty in determining the individual deposition efficiencies. Based on the above Finlay (2001) concluded that the choice of exponent a is not a critical issue in calculating lung deposition.

In calculating airflow in each airway, and the associated particle fluxes, the replenishing nature of the airflow is account for (i.e. the RT is not a flow-through network). Hence, the more distal the airway generation the smaller the total flow that passes in its airways. This is usually done by considering properly the cumulative volume of the preceding airways in determining the successively smaller fraction of inspired air passing through each generation. In general, all other airflow, or air mixing effects, in the different lung generations are accounted for only indirectly, via their reflection on the deposition efficiencies.

Since breathing is a tidal physical process, the natural coordinate system to describe inhalation mechanics is the one moving with the tide. For this reason, as mentioned before, Lagrangian modeling is conceptually more simple, which generally translates to less demanding computational effort in terms of numerical analysis. On the other hand, Lagrangian modeling has also notable drawbacks. A Lagrangian scheme has serious limitations in describing the change in form of the aerosol bolus due to axial dispersion. This limitation is significant when dealing with small particles, characterized by enhanced diffusional mobility. For example, Lagrangian modeling may not be very suitable in calculating the lung deposition of nanoparticles (currently, this is a topic of growing interest due to concerns about the risks emerging from the extensive production and use of nanomaterials). Another limitation of Lagrangial modeling is the little flexibility in describing time-varying flow rates and aerosol concentrations, which may makes it difficult to address time-varying exposure scenarios.

6.7.2 Eulerian Modelling

Eulerian models predict deposition fluxes and concentrations along the respiratory tract with respect to a fixed frame of reference. Several Eulerian models can be found in the literature (Mitsakou et al. 2005; Lazaridis et al. 2001).

In the Eulerian approach one usually employs the so-called trumpet model to describe the geometry of flow path in the lung. Trumpet models portray the human lung as a chamber with a variable cross-sectional area (based on true morphometric data), that take the shape of a trumpet (Taulbee and Yu 1975; Nixon and Egan 1987; Darquenne and Paiva 1994; Broday and Robinson 2003). This approach does not intend to provide a realistic description of the lung, but rather to enable efficient implementation of one-dimensional numerical schemes for solving the spatio-temporal concentration field and obtaining a relatively detailed solution of the (one-dimensional) convection-diffusion-deposition mass balance differential equation. The trumpet model is illustrated in Fig. 6.22. Each cross-section in the trumpet-tube model represents the sum of all individual airways sharing the same axial location along the lung pathway, and the growing cross sectional area represents the increase in total cross section in distal regions of the lung. The time-dependent alveolar volume in each generation is added to the volume of the conducting airways as a shell surrounding the trumpet-tube. This simplified geometric model cannot account for non-uniform air velocities and for non-instantaneous radial diffusion. Moreover, resolution of the deposition at the individual airway level cannot be achieved, because all the airways that are equally distanced from the trachea, in this scheme, implicitly have identical dimensions.

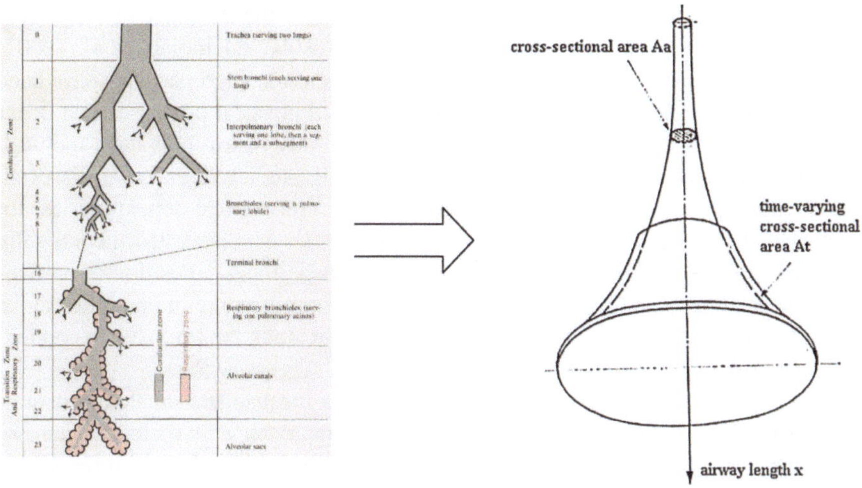

Fig. 6.22 Illustration of the trumpet model: each cross-section in the trumpet-tube corresponds to the sum of all individual airways sharing the same axial location along the lung pathway

6.7.3 *Stochastic Lung Modelling*

From a practical point of view, only stochastic models are capable of describing properly the structure and function of the respiratory tract, and consequently the PM deposition pattern, since variability is an inherent feature of all biological systems (Goo and Kim 2003; Asgharian et al. 2001a,b). Therefore the Weibel's symmetric model, which in general is considered to overestimate the total number of airways and to underestimate the cross-sectional area available for flow (Kaye and Phillips 1997), was argued to consistently underestimate particle deposition (Martonen and Gibby 1982). However, asymmetric models used to test the effect of lung morphology on the estimated deposition pattern (Yu and Diu 1982) revealed no significant differences in total deposition between Weibel's symmetric model and the more complex asymmetric models. Moreover, the total deposition in Weibel's lung model represents the median deposition of numerous realisations of asymmetric lung models (although differences in regional deposition were evident). Stochastic modelling provides not only mean values for the total, regional and differential deposition, but in addition statistics, such as the standard deviation and the skewness, of the deposition pattern. The biological dose is obtained therefore by means of a probability density function rather than just a single mean value.

6.7.4 *Aerosol Dynamics Effects*

The whole-RT mechanistic models that have appeared in the literature deal primarily with the investigation of effects connected with physiological and morphological attributes. For example, Nixon and Egan (1987) studied the influence of ventilation asymmetry to regional deposition of inhaled aerosols. Asgharian et al. (2001) examined the impact of various lung geometries. As a rule, inert particles have been considered. Less focus has been given to the aerosol physical processes that occur in the airways of the respiratory system. Only recently researchers have included aerosol dynamics effects in mechanistic modelling studies (e.g. Lazaridis et al. 2001). Such processes may have substantial impact on particle deposition, as, for instance, shown by Broday and Georgopoulos (2001), who analyzed theoretically the effect of condensation – evaporation on the transport and deposition of particulate matter. It has been also reported that the hygroscopic nature of cigarette smoke particles changes significantly the deposition pattern along the lung due to condensational growth processes (Robinson and Yu 2001; Asgharian 2004).

The purpose of the present work is to develop a mechanistic inhalation dosimetry model based on a one-dimensional (1-D) approach and including aerosol dynamics. A 1-D model is computationally tractable and can deal with the whole airway tree structure. The model is based on a comprehensive description of aerosol processes along the human respiratory tract by considering each process individually, rather than applying empirical global relationships. The inclusion of aerosol dynamics permits to assess respiratory deposition in terms of a variety of metrics, as number, mass or surface. This is of relevance to current toxicological and biological studies, as surface and number represent prominent parameters for describing the interactions between particles and biological fluids, cells and tissues (Kreyling et al. 2002; Oberdörster et al. 2002, Wichmann et al. 2000).

6.8 CFD-Based Mechanistic Modeling

There are many analyses based on CFD which enable a detailed description of particulate deposition in selected elements of the respiratory system in isolation from the rest of the system (e.g. in the nasopharyngeal region, in an alveolated duct, etc.).

Besides the above modelling approaches, exceedingly complicated computational fluid dynamics (CFD) models have been also applied to study in detail deposition patterns in subsections of the human RT. Since this method is computationally very intensive, it cannot be applied to the whole respiratory tract. Instead, analysis focuses on local elements and features of the RT, such as airway-bifurcation induced secondary flow and flow separation and their role on the enhanced mixing and deposition evident there, the unsteadiness in the alveoli deflation-inflation, the cartilaginous rings and their effect on inset of turbulence, and on the complex flow patterns in the oronasal cavity and past the glottis aperture.

Computational fluid dynamics is often used to study the effect of local structural features on the velocity field and pollutant deposition. This approach utilizes advance numerical techniques for accurate calculation of velocity fields within very detailed morphological systems (Kimbell and coworkers 1993, 1997; Subramaniam et al. 1998; Yu et al. 1998; Oldham et al. 2000; Zhang et al. 2002a, b; Moskal and

Gradon 2002; Darquenne 2002 ; Zamankhan et al. 2006). CFD Simulations were also performed for generic geometries, such as single and multi bifurcations, a central aperture within a tube, and a ragged-wall tube (Ferron and Edwards 1996; Martonen et al. 1995; Balashazy and Hofmann 1993a, b; Balashazy 1994, Balashazy and Hofmann 1995, Balashazy et al. 1996; Katz and Martonen 1996). Such studies are motivated by data pointing at specific deposition hot spots and cellular lesion pattern, which were hypothesized to occur as a result of local airflow patterns. For example, downstream of the oral cavity the larynx acts as an orifice that promotes the transition of airflow from laminar to turbulent. This suggests that turbulence normally occurs in the trachea and the largest distal airways (Olson et al. 1972; Chang 1989), which tends to enhance local particle deposition (Chan and Lippmann 1980). Large-Eddy simulations were also performed to study the aerosol deposition in the mouth region (Matida et al. 2006).

6.8.1 *Airflow Within the Respiratory Tract*

6.8.1.1 Airway Ventilation

The tracheobronchial tree is a dynamic structure that expands and contracts during breathing, and therefore experience geometric changes during respiration which results in very complex airflow patterns. Breathing mode and intensity depend upon activity. For example, during intense exercise normal nasal breathing changes into oronasal breathing and in concomitant, both mouth deposition and central to peripheral airway deposition increase (Bennett et al. 1985). Such conditions are further characterized by higher breathing rates, which increase turbulence in the upper airways and alter the regional deposition patterns in the lung (Martonen and Katz 1993).

Lung ventilation is described in terms of few physiological parameters: the tidal volume V_T – the air volume inspired and expired during a breath (500–3,800 ml depending on exercise level), the breathing frequency f_R (12–18 min^{-1} for normal activity), the functional residual capacity *FRC* – the lung volume at the end of normal exhalation ($FRC < 2000$ ml), the total lung capacity *TLC* (about 6,000–7,000 ml), and the anatomic dead space – the volume of non-alveolated airways (~170 ml). The salient point is that after each breath there is a considerable amount of air (and particles) residing in the lungs, which is apt to mix during consequent breaths with fresh inspired air that do not occupy the tracheobronchial anatomic dead-space. This mixing process depends to a large extent on the *FRC*, which itself depends on age and health state (O'Riordan et al. 1992; Smaldone et al. 1993). The ventilation of the lungs is measured by the so-called minute volume, $\dot{V}_E = V_T f_R$. During heavy exercise, the minute volume can be ten-fold its normal range, resulting from an increase in both the tidal volume and the frequency of breathing. Inhalation and exhalation periods, T_{inhl} and T_{exhl}, are usually assumed to equally span 37.5–50% of the respiration cycle, the rest being a pause.

6.8.1.2 Velocity Field

The anatomic complexity of the RT does not allow, in general, a simple solution for the continuity and Navier–Stokes equations. Traditionally, simple velocity fields are considered. In the upper airways, homogeneous turbulent flow with a flat core that occupy (most of) the cross section is oftentimes assumed, whereas Poiseuille profile is presupposed in the lower airways. More complex geometries than a straight solid tube, such as a compliant tube (Ginzburg and Elad 1993) and rhythmically expanding spherical alveoli (Haber and Tsuda 1998) were also studied. Radial gradients and boundary layer effects are usually neglected.

In order to calculate particle transport and fate in the lung using a modeling scheme which is based on a sequence of well mixed chambers, the concept of deposition efficiency is employed. To accomplish this, the airflow within each airway should be known. For a symmetric dichotomous branching network of airways the two daughter airways at any bifurcation are essentially identical. Under such conditions the flow divides symmetrically and the average velocity in each airway is given by the ratio of the flow in the airway and its cross sectional area. Since the area ratio of the two daughter airways to the parent tube is usually greater then unity, continuity requires that the velocity in the daughters is smaller than in the parent airway. In asymmetric dichotomous models calculation of flow division at the bifurcations is according to the area ratio of the two daughter airways (Koblinger and Hofmann 1990; Phillips and Kaye 1997; Andrade et al. 1998).

Due to repeating airway branching, fully developed air flow hardly happen. This conclusion is supported by detailed measurements (Schroter and Sudlow 1969), suggesting that developed flows are not established in most of the airways but rather that extended entrance regions prevail.

6.8.1.3 Mixing of Fresh and Resident Air

Mixing of fresh and residual air is modeled using an effective diffusion coefficient D_{eff} in the diffusion term of the convective–diffusive–removal mass balance equation (Koblinger and Hofmann 1990; Hofmann and Koblinger 1990, 1992). This dispersion mechanism results from simultaneous axial convection and radial diffusion, and prevails only for non-flat velocity profiles. The dependence of the effective diffusion coefficient on the flow regime can be predicted theoretically for simple flows (Taylor 1953; Watson 1983; Dill and Brenner 1982; Broday and Kimmel 1999) or from experimental measurements (Slutsky et al. 1980; Joshi et al. 1983). Scherer et al. (1975) suggested that in terms of the effective diffusion coefficient, axial mixing of gases and inertialess particles is $D(1+1.08Pe)$ during inspiration and $D(1+0.37Pe)$ on expiration, where D is the particle Brownian diffusivity (molecular diffusivity for gases) in quiescent air, Pe is the Peclet number, $Pe = \overline{v}d / D$, and d is the airway's diameter. The linear dependence of D_{eff} on Pe in these expressions (in contrast with the proportion to the square of Pe in the classical

Taylor's dispersion) resembles the relation between the effective diffusivity and steady streaming in unsteady (i.e. periodic) flows (Ultman 1985).

Taulbee and Yu (1975) proposed a dispersion mechanism that is due to asymmetric flow partitioning among airways in the same generation. This mechanism causes the inspired air to proceed in a non-uniform fashion along different pathways. The velocity dispersion resulted from the randomness in geometrical sizes and dynamical properties of the airways was lumped into an apparent axial dispersion coefficient, $D_{eff} = \sigma\overline{v}l/2$, where σ is the standard deviation of the (uniform) velocity distribution among different airways in any given generation, $\overline{v}$ is the average velocity in the airways of that generation, and l is the average length of the airways in the generation considered. This dispersion mechanism is thought to dominate over the axial diffusion process and to account for most of the pulmonary mixing.

Edwards (1994, 1995) applied the macrotransport theory to study particle dispersion in a lung, which was represented by a series of compartments following Weibel's model A, modified to include the distributed alveoli volume in the respiratory generations. Each airway was taken to be a well mixed chamber that possesses its own distinct dispersion coefficient $D_{eff,i}$, a mean velocity, and a combined deposition efficiency. These local properties were spatially averaged according to the details of the macrotransport theory, to obtain global macrotransport coefficients. Strictly speaking, the transformation from a micro- to a macrotransport scale is possible only for sufficiently long times. Yet its predictions compare favourably with experimental observations (Heyder et al. 1986, 1988; Hu et al. 1992; Anderson et al. 1989) and previous theoretical predictions (Soong et al. 1979; Hofmann and Koblinger 1990), which are themselves in good agreement with available deposition data.

6.8.1.4 Particle Deposition in the Respiratory Tract –Extrathoracic models

Aerosol particle deposition in the mouth and throat has been investigated experimentally and numerically since some time now. Most experimental works refer to natural, unperturbed inhalation connected with environmental or occupational exposures (Lippmann and Albert 1969; Foord et al. 1978; Stahlhofen et al. 1980, 1983; Emmett et al. 1982; Bowes and Swift 1989). Based on simple fitting correlations of the experimental data there have been proposed a number of mathematical models for mouth-throat deposition. These models are empirical formulae in the form of algebraic expressions, permitting to express the deposited fraction as function of the aerodynamic diameter and flow rate. The most widely used correlation is that of Rudolf et al. (1994). The latter is employed in ICRP (1994) for estimating the deposition of occupational aerosols during unperturbed inhalation. Recently, the experimental works have been extended to cases in which the aerosol is delivered through special inlets to mimic pharmaceutical inhalers (Lin et al. 2001; DeHaan and Finlay 2001). Based on these measurements there have been proposed accustomed algebraic fittings, permitting to estimate deposition (DeHaan and Finlay 2004) (in the mouth only).

Clearly, the application range of the available mathematical models is strictly limited to the specific conditions for which the model parameters were adjusted, and therefore it is not possible to extrapolate to different conditions, such as different flow-rates, or different particle characteristics (e.g. hygroscopic particles). Therefore, very often one needs to employ complicated modelling approaches, based on computational fluid dynamics (CFD). With this method the flow field is first calculated over a grid that describes in detail the geometry of the mouth and throat, and then deposition is determined with a Lagrangian particle tracking approach (Yu et al. 1998; Stapleton et al. 2000; Matida et al. 2004; Heenan et al. 2003). Needless to say, that this method, although physically sound and straightforward, requires a great deal of effort and skill to implement, as well as considerable computing resources. Also, there is uncertainty on the adequacy of the turbulent models usually employed in CFD to describe turbulent effects in the mouth-throat region (Stapleton et al. 2000). In this respect, Matida et al. (2004) implemented special near-wall corrections in the Lagrangian stochastic eddy-interaction model to obtain agreement between CFD-based simulations and mouth-throat deposition experimental data. Moreover, the accommodation in computational fluid-particle dynamics of key effects, like hygroscopic growth of particles or coagulation, is a challenging modelling task in itself, and particularly intensive computationally. Therefore, such effects have been scarcely addressed in the CFD-based analyses of respiratory deposition in the extrathoracic region. Sarangapani and Wexler (2000) investigated the deposition of hygroscopic particles with a CFD model, but in the nasal route. They concluded that rapid growth takes place there, influencing significantly the deposition characteristics. A similar trend may be expected in the oral route as well, considering that ambient or occupational aerosols are frequently hygroscopic, and also metered-dose inhalers include in general water-soluble compounds.

Furthermore, Mitsakou et al. (2007) developed a model which requires simple computational means, able to predict deposition coupled with hygroscopic growth in the mouth-throat region. The intent was to develop a non-CFD-based model, but still using a fully mechanistic description of the aerosol deposition and dynamics, avoiding the use of empirical correlations. The developed mathematical model is based on a one-dimensional representation of the aerosol flow along the mouth-throat extrathoracic flow-path. This requires the use of an idealised description of the geometry of the flow-path. Geometrical models for the mouth and throat region have been implemented by several investigators with various degrees of complexity. There exist complicated descriptions based on detailed replicas of the extrathoracic airways (Yu et al. 1998; Li et al. 1996), while on the other extreme one may find extremely idealized descriptions consisting of simple U-shaped (Cheng et al. 1999) or L-shaped (Zhang et al. 2004; USP 1995, 2000) tube bends. Such highly idealized geometrical models may be inadequate in reproducing the filtering characteristics of the human mouth and throat, as for instance shown by Zhang et al. (2004). On the other hand, detailed three-dimensional geometries are not suitable for our purposes, but rather for laborious CFD-based models. Stapleton et al. (2000) proposed an idealized mouth-throat geometry of intermediate complexity, using information from the literature, supplemented with separate measurements from Computed Tomography (CT), Magnetic Resonance Imaging (MRI) scans and observations of living subjects. The model of Stapleton et al. (2000) has been shown to reproduce successfully average filtering efficiencies in vivo (DeHaan and Finlay 2001), and can be therefore considered as a practical compromise with respect to the required degree of geometrical complexity.

References

ACGIH. (1997). *Threshold limit values and biological exposure indices*. Cincinnati, OH: ACGIH.

Anderson, P. J., Blanchard, J. D., Brain, J. D., Feldman, H. A., McNamara, J. J., & Heyder, J. (1989). Effect of cystic fibrosis on inhaled aerosol boluses. *American Review of Respiratory Disease, 140*, 1317–1324.

Andrade, J. S., Alencar, A. M., Almeida, M. P., Mendes, F. J., Buldyrev, S. V., Zapperi, S., et al. (1998). Asymmetric flow in symmetric branched structures. *Physical Review Letters, 81*, 926–929.

Anjilvel, S., & Asgharian, B. (1995). A multiple-path model of particle deposition in the rat lung. *Fundamental and Applied Toxicology, 28*, 41–50.

Asgharian, B. (2004). A model of deposition of hygroscopic particles in the human lung. *Aerosol Science and Technology, 38*, 938–947.

Asgharian, B., & Anjilvel, S. (1994). A Monte Carlo calculation of the deposition efficiency of inhaled particles in lower airways. *Journal of Aerosol Science, 25*(4), 711–721.

Asgharian, B., Hofmann, W., & Bergmann, R. (2001b). Particle deposition in a multiple-path model of the human lung. *Aerosol Science and Technology, 34*, 332–339.

Asgharian, B., Hofmann, W., & Miller, F. J. (2001a). Mucociliary clearance of insoluble particles from the tracheobronchial airways of the human lung. *Journal of Aerosol Science, 32*, 817–832.

Balashazy, I. (1994). Simulation of particle trajectories in bifurcating tubes. *Journal of Computational Physics, 110*, 11–22.

Balashazy, I., Balint, A., Molnar, A. J., Hofmann, W., Szoke, I. A., & Kiss, E. (2007). Aerosol drug delivery optimization by computational methods for the characterization of total and regional deposition of therapeutic aerosols in the respiratory system. *Current Computer-Aided Drug Design, 3*, 13–32.

Balashazy, I., Heistracher, T., & Hofmann, W. (1996). Air flow and particle deposition patterns in bronchial airway bifurcations – The effect of different CFD models and bifurcation geometries. *Journal of Aerosol Medicine, 9*(3), 287–301.

Balashazy, I., & Hofmann, W. (1993a). Particle deposition in airway bifurcations – I. Inspiratory flow. *Journal of Aerosol Science, 24*(6), 745–772.

Balashazy, I., & Hofmann, W. (1993b). Particle deposition in airway bifurcations – II. Expiratory flow. *Journal of Aerosol Science, 24*(6), 773–786.

Balashazy, I., & Hofmann, W. (1995). Deposition of aerosols in asymmetric airway bifurcations. *Journal of Aerosol Science, 26*(2), 273–292.

Balashazy, I., & Hofmann, W. (2000). Quantification of local deposition patterns of inhaled radon decay products in human bronchial airway bifurcations. *Health Physics, 78*(2), 147–158.

Balashazy, I., Hofmann, W., & Lorine, M. (1999). Relative contributions of individual deposition mechanisms to total aerosol deposition in human airways. *Journal of Aerosol Science, 30*, S729–S730.

Balashazy, I., Martonen, T. B., & Hofmann, W. (1990). Inertial impaction and gravitational deposition of aerosols in curved tubes and airway bifurcations. *Aerosol Science and Technology, 13*, 303–321.

Bennett, W. D., Messina, M. S., & Smaldone, G. C. (1985). Effect of exercise on deposition and subsequent retention of inhaled particles. *Journal of Applied Physiology, 59*(4), 1046–1054.

Bowes, S. M., & Swift, D. L. (1989). Deposition of inhaled particles in the oral airway during oronasal breathing. *Aerosol Science and Technology, 11*, 157–167.

Broday, D. M. (2004). Deposition of ultrafine particles at carinal ridges of the upper bronchial airways. *Aerosol Science and Technology, 38*, 991–1000.

Broday, D. M., & Georgopoulos, P. G. (2001). Growth and deposition of hygroscopic particulate matter in the human lungs. *Aerosol Science and Technology, 34*(1), 144–159.

Broday, D. M., & Kimmel, E. (1999). On the axial dispersion induced by vibrations of a flexible tube. *IJES, 37*(7), 863–881.

Broday, D. M., & Robinson, R. (2003). Application of cloud dynamics to dosimetry of cigarette smoke particles in the lungs. *Aerosol Science and Technology, 37*(6), 510–527.

CAFE. (2004). *Second position paper on particulate matter*. CAFE working group on particulate matter. Retrieved from http://ec.europa.eu/environment/air/cafe/pdf/working_groups/2nd_position_paper_pm.pdf

Chan, T. L., & Lippmann, M. (1980). Experimental measurements and empirical modelling of the regional deposition of inhaled particles in humans. *American Industrial Hygiene Association Journal, 41*, 399–409.

Chang, H. K. (1989). Flow dynamics in the respiratory tract. In H. K. Chang & M. Paiva (Eds.), *Respiratory physiology: An analytical approach* (pp. 57–138). New York: Marcel Dekker.

Chang, I., Griffith, W., Shyr, L., Yeh, H., Cuddihy, G., & Seiler, A. (1991). Software for the draft NCRP respiratory tract dosimetry model. *Radiation Protection Dosimetry, 38*(1/3), 193–199.

Cheng, Y. S., Zhou, Y., & Chen, B. T. (1999). Particle deposition in a cast of human oral airways. *Aerosol Science and Technology, 31*, 286–300.

Coulson, G., Bartonova, A., Bohler, T., Broday, D. M., Colbeck, I., Floisand, I., et al. (2005). Exposure risks form pollutants in domestic environments. *Indoor and Built Environment, 14*, 209–213.

Darquenne, C. (2002). Heterogeneity of aerosol deposition in a two-dimensional model of human alveolated ducts. *Journal of Aerosol Science, 33*, 1261–1278.

Darquenne, C., & Paiva, M. (1994). One-dimensional simulation of aerosol transport and deposition in the human lung. *Journal of Applied Physiology, 77*, 2889–2898.

Davies, C. N. (1982). Deposition of particles in the human lungs as a function of particle size and breathing pattern: An empirical model. *Annals of Occupational Hygiene, 26*(1–4), 119–135.

DeHaan, W. H., & Finlay, W. H. (2001). In vitro monodisperse aerosol deposition in a mouth and throat with six different inhalation devices. *Journal of Aerosol Medicine, 14*, 361–367.

DeHaan, W. H., & Finlay, W. H. (2004). Predicting extrathoracic deposition from dry powder inhalers. *Journal of Aerosol Science, 35*, 309–331.

Dill, H. L., & Brenner, H. (1982). *PCH PhysicoChemical Hydrodynamics, 3*, 267–292.

Dockery, D. W., Pope III, C. A., Xu, X., Spengler, J. D., Ware, J. H., Fay, M. E., Ferris Jr., B. G., & Speizer, F. E. (1993). An association between air pollution and mortality in six U.S. cities. *New England Journal of Medicine, 329*(24), 1753–1759.

Donaldson, K., Stone, V., Gilmour, P. S., Brown, D. M., & Macnee, W. (2000). Ultrafine particles: Mechanisms of lung injury. *Philosophical Transactions of the Royal Society of London, Series A, 358*, 2741–2749.

Drossinos, Y. and Housiadas, C., Aerosol Flows, In: *The Multiphase Flow Handbook* (ed. C. Crowe), Chapter 6, CRC Press – Taylor & Francis, Boca Raton, FL, 2006.

Edwards, D. A. (1994). A general theory of the macrotransport of nondepositing particles in the lung by convective dispersion. *Journal of Aerosol Science, 25*(3), 543–565.

Edwards, D. A. (1995). The macrotransport of aerosol particles in the lung: Aerosol deposition phenomena. *Journal of Aerosol Science, 26*(2), 293–317.

Egan, M. J., & Nixon, W. (1985). A model of aerosol deposition in the lung for use in inhalation dose assessments. *Radiation Protection Dosimetry, 11*(1), 5–17.

Emmett, P. C., Aitken, R. J., & Hannan, W. J. (1982). Measurements of the total and regional deposition of inhaled particles in the human respiratory tract. *Journal of Aerosol Science, 13*, 549–560.

Ferron, G. A., & Edwards, D. A. (1996). Numerical simulation of air and particle transport in the conducting airways. *Journal of Aerosol Medicine, 9*(3), 303–316.

Findeisen, W. (1935). Über das Absetzen kleiner, in der Luft suspendierter Teilchen in der menschlichen Lunge. *Pflüger Archives Fd ges Physiology, 236*, 367–379.

Finlay, W. H. (2001). *The mechanics of inhaled pharmaceutical aerosols*. London: Academic.

Finlay, W. H., & Martin, A. R. (2008). Recent advances in predictive understanding of respiratory tract deposition. *Journal of Aerosol Medicine and Pulmonary Drug Delivery, 21*, 189–205.

Finlay, W. H., & Stapleton, K. W. (1995). The effect on regional lung deposition of coupled heat and mass transfer between hygroscopic droplets and their surrounding phase. *Journal of Aerosol Science, 26*, 655–670.

Finlay, W. H., Stapleton, K. W., & Yokota, J. (1996). On the use of computational fluid dynamics for simulating flow and particle deposition in the human respiratory tract. *Journal of Aerosol Medicine: Deposition, Clearance, and Effects in the Lung, 9*(3), 329–341.

Foord, N., Black, A., & Walsh, M. (1978). Regional deposition of 2.5–7.5 μm diameter inhaled particles in healthy male non-smokers. *Journal of Aerosol Science, 9*, 343–357.

Freijer, J. I., Cassee, F. R., & van Bree, L. (1997). *Modeling of particulate matter deposition in the human airways* (Report No. 624029 001). Bilthoven, The Netherlands: National Institute of Public Health and the Environment.

Fung, Y. C. (1990). *Biomechanisms. Motion, flow, stress, and growth* (569 pp.). New York: Springer.

Georgopoulos, P. G., Wang, S.-W., Vyas, V. M., Sun, Q., Burke, J., Vedantham, R., McCurdy, T., & Özkaynak, H. (2005). A source-to-dose assessment of population exposures to fine PM and ozone in Philadelphia, PA, during a summer 1999 episode. *Journal of Exposure Analysis and Environmental Epidemiology, 15*(5), 439–457.

Ginzburg, I., & Elad, D. (1993). Dynamic model of the bronchial tree. *Journal of Biomechanical Engineering, 15*(4), 283–288.

Goo, J., & Kim, C. S. (2003). Theoretical analysis of particle deposition in human lungs considering stochastic variations of airway morphology. *Journal of Aerosol Science, 34*, 585–602.

Gradon, L., Pratsinis, E., Prodgorski, A., Scott, S., & Panda, S. (1996). Modeling retention of inhaled particles in rat lungs including toxic and overloading effects. *Journal of Aerosol Science, 27*(3), 487–503.

Haber, S., & Tsuda, A. (1998). The effect of flow generated by rhythmically expanding pulmonary acinus on aerosol dynamics. *Journal of Aerosol Science, 29*(3), 309–322.

Heenan, A. F., Matida, E., Pollard, A., & Finlay, W. H. (2003). Experimental measurements and computational modeling of the flow field in an idealized oropharynx. *Experiments in Fluids, 35*, 70–84.

HEI. (2001). *Health Effects Institute. Airborne particles and health: HEI Epidemiologic evidence*. Cambridge, MA: HEI. Retrieved from http://healtheffects.org/pubs-special.htm

Heyder, J., Blanchard, J. D., Feldman, H. A., & Brain, J. D. (1988). Convective mixing in human respiratory tract: Estimates with aerosol boli. *Journal of Applied Physiology, 64*, 1273–1278.

Heyder, J., Gebhart, J., Rudolf, G., Schiller, C. F., & Stahlhofen, W. (1986). Deposition of particles in the human respiratory tract in the size range 0.005–15 micron. *Journal of Aerosol Science, 17*, 811–825.

Hinds, W. C. (1999). *Aerosol technology: Properties, behavior, and measurement of airborne particles* (2nd ed.). New York: Wiley.

Hofmann, W., & Koblinger, L. (1990). Monte-Carlo modeling of aerosol deposition in human lungs. Part II: Deposition fractions and their sensitivity to parameter variation. *Journal of Aerosol Science, 21*(5), 675–688.

Hofmann, W., & Koblinger, L. (1992). Monte-Carlo modeling of aerosol deposition in human lungs. Part III: Comparison with experimental data. *Journal of Aerosol Science, 23*(1), 51–63.

Hu, S. L., Ben-Jebria, A., & Ultman, J. S. (1992). Longitudinal distribution of ozone absorption in the lung: Quiet respiration in healthy subjects. *Journal of Applied Physiology, 73*, 1655–1661.

Hughes, W. W. (1996). *Essentials of environmental toxicology: The effects of environmentally hazardous substances on human health* (176 pp.). Levittown: Taylor & Francis.

International Commission on Radiological Protection (ICRP). (1994). Human respiratory tract model for radiological protection, ICRP Publication 66. *Annals of the ICRP, 24*(4). Oxford: Pergamon, 482 pp.

International Commission on Radiological Protection (ICRP) (2002). Guide for the Practical Application of the ICRP Human Respiratory Tract Model. *Annals of the ICRP, 32*(1–2), 1–312. Oxford: Pergamon.

Isaacs, K., Rosati, J. A., & Martonen, T. B. (2005). Mechanisms of particle deposition. In L. S. Ruzer, & N. H. Harley (Eds.), *Aerosols handbook*. Boca Raton, FL: CRC Press.

Joshi, C. H., Kamm, R. D., Drazen, J. M., & Slutsky, A. S. (1983). An experimental study of gas exchange in laminar oscillatory flow. *Journal of Fluid Mechanics, 133*, 245–254.

Katz, I. M., & Martonen, T. B. (1996). Flow patterns in three-dimensional laryngeal models. *Journal of Aerosol Medicine, 9*(4), 501–511.

Kaye, S. R., & Phillips, C. G. (1997). The influence of the branching pattern of the conducting airways on flow and aerosol deposition parameters in the human, dog, rat and hamster. *Journal of Aerosol Science, 28*(7), 1291–1300.

Kimbell, J. S., Godo, M. N., Gross, E. A., Joyner, D. R., Richardson, R. B., & Morgan, K. T. (1997). Computer simulation of inspiratory airflow in all regions of the F344 rat nasal passages. *Toxicology and Applied Pharmacology, 145*, 388–398.

Kimbell, J. S., Gross, E. A., Joyner, D. R., Godo, M. N., & Morgan, K. T. (1993). Application of computational fluid dynamics to regional dosimetry of inhaled chemicals in the upper respiratory tract of the rat. *Toxicology and Applied Pharmacology, 121*, 253–263.

Koblinger, L., & Hofmann, W. (1990). Monte Carlo modeling of aerosol deposition in human lungs. Part I: Simulation of particle transport in a stochastic lung structure. *Journal of Aerosol Science, 21*(5), 661–674.

Kremer, D. M., & Hancock, B. C. (2006). Process simulation in the pharmaceutical industry: A review of some basic principle models. *Journal of Pharmateutical Sciences, 95*(3), 517–529.

Kreyling, W. G., Semmler, M., Erbe, F., Mayer, P., Takenaka, S., & Schulz, H. (2002). Translocation of ultrafine insoluble iridium particles from lung epithelium to extrapulmonary organs is size dependent but very low. *Journal of Toxicology and Environmental Health, Part A, 65*, 1513–1530.

Lazaridis, M., Broday, D. M., Hov, Ø., & Georgopoulos, P. G. (2001). Integrated exposure and dose modeling and analysis system. 3. Deposition of inhaled particles in the human respiratory tract. *Environmental Science Technology, 35*, 3727–3734.

Li, W., Perzl, M., Heyder, J., Langer, R., Brain, J. D., Englmeier, K. H., et al. (1996). Aerodynamics and aerosol particle deaggregation phenomena in model oral-pharyngeal cavities. *Journal of Aerosol Science, 27*, 1269–1286.

Lin, T. C., Breysse, P. N., Laube, B. L., & Swift, D. L. (2001). Mouthpiece diameter affects deposition efficiency in cast models of the human oral airways. *Journal of Aerosol Medicine, 14*, 335–341.

Lippmann, M. (1977). Regional deposition of particles in the human respiratory tract. In D. H. K. Lee, H. L. Falk, S. D. Murphy, & S. R. Geiger (Eds.), *Handbook of physiology* (pp. 213–232). Philadelphia, PA: American Physiological Society.

Lippmann, M., & Albert, R. E. (1969). The effect of particle size on the regional deposition of inhaled aerosols in the human respiratory tract. *American Industrial Hygiene Association Journal, 30*, 257–275.

Londahl, J., Massling, A., Pagels, J., Swietlicki, E., Vaclavik, E. A., & Loft, S. (2007). Size-resolved respiratory-tract deposition of fine and ultrafine hydrophobic and hygroscopic aerosol particles during rest and exercise. *Inhalation Toxicology, 19*, 109–116.

Martonen, T. B., & Gibby, D. (1982). Computer models of aerosol deposition in two human tracheobronchial geometries. *Computers and Biomedical Research, 15*, 425–433.

Martonen, T. B., & Katz, I. (1993). Deposition patterns of polydisperse aerosols within human lungs. *Journal of Aerosol Medicine, 6*(4), 251–274.

Martonen, T. B., Rosati, J. A., & Isaacs, K. K. (2005). Modeling deposition of inhaled particles. In L. S. Ruzer, & N. H. Harley (Eds.), *Aerosols hanbook.* Boca Raton, FL: CRC Press.

Martonen, T. B., Yang, Y., Hwang, D., & Fleming, J. S. (1995). Computer model of human lung morphology to complement SPECT analyses. *International Journal of Bio-Medical Computing, 40*(1), 5–16.

Matida, E. A., Finlay, W. H., Breuer, M., & Lange, C. F. (2006). Improving prediction of aerosol deposition in an idealized mouth using large-eddy simulation. *Journal of Aerosol Medicine, 19*, 290–300.

Matida, E. A., Finlay, W. H., Lange, C. F., & Grgic, B. (2004). Improved numerical simulation of aerosol deposition in an idealized mouth – throat. *Journal of Aerosol Science, 35*, 1–19.

Mauderly, J. L., & Chow, J. C. (2008). Health effects of organic aerosols. *Inhalation Toxicology, 20*, 257–288.

Mitsakou, C., Helmis, C., & Housiadas, C. (2005). Eulerian modelling of lung deposition with sectional representation of aerosol dynamics. *Journal of Aerosol Science, 36*, 75–94.

Mitsakou, C., Mitrakos, D., Neofytou, P., & Housiadas, C. (2007). A simple mechanistic model of deposition of water-soluble aerosol particles in the mouth and throat. *Journal of Aerosol Medicine – Deposition Clearance in the Lung, 20*, 519–529.

Morawska, L., Homann, W., Loveday, J. H., Swanson, C., & Mengersen, K. (2005). Experimental study of the deposition of combustion aerosols in the human respiratory tract. *Journal of Aerosol Science, 36*, 939–957.

Moskal, A., & Gradon, L. (2002). Temporary and spatial deposition of aerosol particles in the upper human airways during breathing cycle. *Journal of Aerosol Science, 33*, 1525–1539.

National Council on Radiation Protection and Measurement (NCRP). (1994). *Deposition, retention and dosimetry of inhaled radioactive substances* (Report S. C. 57-2). Bethesda, MD: NCRP.

Neuberger, M., Schimekc, M. G., Horak, F., Jr., Moshammerb, H., Kundib, M., Frischerd, T., et al. (2004). Acute effects of particulate matter on respiratory diseases, symptoms and functions: Epidemiological results of the Austrian Project on Health Effects of Particulate Matter (AUPHEP). *Atmospheric Environment, 38*, 3971–3981.

Nixon, W., & Egan, M. J. (1987). Modeling study of regional deposition of inhaled aerosols with special reference to effects of ventilation asymmetry. *Journal of Aerosol Science, 18*(5), 563–579.

O'Riordan, T. G., Zwang, J., & Smaldone, G. C. (1992). Mucociliary clearance in adult asthma. *American Review of Respiratory Disease, 146*(3), 598–603.

Oberdörster, G., Sharp, Z., Atudorei, V., Elder, A., Gelein, R., Kreyling, W., et al. (2004). Translocation of inhaled ultrafine particles to the brain. *Inhalation Toxicology, 16*, 437–445.

Oberdörster, G., Sharp, Z., Atudorei, V., Elder, A., Gelein, R., Lunts, A., et al. (2002). Extrapulmonary translocation of ultrafine carbon particles following whole-body inhalation exposure of rats. *Journal of Toxicology and Environmental Health, Part A, 65*, 1531–1543.

Oldham, M. J., Phalen, R. F., & Heistracher, T. (2000). Computational fluid dynamic predictions and experimental results for particle deposition in an airway model. *Aerosol Science and Technology, 32*, 61–71.

Olson, D. E., Iliff, L. D., & Sudlow, M. F. (1972). Some aspects of the physics of flow in the central airways. *Bulletin of Physiopathology Respiration, 8*, 391–408.

Park, S. S., & Wex, A. S. (2008). Size-dependent deposition of particles in the human lung at steady-state breathing. *Journal of Aerosol Science, 39*, 266–276.

Phalen, R. F., Cuddihy, R. G., Fisher, G. L., Moss, O. R., Schlesinger, R. B., Swift, D. L., & Yeh, H.-C. (1991). Main features of the proposed NCRP respiratory tract model. *Radiation Protection Dosimetry, 38*(1–3), 179–184.

Phalen, R. F., Yeh, H. C., Raabe, O. G., Hulbert, A. J., & Crain, C. R. (1974). Comparative airway anatomy in human, beagle dog, rat and hamster and implications in inhaled particles. *Health Physics, 27*, 634–635.

Phillips, C. G., & Kaye, S. R. (1997). On the asymmetry of bifurcations in the bronchial tree. *Respiration Physiology, 107*, 85–98.

Pope, C. A., Burnett, R. T., Thun, M. J., Calle, E. E., Krewski, D., Ito, K., et al. (2002). Lung cancer, cardiopulmonary mortality and long-term exposure to fine particulate matter pollution. *Journal of the American Medical Association, 287*, 1132–1141.

Pope, C. A., Thun, M. J., Namboodiri, M. M., Dockery, D. W., Evans, J. S., Speizer, F. E., et al. (1995). Particulate air pollution as a predictor of mortality in a prospective study of US adults. *American Journal of Respiratory Critical Care Medicine, 151*, 669–674.

Raabe, O. G., Boyd, H. A., Kanapilly, G. M., Wilkinson, C. J., & Newton, G. J. (1975). Developing and use of a system for routine production of monodisperse particles of PU-238O2 and evaluation of gamma emitting labels. *Health Physics, 28*, 655.

Robinson, R. J., & Yu, C. P. (2001). Deposition of cigarette smoke particles in the human respiratory tract. *Aerosol Science and Technology, 34*(2), 202–215.

Rudolf, G., Gebhart, J., Heyder, J., Schiller, C. F., & Stahlhofen, W. (1986). An empirical formula describing aerosol deposition in man for any particle size. *Journal of Aerosol Science, 17*, 350–355.

Rudolf, G., Köbrich, R., Stahlhofen, W., & James, A. C. (1994). Regional airway deposition in man – a statistical and algebraic model. In J. Dodgson & R. L. McCallum (Eds.), *Inhaled particles VII*. Proceedings of an International Symposium on Inhaled Particles Organized by the British Occupational Hygiene Society, Elsevier Science, Oxford.

Sarangapani, R., & Wexler, A. S. (2000). Modeling particle deposition in the extrathoracic airways. *Aerosol Science and Technology, 31*, 72–89.

Scherer, P. W., Shendalman, L. H., Greene, N. M., & Bouhuys, A. (1975). Measurement of axial diffusivities in a model of the bronchial airways. *Journal of Applied Physiology, 38*(4), 719–723.

Schlesinger, R. (1988). Biological disposition of airborne particles: Basic principles and application to vehicular emissions. In Ann Y. Watson, Richard R. Bates, & D. Kennedy (Eds.), *Air pollution, the automobile, and public health* (704 pp.). Washington, DC: National Academy Press.

Schlesinger, R. B. (1995). Toxicological evidence for health effects from inhaled particulate pollution: Does it support the human experience? *Inhalation Toxicology, 7*, 99–109.

Schlesinger, R. B., & Cassee, F. R. (2003). Atmospheric secondary inorganic particulate matter: The toxicological perspective as a basis for health effects risk assessment. *Inhalation Toxicology, 15*, 197–235.

Schroter, R. C., & Sudlow, M. F. (1969). Flow patterns in models of the human bronchial airways. *Respiration Physiology, 7*, 341–355.

Schwartz, J. (1994). What are people dying of an air pollution days? *Fundamental and Applied Toxicology, 64*, 26–35.

Slutsky, A. S., Drazen, J. M., Ingram, R. H., Kamm, R. D., Shapiro, A. H., Fredberg, J. J., et al. (1980). Effective pulmonary ventilation with small-volume oscillations at high frequency. *Science, 209*, 609–611.

Smaldone, G. C., Foster, W. M., O'Riordan, T. G., Messina, M. S., Perry, R. J., & Langenback, E. G. (1993). Regional impairment of mucociliary clearance in chronic obstructive pulmonary disease. *Chest, 103*(5), 1390–1396.

Smith, J. R. H., Bailey, M. R., Etherington, G., Shutt, A. L., & Youngman, M. J. (2008). Effect of particle size on slow particle clearance from the bronchial tree. *Experimental Lung Research, 34*, 287–312.

Soong, T. T., Nicolaides, P., Yu, C. P., & Soong, S. C. (1979). A statistical description of the human tracheobronchial tree gepmetry. *Respiration Physiology, 37*, 161–172.

Stahlhofen, W., Gebhart, J., & Heyder, J. (1980). Experimental determination of the regional deposition of aerosol particles in the human respiratory tract. *American Industrial Hygiene Association Journal, 41*, 385–398.

Stahlhofen, W., Gebhart, J., Heyder, J., & Scheuch, G. (1983). New regional deposition data of the human respiratory tract. *Journal of Aerosol Science, 14*, 186–188.

Stahlhofen, W., Gebhart, J., Heyder, J., & Scheuch, G. J. (1986). Measurement of lung clearance with pulses of radioactively labeled aerosols. *Journal of Aerosol Science, 17*, 333–336.

Stapleton, K. W., Guentsch, E., Hoskinson, M. K., & Finlay, W. H. (2000). On the suitability of k-ω turbulence modeling for aerosol deposition in the mouth and throat: A comparison with experiment. *Journal of Aerosol Science, 31*, 739–749.

Su, W. C., Wu, J., Marijnissen, C. M., & Chengi, Y. S. (2008). Deposition of man-made fibers in a human nasal airway. *Aerosol Science and Technology, 42*, 173–181.

Subramaniam, R. P., Richardson, R. B., Morgan, K. T., Kimbell, J. S., & Guilmette, R. A. (1998). Computational fluid dynamics simulations of inspiratory airflow in the human nose and nasopharynx. *Inhalation Toxicology, 10*, 473–502.

Taulbee, D. B., & Yu, C. P. (1975). A theory of aerosol deposition in the human respiratory tract. *Journal of Applied Physiology, 38*(1), 77–85.

Taylor, G. I. (1953). Dispersion of soluble matter in solvent flowing slowly through a tube. *Proceedings of the Royal Society London, A219*, 186–203.

Tran, C., Jones, A., & Cullen, R. (1999a). Mathematical modelling of the retention and clearance of low-toxicity particles in the lung. *Inhalation Toxicology, 11*, 1059–1076.

Tran, C., Jones, A., & Cullen, R. (1999b). Exploration of the mechanisms of retention and clearance of low-toxicity particles in the rat lung using a mathematical model. *Inhalation Toxicology, 11*, 1077–1108.

Ultman, J. S. (1985). Gas transport in the conducting airways. In L. A. Engel & M. Paiva (Eds.), *Gas mixing and distribution in the lung* (vol. 25, pp. 63–136). New York: Marcel Dekker.

United States Pharmacopeia. (1995). USP (United States Pharmacopeia), *23*, 1765–1766.

United States Pharmacopeia. (2000). USP (United States Pharmacopeia), *24*, 2678–2679.

USEPA. (2004). *Air quality criteria for particulate matter* (Report EPA/600/P-99/002aF and bF). Research Triangle Park, NC: USEPA.

Vincent, J., Mark, D., Miller, B. G., Armbruster, L., & Ogden, J. L. (1990). Aerosol inhalability at higher windspeeds. *Journal of Aerosol Science, 21*, 577–586.

Watson, E. J. (1983). Diffusion in oscillatory pipe flow. *Journal of Fluid Mechanics, 133*, 233–244.

Weibel, E. R. (1963). *Morphometry of the human lung*. New York: Academic.

Weibel, E. R. (1991). Fractal geometry – a design principle for living organisms. *American Journal of Physiology, 261*(6), L361–L369.

Wichmann, H. E., Spix, C., Tuch, T., Wölke, G., Peters, A., Heinrich, J., et al. (2000). *Daily mortality and fine and ultrafine particles in Erfurt, Germany, part I: Role of particle number and particle mass* (Health Effects Institute Report No. 98). Boston, MA: HEI.

Yang, W., Peters, J. I., & Williams, R. O. (2008). Inhaled nanoparticles – a current review. *International Journal of Pharmaceutics, 356*, 239–247.

Yeh, H. C., & Schum, G. M. (1980). Models of human lung airways and their application to inhaled particle deposition. *Bulletin of Mathematical Biology, 42*, 461–480.

Yu, C. P., & Diu, C. K. (1982). A comparative study of aerosol deposition in different lung models. *American Industrial Hygiene Association Journal, 43*, 54–65.

Yu, C. P., & Diu, C. K. (1983). Total and regional deposition of inhaled aerosols in humans. *Journal of Aerosol Science, 14*(5), 599–609.

Yu, C. P., Zhang, L., Becquemin, M. H., Roy, M., & Bouchikhi, A. (1992). Algebraic modeling of total and regional deposition of inhaled particles in the human lung of various ages. *Journal of Aerosol Science, 23*(1), 73–79.

Yu, G., Zhang, Z., & Lessmann, R. (1998). Fluid flow field and particle diffusion in the human upper respiratory system. *Aerosol Science and Technology, 28*(2), 146–158.

Zamankhan, P., Ahmadi, G., Wang, Z., Hopje, P. K., Cheng, Y.-S., Su, W. C., et al. (2006). Airflow and deposition of nano-particles in a human nasal cavity. *Aerosol Science and Technology, 40*, 463–476.

Zhang, Y., & Finlay, W. H. (2005). Measurement of the effect of cartilaginous rings on particle deposition in a proximal lung bifurcation model. *Aerosol Science and Technology, 39*, 394–399.

Zhang, Y., Finlay, W. H., & Matida, E. A. (2004). Particle deposition measurements and numerical simulation in a highly idealized mouth – throat. *Journal of Aerosol Science, 35*, 789–803.

Zhang, Z., & Kleinstreuer, C. (2001). Effect of particle inlet distributions on deposition in a triple bifurcation lung airway model. *Journal of Aerosol Medicine, 14*, 13–29.

Zhang, Z., Kleinstreuer, C., & Kim, C. S. (2002a). Cyclic micron-size particle inhalation and deposition in a triple bifurcation lung airway model. *Journal of Aerosol Science, 33*, 257–281.

Zhang, Z., Kleinstreuer, C., & Kim, C. S. (2002b). Aerosol deposition efficiencies and upstream release positions for different inhalation modes in the upper bronchial airway model. *Aerosol Science and Technology, 36*, 828–844.

Zhou, Y., & Cheng, Y.-S. (2005). Particle deposition in a cast of human tracheobronchial airways. *Aerosol Science and Technology, 39*, 492–500.

Chapter 7
Dermal Absorption Modelling

Jim E. Riviere

Abstract The focus of this chapter is on modeling the absorption of chemical pollutants into and across human skin. The skin is a primary route of systemic exposure to a number of environmental pollutants either directly as neat chemicals, in aqueous solution when exposure is to polluted water, or in complex solvents when exposure occurs secondary to contact at industrial waste sites. These differences in exposure modalities may significantly modulate the extent of dermal absorption seen and thus should be factored into any risk assessment.

7.1 Introduction

There are a number of considerations needed to assess dermal absorption of chemicals from both the perspective of the interaction of the absorbing drug's chemistry with the biology of the skin involved, the experimental approaches used to assess absorption as well as the mathematical endpoints employed to quantitate it. All of these factors must be taken into account when any assessment of dermal absorption is made. Of equal importance is that these methodological assumptions must also be considered when dermal absorption or permeability estimates are subsequently used in the risk assessment process. This chapter will provide an overview of these basic concepts.

J.E. Riviere (✉)
Center for Chemical Toxicology Research and Pharmacokinetics, North Carolina State University, Raleigh, North Carolina 27606, USA
e-mail: jim_riviere@ncsu.edu

M. Lazaridis and I. Colbeck (eds.), *Human Exposure to Pollutants via Dermal Absorption and Inhalation*, Environmental Pollution 17, DOI 10.1007/978-90-481-8663-1_7,

7.2 The Dermal Absorption Barrier

The skin is a relatively complex organ that serves a number of physiological functions, which in addition to serving as an environmental barrier that is the focus of this chapter, also includes temperature regulation in mammals, and endocrine, immunological and metabolic activities. The skin provides the structure upon which animals project their identity to other individuals of the same and different species, a function that is the target of the cosmetic industry in humans.

Topically applied chemicals may either *penetrate* the skin to have a local effect and/or be *absorbed* across the skin and result in systemic absorption and distribution throughout the body. It is important to keep these terms separate since both are important parameters with different experimental endpoints. A *barrier* is defined as something that obstructs or impedes. From our perspective this obstruction refers to the movement of chemicals across or into the skin. This barrier can be viewed as restricting diffusion, but also could be viewed as impeding absorption of intact drug through metabolism in the skin.

7.2.1 General Structure of Skin

The skin is a multifunctional organ responsible for providing a barrier to the environment, thermoregulation, metabolism, immune recognition, sensation and some endocrine functions. Skin is composed of two primary cell layers, the outermost epidermis containing keratinocytes and the inner vascularized dermis (Fig. 7.1) (Monteiro-Riviere 1991, 2006). The layer that interfaces with the environment, the

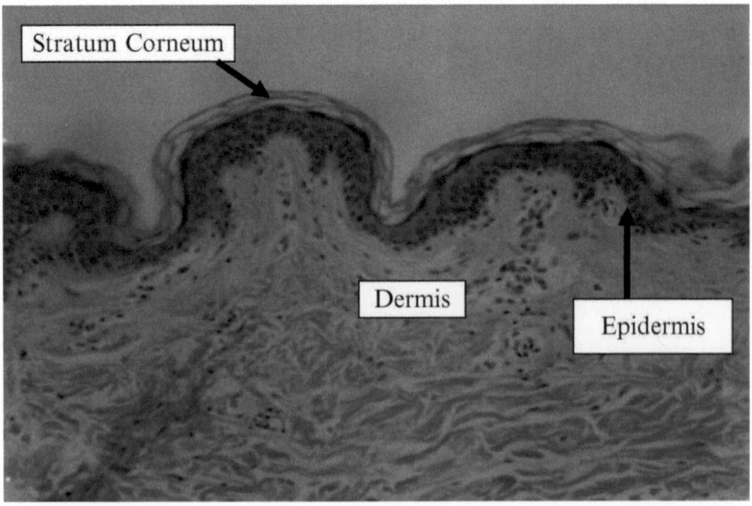

Fig. 7.1 Histogram of skin depicting cellular layers

stratum corneum, is the primary barrier to insensible water loss to the environment and to absorption for most exogenous chemicals. Removal of the stratum barrier by tape stripping increase absorption of most compounds.

This unique barrier layer is composed of dead protein-filled cells that are embedded in a heterogeneous complex lipid matrix whose biophysical properties are related to compound permeability (Elias 1981; Monteiro-Riviere 2006). This structure is often described using a "brick and mortar" analogy where the dead corneocytes are the "bricks" and the intercellular lipids the "mortar." There is a consensus that chemicals are absorbed through skin via this intercellular pathway, with partitioning into and diffusion through the lipid predictive of compound absorption. The bottom layer of these basal keratinocytes sits on the epidermal-dermal junction and contains the only viable cells of the epidermis. As the cells move outward, the nucleus disappears as they manufacture the lipids that are later extruded into the intercellular space. Interspersed between the keratinocytes are other cell types including the melanocytes responsible for pigmentation, Merkel cells that interact with the nervous system, and roaming Langerhans cells that function in immune recognition and antigen processing.

The dermis is composed primarily of collagen fiber networks, fibroblasts, immune cells and lymphatic and vascular endothelium making up the vast capillary network that perfuses the skin. Nutrients for maintenance of both the epidermis and dermis are supplied via this network, whose primary function also includes thermoregulation in mammals via regulation of blood flow to dissipate or conserve heat. The lymphatic spaces drain to regional lymph nodes and are involved in trafficking activated Langerhans cells, as well as some absorbed lipophilic chemicals or particles, to the lymph nodes for modulation of immune function. The dermis provides structure to skin and is the origin of appendages that penetrate the epidermis to the surface. These include hair follicles and various types of sweat ducts. Although appendages are important for other functions attributed to skin, they are not primary pathways for chemical dermal absorption. Finally, it should be noted that the adjective dermal is often applied to absorption. In this context of this chapter, *dermal* refers to the entire skin and is not restricted to the dermis.

7.2.2 Other "Barriers" to Absorption

The stratum corneum is properly referred to as a diffusional barrier. Most approaches to predict dermal absorption and penetration are based upon chemical movement through this intercellular lipid pathway. However, the skin is also a very metabolically active organ, responsible for generating active forms of Vitamin D from cholesterol. Most drug metabolizing enzymes are present in the epidermal cell layer and biotransformation of absorbed drugs has been demonstrated (Riviere 2006). There are two aspects of how such a process could interact with absorption. First, substances that penetrate the stratum corneum barrier could be metabolized before systemic absorption. If an active form of the drug penetrates the stratum corneum, epidermal metabolism could result in compound deactivation and thus function as an absorption first-pass metabolic barrier. In contrast, if the chemical is inactive before metabolism, these enzymes may activate the compound. This is a strategy to deliver normally non-permeable drugs using lipid soluble pro-drugs (e.g. esters of the active drug). If the metabolized chemical is activated to a toxic compound (e.g. benzo-a-pyrene), adverse effects may occur. This can be a mechanism for carcinogenesis or topically applied compounds.

The vasculature could also be viewed as a barrier to absorption of compounds that either cause vasoconstriction, or are delivered by high flux transport approaches (e.g. iontophoresis, microneedle delivery) where vascular uptake becomes the rate limiting step. This effect is compound and formulation/delivery strategy specific.

7.3 Animal Models

The focus of dermal absorption studies is to predict chemical behavior in humans. For many drugs and cosmetics, studies can be done directly in humans. However, for screening of unknown molecules, or to serve the needs of traditional toxicology testing, surrogate animal species are used. The selection of an appropriate species to study dermal absorption is largely based on the use of the data that will be generated. These objectives include assessing absorption and systemic exposure after topical application, dermal sensitivity testing, carcinogenesis, and potential for direct toxicity to skin. Anatomical factors which play a role in selection of a model species include skin thickness, hair follicle density and blood flow (Bronaugh et al. 1982; Monteiro-Riviere et al. 1990; Monteiro-Riviere 2006).

There is a large literature that compares percutaneous absorption of a wide variety of substances in humans and different animal species (Feldmann and Maibach 1974; Maibach and Feldmann 1974; Bronaugh et al. 1982; Scott et al. 1991; Wester et al. 1998; Riviere 2006). To compile such comparative data, factors such as applied dose, surface area, use of occlusive dressings and vehicle or formulation must be controlled. Since rodents are the primary laboratory animal used to assess safety of numerous drugs and chemicals, their use in dermal absorption studies is required for making route-to-route extrapolations. Mice and rats are also employed for carcinogenicity "skin-painting" studies. In general, rodents (mice, rats) are more permeable to chemicals than are humans, making them suitable for defining worst-case absorption scenarios for toxicological endpoints. Rabbits and guinea pigs have been utilized in dermal immunology protocols assessing issues such as contact dermatitis and sensitization. Rabbits have also been used to assess dermal absorption, with most studies showing increased absorption compared to humans.

When the goal of the study is to predict the rate and extent of chemical absorption in humans, animal species with minimal hair or fur are used. As hair follicle density increases to the level seen in furred rats ($\approx 300/cm^2$) or mice ($\approx 650/cm^2$), the interfollicular skin becomes thinner, thereby increasing absorption In contrast, hair follicle density in humans is only $11/cm^2$ which is the same in pigs.

So called hairless rodents also have a lower follicular density (≈75/cm^2), the term hairless being applied to their lack of hair shafts and not follicles (Bronaugh et al. 1982). Apes are often selected due to their evolutionary closeness to humans, however regions of the body with minimal hair (ventral abdomen) must be used.

The pig is widely touted as an appropriate animal model (Feldmann and Maibach 1974; Maibach and Feldmann 1974; Scott et al. 1991; Wester et al. 1998). In addition to similarities in hair follicle density, number of epidermal cell layers, thickness and cutaneous blood flow, the biochemistry and biophysics of the stratum corneum lipids are comparable to humans. Since body mass/surface area ratios are also like humans, extrapolations of systemic exposure are facilitated. In whatever species are employed, one should be cognizant of the fact that regional differences in skin anatomy exist making the crucial animal descriptor both species and body site.

7.4 Experimental Approaches

Once an appropriate species is selected, the decision must be made between using intact animals in vivo versus numerous in vitro approaches. The "gold-standard" for absorption studies are in vivo, however the extent of variability inherent to this work often precludes intact animals for being used for detailed probing of mechanisms involved. Additionally, in vivo approaches are expensive, require special facilities, and are not accepted by some regulatory authorities for cosmetic screening. These approaches are extensively reviewed elsewhere (Riviere 2006).

7.4.1 In Vivo Approaches

The classic approach to assessing absorption of any compound exposed to the body is by measuring the amount of compound excreted into urine and feces compared to that excreted after intravenous administration of an equivalent dose. The ratio of the total amount excreted after dermal to intravenous dosing is termed the *bioavailability*, and is referred to as F. This ratio can also be calculated by collecting plasma concentrations and comparing the ratio of their areas under the concentration-time profile (AUC). The reason F has to be determined relative to a parenteral route such as intravenous is that differences in metabolism between chemicals may change the rate and or extent of excretion or the percentage of a drug present in the central plasma compartment. Once the fraction of the dose eliminated in urine or feces is known, then one need only collect that excreta to get future estimates of absorption. In the case of urine monitoring, if creatinine concentrations are also collected (marker of renal function), and drug mass normalized by it, then monitoring of only urine and this excretion marker is sufficient to make estimates of relative systemic exposure or absorption after different treatments.

These studies require timed samples and careful analysis. Experiments should only be ended when the majority of the compound has been eliminated (e.g. approximately 80%). For dermal applications with very slow rates of absorption, the length of the study required to insure systemic distribution and elimination are complete can be lengthy. Truncating studies earlier may lead to erroneous conclusions. When conducting such studies, care must be taken to insure the dose is completely covering the marked exposure area and that this surface area is measured.

Additional approaches have recently been developed which attempt to assess in vivo absorption by monitoring the drug absorption process from the perspective of concentrations in stratum corneum tape strips (Rougier et al. 1985; Nylander-French 2000). This approach termed *dermatopharmacokinetics*, analyzes the amount of drug in skin of the application site by sequentially removing and analyzing compound in layers of stratum corneum obtained by repeated adhesive tape strips. Surface nonabsorbed drug must first be removed by gentle washing and swabbing. Various approaches are then employed to model the diffusion gradient reflected in the stratum corneum depth profiles. These techniques must first be calibrated to normal human absorption endpoints before reliable predictions can be made.

7.4.2 In Vitro Approaches

One of the most common techniques used to study dermal absorption from both humans and animals are to mount skin samples into a diffusion cell and measure chemical flux into a perfusate bathing the dermal side of the ski sample. Specific protocols have been established relative to membrane preparation and perfusate composition (Bronaugh and Stewart 1984, 1985). Either full-thickness, heat separated epidermal membranes or dermatomed skin (skin sliced into disc of predetermined thickness) may be used. If furred animal skin is employed, the skin section must be sufficiently thick to avoid cutting off the base of the hair follicles in the dermis. The skin is then clamped between two chambers, one of which contains a vehicle from which absorbed chemical will be sampled. *Static* diffusion cells sample this chamber and replace with new perfusate at each time point. *Flow-through* cells use a pump to pass perfusate through the receptor chamber and collect flux by repeatedly collecting perfusate.

In some systems, the skin disc is first fully hydrated and allowed to equilibrate. Receptor fluid is usually saline for pharmaceutical drug studies or a media containing albumin or a surfactant/solvent for studies of organic chemicals where some degree of lipid solubility in the receptor fluid is required. Perfusate should be heated to 32°C or 37° C according to experimental guidelines. If drug is dissolved in water and dosed in the donor chamber at a dose that far exceeds amount absorbed, this experimental is termed an *infinite dose* experiment. In contrast, if the dose is applied at typical exposure situations and exposed to ambient air or covered (occluded), the experiment is termed *finite dose*. In this case, relative humidity of the dosing environment may be important.

Skin source is an important consideration. In human studies, skin may either be obtained fresh from reconstructive surgical procedures, or dead from cadaver sources. In the later case, skin is often frozen and then thawed before use. In these cases, a membrane integrity test is conducted using ^3H water absorption to screen skin discs for lack of barrier integrity. Fresh skin can be used to insure an intact barrier and live tissue, a factor important for metabolism studies. These skin samples are often perfused with oxygenated media to maintain metabolic functions. Artificial skin grown using air-liquid interface cultures are also available, although the permeability through such systems are presently much greater than that of human skin. Such artificial systems are monocultures of keratinocytes and do not contain other cell types. When only the stratum corneum is to be studied, some workers have used shed snake skin as absorption properties have been reported to be comparable to that of humans.

In vitro models are the predominant approach used to assess dermal absorption today. Numerous protocols have been proposed for specific purposes, that is for assessing absorption of lipophilic pesticides versus transdermal delivery of more hydrophilic drugs. Differences between such protocols often relate to surface decontamination methods (swabs, wash, etc), perfusate composition (addition of vehicle to penetrate solubility of penetrant) and length of experiment. If penetration is to be assessed, chemical concentrations in the remaining skin disc after surface drug removal must be measured. Finally, specific types of experimental designs must be used to provide data for specific mathematical modeling approaches, considerations which often dictate specific approaches.

7.5 Mathematical Modeling Considerations

No matter what the experimental system selected, data must be analyzed to arrive at a quantitative estimate of dermal absorption or transdermal flux for estimates of systemic exposure. Again, the nature of how the question is poised will determine the nature of analysis needed. The major endpoints are either permeability constants for a chemical in question, or estimates of the rate and extent of absorption after topical dosing of a specific concentration and formulation of a chemical of interest. Dosing conditions could also be designed to mimic occupational or environmental exposure scenarios, and often are conducted for specific regulatory issues. These approaches are fully developed and discussed elsewhere (Cleek and Bunge 1993; Bunge and Cleek 1995; Riviere 2006).

The movement of chemicals across the stratum corneum barrier into the epidermis occurs primarily by passive diffusion driven by the thermodynamic force generated by the applied concentration of drug on the surface of the skin. This is best expressed using Fick's Law of Diffusion that states that the steady state of drug flux across a membrane can be expressed as:

$$\text{Flux}\left(\text{J}\right)=\frac{\text{DP}}{\text{h}}\left(\text{Concentration Gradient}\right)\left(\text{Surface Area}\right)$$

where D is the diffusion coefficient or diffusivity of the drug in the intercellular lipids of the stratum corneum, P is the partition coefficient for the drug between the stratum corneum and the dosing medium on the skin surface, and h is the skin thickness or actual path-length through which the drug diffuses across the diffusion barrier. Because of the aforementioned "brick and mortar" structure of the stratum corneum, the diffusion path length is tortuous and much greater than the actual thickness. The driving force for this thermodynamic process is the concentration gradient that exists between the applied dose and the dermal perfusate or blood.

The term DP/H is termed the permeability coefficient, Kp. Kinetically, this is a first-order rate constant that forms the basis for the absorption rate constant (Ka) used in pharmacokinetic models of transdermal drug delivery studies. Transdermal flux should be expressed in terms of the surface area of the dosing site, making the two important properties of dosage after topical exposure the concentration of drug applied *and* the surface area of application. Finally, Fick's law expresses the steady state flux of drug that occurs when this rate becomes constant. In skin diffusion studies, this occurs after passage of a *lag time* that is a function of the drug "loading" the stratum corneum and dermis, diffusivity, and thickness of the skin. Lag time can be calculated as $(h^2)/6D$. For molecules with small diffusivities, initial absorption rates may be very low.

The diffusivity of a chemical is a function of the molecular weight, molecular size, molecular interactions with skin constituents (e.g. hydrogen bonding, hydrophobic interactions, etc.), the drug's solubility in the membrane milieu, and the degree of ionization. These properties form the basis of the quantitative structure activity relationship (QSAR) equations presented below. Large molecular weight drugs (approximately greater than 500 Da, e.g. proteins) have extremely low diffusivities, thus effectively preventing them from being absorbed across the skin barrier. If diffusivity is very low, than an exceedingly long lag-time could occur due to its increased dependency on this parameter. Note however that this rule of thumb applies relative to the ability to generate significant transdermal fluxes for pharmacological activities, and not necessarily absorption of small quantities of chemicals that could exert a local effect. For compounds that are partially ionized, diffusivity is decreased, indicating that only the non-ionized fraction of a weak acid or base is available for diffusion across the stratum corneum. This is a function both of the pH of the dosing medium as well as the pH of the skin.

The partition coefficient determines the ability of the drug to gain access to the diffusion pathway. Partition coefficient is usually determined in experimental systems by measuring octanol/water or lipid/water partitioning. The higher the ratio, the greater the lipophilicity. The penetrating drug must have some propensity to partition into the intercellular lipids of the stratum corneum. It is generally accepted that the optimal log octanol/water partition coefficient for a drug to penetrate the stratum corneum is approximately two. In other words, the drug is partitioned in the lipid phase approximately 100-fold. For hydrophilic drugs with low partition coefficients, pro-drugs could be formulated (e.g. by esterification) which increases the drug's permeability across the stratum corneum. The lipophilic moiety (e.g. ester) is then cleaved in the epidermis, dermis or even plasma and active parent drug is then distributed throughout the systemic circulation. If the partition coefficient is too great, drug may have a tendency to sequester into the stratum corneum and not enter the more aqueous dermis, thereby decreasing systemic delivery. If it does penetrate into the dermis, the high lipid partition coefficient may favor formation of a dermal depot. It must be noted, that the drug must also have partitioning properties that are favorable for entering into solution into the aqueous plasma, or be able to bind to plasma proteins for systemic absorption to occur. Finally, the tendency to partition is counterbalanced by solubility in the dosing vehicle. For a compound to be absorbed, it must have sufficient solubility to be in solution while still possessing sufficient lipid solubility to partition into the stratum corneum lipids. Optimization of these two independent properties may be required to obtain significant dermal absorption.

There are other factors that also come to play in assessing absorption. After lag time has passed, chemical flux across skin is at steady state. It is this value that should be used to calculate Kp. In an infinite dose experiment, a plot of cumulative flux will easily allow precise measurement of flux. However, in a finite dose experiment, the flux after the lag time has passed will ultimately decrease as chemical is depleted from the dosing solution. The slope of the initial portion of this flux profile is thus termed *pseudo-steady state*.

Data can also be reported as maximum flux (J_{max}) and time to this flux (T_{max}). This term is often used as the parameter to quantify the absorption properties for toxicological endpoints since there are no assumptions inherent to its determination. J_{max} for a specific chemical can be determined from a saturated solution of the compound and will give an estimate of the worst-case dermal exposure for a chemical. However, if a super-saturated solution is applied, J_{max} may be even greater, its value being a function of thermodynamic activity in the dosing vehicle countered by the extent of interaction between the vehicle and the skin. Depending on the circumstances and the nature of the chemical and vehicle involved, Kp or J_{max} may be the optimal parameters to characterize exposure.

7.5.1 QSAR Approach

A great deal of research has been focused on relating the physical chemical properties of the penetrant to its permeability through skin (Moss et al. 2002). The first such relationship widely used to assess chemical absorption is that of Potts and Guy (1992)

$$\text{Log Kp} = 0.71 \log \text{PC}_{\text{octanol/water}} - 0.0061\text{MW} - 6.3\left(R^2 = 0.67\right)$$

where MW is the molecular weight. This equation was subsequently modified (Potts and Guy 1995) to relate Kp to molecular properties of the penetrants as:

$$\mathrm{Log\,Kp} = 0.0256\,\mathrm{MV} - 1.72\sum \alpha_2^{\mathrm{H}} - 3.93\sum \beta_2^{\mathrm{H}} - 4.85\left(\mathrm{R}^2 = 0.94\right)$$

where MV is molecular volume, $\Sigma\alpha_2{}^{H}$ is the hydrogen-bond donor acidity, and $\Sigma\beta_2{}^{H}$ is the hydrogen-bond acceptor basicity.

The most promising approach is to further extend this rational using linear free energy relationships (LFER) to relate permeability to the physical properties of the penetrant under defined experimental conditions (dose, membrane selection, vehicle). Geinoz et al. (2004) critically reviewed most such quantitative structure permeability relationships (QSPeR) applied to dermal absorption and should be consulted. Abraham's LFER model is representative of the dermal QSPeR approaches presently available (Abraham et al. 1999). This model was selected since it is broadly accepted by the scientific community as being descriptive of the key molecular/physiochemical parameters relevant to solute absorption across skin. This basic model can be written as:

$$\log k_{\mathrm{p}} = \mathrm{c} + \mathrm{a}\sum \alpha_2^{\mathrm{H}} + \mathrm{b}\sum \beta_2^{\mathrm{H}} + s\pi_2^{\mathrm{H}} + \mathrm{r}R_2 + \mathrm{v}V_{\mathrm{x}}$$

where $\pi_2{}^{H}$ is the dipolarity/polarizability, R_2 represents the excess molar refractivity, V_x is the McGowan volume and the other parameters are as described earlier. The variables c, a, b, s, r, and v are strength coefficients coupling the molecular descriptors to skin permeability in the specific experimental system studied.

7.5.2 Mixture Exposure

Dermal formulations may significantly modify percutaneous absorption by modulating stratum corneum partitioning and solubility. For a lipid soluble drug, a lipid base formulation would tend to decrease absorption by retaining applied drug at the skin surface. In contrast, an aqueous base would promote absorption solely by this partitioning phenomenon that would favor drug movement out of the formulation into the more favorable lipid environment of the stratum corneum. The actual solubility of the compound in the vehicle further modifies this relationship. The reverse scenario would be operative for a hydrophilic drug.

Formulations and solvents have long been known to alter a chemical penetrant's solubility and partition coefficient into the stratum corneum, the two factors which determine dermal absorption. Our laboratory has focused significant research on the effects of chemical mixtures on dermal absorption of penetrant compounds (Riviere and Brooks 2005, 2007). In order to incorporate mixture effects, we have been exploring using an additional term, referred to as the mixture factor (MF), to modify a QSPeR equation to better predict absorption from mixtures:

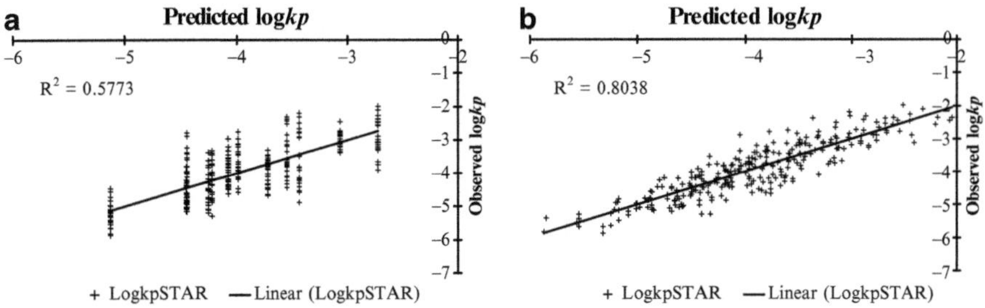

Fig. 7.2 Ability of mixture factor (MF) to predict dermal absorption from varied vehicles. (**a**) No MF, (**b**) MF (see text for full description)

$$\log k_p = c + mMF + a\sum \alpha_2^H + b\sum \beta_2^H + s\pi_2^H + rR_2 + vV_x$$

The nature of the MF is determined by examining the residual plot (actual – predicted log k_p) generated from the base LFER equation based on molecular descriptors of the permeants, against a function of the physical chemical properties of the mixture/solvents in which they were dosed (Riviere 2006; Riviere and Brooks 2005).

The selection of an appropriate MF is based on its correlation to the residuals of the normal QSPeR model. The MF seems to improve prediction of permeability and absorption across different LFER models (e.g. Potts and Guy as well as Abraham approaches). Figure 7.2 illustrates thus approach using flow-through diffusion cell data into pig skin.

7.6 Conclusion

As can be appreciated from this discussion, assessing the absorption of a chemical after topical exposure to skin is in one breath a relatively simple problem, yet in another is affected by experimental and biological nuances that confounds simple solutions. The variability that this biological and experimental diversity brings to mathematically modeling dermal penetration and absorption interferes with accurately quantitating these processes. Models that have worked are structured so that the rate-limiting process under different exposure scenarios is captured. For any exposure scenario, effort should be expended to define this process (e.g. partitioning out of exposure vehicle – soil, occupational mixture, repeated environmental exposures). When these conditions are understood, the appropriate experimental design can be selected to that models and parameters can then appropriately estimated. All studies should aim to precisely define experimental conditions so that assumptions can be realistically evaluated.

Modeling dermal absorption processes is an active field of research in occupational, pharmaceutical and environmental sciences. Each field has its own unique history and accepted practices that have evolved as a function of compounds studied and likely exposure scenarios. Recent work has started to cross these disciplinary boundaries that has enriched the filled and made some approaches more generally applicable. These trends should continue and lead to more generally applicable dermal modeling approaches in the future.

References

Abraham, M. H., Chadha, H. S., Martins, F., Mitchell, R. C., Bradbury, M. W., & Gratton, J. A. (1999). A review of the correlation and prediction of transport properties by an LFER method: Physiochemical properties, brain penetration and skin permeability. *Pesticide Science, 55*, 78–88.

Bronaugh, R. L., & Stewart, R. F. (1984). Methods for in vitro percutaneous absorption studies. III. Hydrophobic compounds. *Journal of Pharmaceutical Sciences, 73*, 1255–1258.

Bronaugh, R. L., & Stewart, R. F. (1985). Methods for in vitro percutaneous absorption studies. IV. The flow-through diffusion cell. *Journal of Pharmaceutical Sciences, 74*, 64–67.

Bronaugh, R. L., Stewart, R. F., & Congdon, E. R. (1982). Methods for in vitro percutaneous absorption studies. II. Animal models for human skin. *Toxicology and Applied Pharmacology, 62*, 481–488.

Bunge, A. L., & Cleek, R. L. (1995). A new method for estimating dermal absorption from chemical exposure: 2. Effect of molecular weight and octanol-water partitioning. *Pharmaceutical Research, 12*, 88–95.

Cleek, R. L., & Bunge, A. L. (1993). A new method for estimating dermal absorption from chemical exposure. 1. General approach. *Pharmaceutical Research, 10*, 497–506.

Elias, P. M. (1981). Epidermal lipids, membranes, and keritinization. *International Journal of Dermatology, 20*, 1–19.

Feldmann, R. J., & Maibach, H. I. (1974). Percutaneous penetration of some pesticides and herbicides in man. *Toxicology and Applied Pharmacology, 28*, 126–132.

Geinoz, S., Guy, R. H., Testa, B., & Carrupt, P. A. (2004). Quantitative structure-permeation relationships (QSPeRs) to predict skin permeation: A critical evaluation. *Pharmaceutical Research, 21*, 83–92.

Maibach, H., & Feldmann, R. (1974). *Systemic absorption of pesticides through skin of man* (pp. 120–127). Occupational Exposure to Pesticides: Report to the Federal Working Group on Pest Management for the Task Group. Washington, DC: Task Group on Occupational Exposure to Pesticides.

Monteiro-Riviere, N. A. (1991). Comparative anatomy, physiology, and biochemistry of mammalian skin. In D. W. Hobson (Ed.), *Dermal and ocular toxicology: fundamentals and methods* (pp. 3–71). Boca Raton, FL: CRC Press.

Monteiro-Riviere, N. A. (2006). Structure and function of skin. In J. E. Riviere (Ed.), *Dermal absorption models in toxicology and pharmacology* (pp. 1–27). Boca Raton, FL: CRC Press.

Monteiro-Riviere, N. A., Bristol, D. G., Manning, T. O., & Riviere, J. E. (1990). Interspecies and interregional analysis of the comparative histological thickness and laser Doppler blood flow measurements at five cutaneous sites in nine species. *Journal of Investigative Dermatology, 95*, 582–586.

Moss, G. P., Dearden, J. C., Patel, H., & Cronin, M. T. D. (2002). Quantitative structure-permeability relationships (QSPRs) for percutaneous absorption. *Toxicology In Vitro, 16*, 299–317.

Nylander-French, L. A. (2000). A tape-stripping method for measuring dermal exposure to multifunctional acrylates. *Annals of Occupational Hygiene, 44*, 645–651.

Potts, R. O., & Guy, R. H. (1992). Predicting skin permeability. *Pharmaceutical Research, 9*, 663–669.

Potts, R. O., & Guy, R. H. (1995). A predictive algorithm for skin permeability: the effects of molecular size and hydrogen bond activity. *Pharmaceutical Research, 12*, 1628–1633.

Riviere, J. E. (Ed.). (2006). *Dermal absorption models in toxicology and pharmacology.* Boca Raton, FL: CRC Press.

Riviere, J. E., & Brooks, J. D. (2005). Predicting skin permeability from complex chemical mixtures. *Toxicology and Applied Pharmacology, 208*, 99–110.

Riviere, J. E., & Brooks, J. D. (2007). Prediction of dermal absorption from complex chemical mixtures: Incorporation of vehicle effects and interactions into a QSPR framework. *SAR and QSAR in Environmental Research, 18*, 31–44.

Rougier, A., Dupuis, D., Lotte, C., & Roguet, R. (1985). The measurement of stratum corneum reservoir, a predictive method for in vivo percutaneous absorption studies: influence of application time. *Journal of Investigative Dermatology, 84*, 66–68.

Scott, R. C., Corrigan, M. A., Smith, F., & Mason, H. (1991). The influence of skin structure on permeability: An intersite and interspecies comparison with hydrophilic penetrants. *Journal of Investigative Dermatology, 96*, 921–925.

Wester, R. C., Melendres, J., Sedik, L., & Maibach, H. I. (1998). Percutaneous absorption of salicylic acid, theophylline, 2, 4-dimethylamine, diethyl Hexylphthalic acid, and ρ-aminobenzoic acid in the isolated perfused procine skin flap compared to man. *Toxicology and Applied Pharmacology, 151*, 159–165.

Chapter 8
Micro-environmental Modelling

Tareq Hussein and Markku Kulmala

Abstract "Indoor air quality" is a wide subject with different social, economic, and health aspects. In developed countries, people spend more than 80% of their time indoors where they are exposed to many kinds of air pollutants either from outdoor origin or produced indoors. An air pollutant can be a gas or an aerosol particle (solid, liquid, radioactive, bio-aerosols, etc.). Indoor air pollutants are transported from the outdoor air by means of mechanical ventilation systems or across the building shell as a result of natural ventilation. In many aspects, the indoor-to-outdoor relationship of air pollutants, as well as, the dynamic behavior of air pollutants can be addressed and investigated by means of mathematical models. However, the accuracy of such mathematical models depends on many factors including, most importantly, the confidence in the input parameters, validity of the assumptions, description of the processes, and user influence.

8.1 Introduction

Indoor air quality can be determined experimentally by directly measuring air pollutant concentrations. The indoor air quality can be predicted by using Indoor Air Models that describes the dynamic behaviour of indoor aerosols after understanding the relationship between the indoor and the outdoor air pollutants and there are no indoor sources of pollutants. Indoor air models can be utilized to predict the indoor air quality when measurements are expensive or even impossible to be performed. Indoor air models are also useful from the engineering point of view to maintain an acceptable indoor air quality in a building to be built in the future.

Generally speaking, there and two major types of indoor air models: ***single-compartment*** (single-zone) and ***multiple-compartments*** (multiple-zone). Single compartment indoor air models (SC-IAM) are typically applied when the indoor air is well mixed so

T. Hussein and M. Kulmala (✉)
Department of Physics, University of Helsinki, P.O. Box 64, FI–00014, UHEL, Finland
e-mail: markku.kulmala@helsinki.fi

M. Lazaridis and I. Colbeck (eds.), *Human Exposure to Pollutants via Dermal Absorption and Inhalation*, Environmental Pollution 17, DOI 10.1007/978-90-481-8663-1_8,

that the pollutant concentrations do not have spatial gradients indoors. If the indoor pollutants show spatial gradients or if the indoor space is physically divided into different compartment, there is a need for multiple compartment indoor air models (MC-IAM). Additionally, if an indoor air model describes the dynamic behaviour of only one pollutant, it is called ***single-component***; otherwise it is called ***multiple-component***.

In this chapter we will present and discuss the basic principles of different types of indoor air models. Our discussion will be focused on indoor aerosol particles in specific and indoor air pollutants in general.

8.2 Simple Indoor Air Models

In general, the change rate of an indoor air pollutant concentration is governed by sources and sinks. Typically, pollutants are exchanged between the indoor and outdoor air across the building shell or through the mechanical ventilation system, and thus, the main source of indoor pollutants is from outdoor origin. Pollutant concentrations are reduced by filtration and loss mechanisms (e.g. deposition onto duct lines, building shell, and indoor surfaces) while transported from the outdoor air into the indoor air. On the other hand, re-suspension and emissions of pollutants increase their concentrations indoors.

A *simple* indoor air model must at least incorporate the air exchange rate between the indoor and outdoor air; that is ventilation rate. Additionally, penetration and filtration processes of air pollutants while transported into the indoor air must be taken into account in a simple indoor air model. A simple indoor air model may also, incorporate deposition process onto indoor surfaces, re-suspension from accumulated material on indoor surfaces, and emission processes (Kulmala et al. 1999). A well mixed indoor air profile is an important assumption for a simple indoor air model to be valid. Here, "mixing" refers to the turbulent mixing of indoor air by mechanical means (fan, mechanical ventilation, etc.), flow through an open window (natural ventilation), or movement of people within a room. Mathematical formulation of the above mentioned processes outlines the basic principle of a simple indoor air model.

Consider a room with a well mixed indoor air profile and assume the room is ventilated with a mechanical ventilation system. The change rate of a pollutant concentration (e.g. total particle number or mass concentration) in the indoor air is then given by the balance equation

$$\frac{d}{dt}N_{in} = P\lambda N_{out} - \lambda N_{in} - \lambda_d N_{in} + S_{in}, \tag{8.1}$$

where N_{in} and N_{out} (in units of m^{-3} for aerosol particle number concentrations) are respectively the indoor and outdoor pollutant concentrations, P is the penetration factor of pollutants through the standard filter installed in the fresh air section of the mechanical ventilation, λ (s^{-1}) is the ventilation rate, λ_d (s^{-1}) is the deposition rate of the indoor pollutant onto available indoor surfaces, and S_{in}

(in units of m^{-3} s^{-1} for aerosol particles emissions) denotes emission rates of pollutants from indoor sources. For the time being, re-suspension processes can be included in the source term S_{in}.

The indoor air model described in Eq. (8.1) illustrates the most relevant processes (penetration – P, ventilation – λ, and deposition – λ_d) affecting the indoor-to-outdoor relationship of aerosol particles. Therefore, we will recall the parameters describing these processes as the ***controlling parameters*** and we will discuss them in more details in the following sections.

This kind of simple indoor air models is the base-block for more complicated models and they have been commonly used to predict and simulate total particle number or mass concentrations, radioactivity of radon gas, the fate of bio-aerosols, etc. (e.g. Jamriska et al. 2003; Riley et al. 2002; Kulmala et al. 1999; Raunemaa et al. 1989; Alzona et al. 1979; Lum and Graedel 1973). The accuracy of such indoor air models is limited to the extent of indoor air mixing. In case of aerosol particles, this accuracy is also limited to the dynamic behaviour of aerosol particles and their size distribution. For example, the penetration and deposition processes are particle-size dependent that a single value of the penetration factor (P) or the deposition rate (λ_d) is not valid for a wide size-range of aerosol particles.

8.2.1 Indoor-Outdoor Air Exchange Process: Ventilation Rate

The first term, $P\lambda C_{out}$, on the right-hand-side of Eq. (8.1) is a source term and it represents the accumulation rate of an indoor pollutant that penetrates from the outdoor air via a mechanical ventilation system. Consequently, indoor pollutants are removed from the indoor air via the outgoing air to the outdoor air; this is represented by the second term λC_{in}. These two terms illustrate the indoor-outdoor air exchange process; that is ventilation. The ventilation rate, λ, represents the number of times the indoor air is changed in a certain time period, per hour, for example, and it is simply the ratio between the indoor-outdoor air flow rate, $Q_{0,in}$ (m^3 s^{-1}), to the indoor volume, V (m^3); that is

$$\lambda = \frac{Q_{0,in}}{V}. \tag{8.2}$$

In Eq. (8.1) we assumed that there are no other paths of air exchange processes except for the ventilation; that is $Q_{0,in} = Q_{0,ex}$.

There are two major types of ventilation: mechanical and natural. Mechanical ventilation have many scenarios and all of them produce a well defined air mixing and velocity fields indoors. Contrary to mechanical ventilation, the air exchange rate between the indoor and the outdoor air and the penetration factor are not well defined during natural ventilation. Air exchange rate during natural ventilation occurs through an open window, door, and air leak across the building shell via wall-cracks or around a window shell.

The ventilation process is one of the most important processes in indoor air quality. For instance, when the source of pollutants is often indoors, a high ventilation rate is recommended to quickly reduce the pollutants concentrations indoors. However, in case of industrial places where typically huge amounts of air pollutants are produced indoors, it is a must to install filtration system on the exhaust air system before releasing the outgoing air into the outdoor air.

8.2.2 Penetration Process

The penetration factor is the fraction of incoming outdoor pollutants as a result of filtration and particle losses across the building shell (e.g. Riley et al. 2002). For a standard air filter (Fig. 8.1), which are commonly used in mechanical ventilation systems, the penetration factor can be derived from its filtration efficiency (FE) according to

$$P = 1 - \frac{FE\%}{100\%}. \tag{8.3}$$

The filtration efficiency is defined as the fraction of entering pollutants retained by the filter. Penetration factor of a stander filters depends on filter type (including the internal structure and design), air flow rate across the filter, and dust loading on

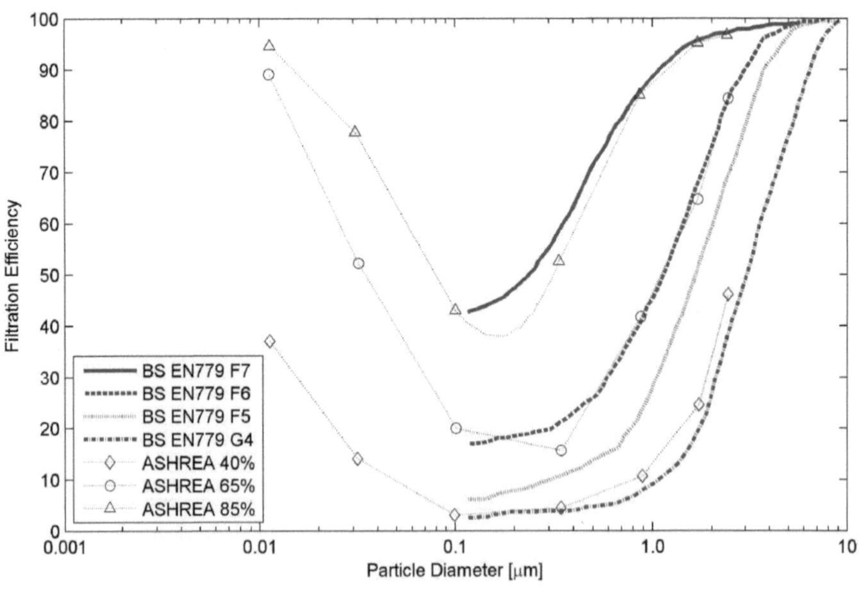

Fig. 8.1 Filtration efficiencies of standard class filters: ASHRAE standards are adopted from Hanley et al. (1994) at 1.3 m/s face velocity, and BS EN 779 standards are adopted from Goodfellow and Tähti (2001) at the minimum efficiency

the filter (Goodfellow and Tähti 2001; Hanley et al. 1994). Across the building shell, that is natural ventilation, the penetration factor varies with the building geometry, surface materials, and pressure drop along the leakage path (e.g. Liu and Nazaroff 2001).

In general, the penetration factor of a certain particle size can have a value between 0 and 1 and it is a function of the particle size. A maximum in the penetration factor curve is often observed in the particle diameter range 0.1–1.0 μm (Hinds 1999), where neither diffusion nor inertial impaction are efficient filtering mechanisms. In the case of simple indoor air models, Eq. (8.1), an average value over the whole particle size-range is assumed for the penetration factor, which in turn limits the accuracy of such simple models.

8.2.3 Deposition Process on Indoor Surfaces

The third term, $\lambda_d C_{in}$, on the right-hand-side of Eq. (8.1) is a sink term (removal process) and it illustrates the change rate due to deposition onto indoor surfaces. Deposition occurs as a result of two main mechanisms: (1) Advection and turbulent diffusion that rapidly transport particles from the room core to the boundary layer nearby the deposition surface and (2) transport of particles through the boundary layer towards the surface. It is likely that the second process is the one that controls the deposition and it occurs through several, processes most importantly to mention: Brownian and Eddy diffusion, inertial impaction, and gravitational settling. Diffusion is the dominant mechanism for ultrafine particles (diameter < 0.01 μm) whereas gravitational settling and impaction are the dominant processes for coarse particles (diameter > 1 μm) as illustrated in Fig. 8.2. Similar to the penetration factor, the deposition velocity is a function of the particle size. An average value for the whole particle size-range is typically assumed for the deposition rate in a simple indoor air model.

Usually, the deposition rate of aerosol particles onto indoor surface is derived from the deposition velocity, the room volume, and the deposition surface area (e.g. Fogh et al. 1997; Porstendörfer and Reineking 1992; Nazaroff and Cass 1989). The deposition surfaces are distinguished into three orientations: vertical such as walls, horizontal facing up such as floor, and horizontal facing down such as ceiling. The total deposition rate is then calculated according to

$$\lambda_d = \frac{1}{V}\left[A_{vertical} v_{d,vertical} + A_{down} v_{d,down} + A_{up} v_{d,up}\right], \tag{8.4}$$

where A (m^2) is the surface area of an indoor surface with a certain orientation (vertical, facing up, or facing down) and v_d (m/s) is the deposition velocity towards that surface. V is again the room volume (m^3).

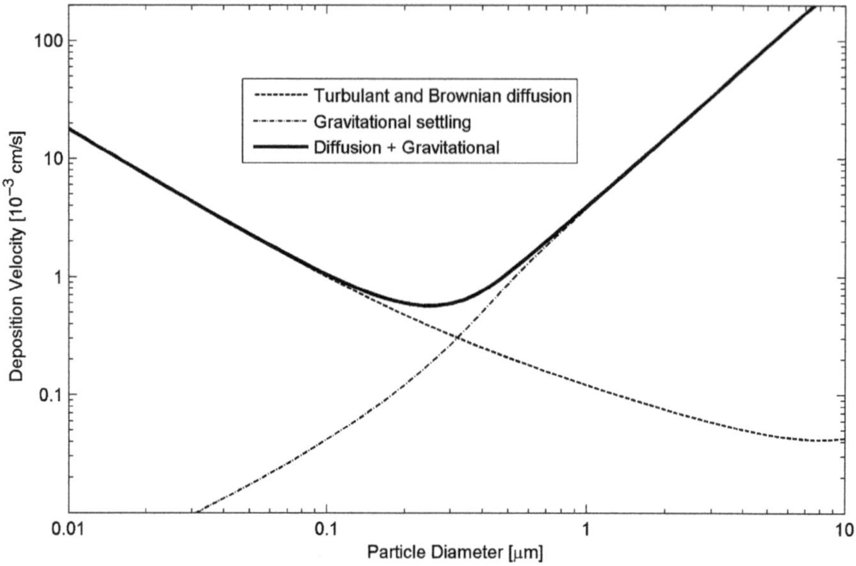

Fig. 8.2 Deposition velocity of indoor aerosol particles onto a smooth surface horizontally oriented facing up according to Lai and Nazaroff (2000) with friction velocity u* = 10 cm/s

Thermophoresis due to temperature gradients or electrostatic drifting due to charged particles or electric fields above charged surfaces can also influence deposition velocity of aerosol particles (e.g. Lai 2006; Lai and Nazaroff 2000; Vanmarcke et al. 1991; Nazaroff and Cass 1989; Shimada et al. 1989; McMurry and Radar 1985). During inhomogeneous turbulence, the gradient in turbulent fluctuating velocity components nearby the deposition surface gives rise to turbophoresis (Fig. 8.3). Turbophoresis is a particle transport mechanism that is distinct from Brownian diffusion and it is expected to be a significant deposition process for particles with large inertia (e.g. Zhao and Wu 2006a; Guha 1997); generally speaking particles with diameters > 5 μm (e.g. Hussein et al. 2009a).

The deposition rate of indoor aerosol particles increases with surface roughness (e.g. Hussein et al. 2009b; Zhao and Wu 2006b; Lai and Nazaroff 2000; Lai et al. 2001, 2002; Thatcher et al. 2002; Abadie et al. 2001; Fogh et al. 1997; Guha 1997). This increased deposition rate is assumed to be due to modifications in the boundary layer over the rough surface: in analogy with heat transfer experiments the thin boundary layer over a rough surface is completely destroyed or at least made thinner due to the surface roughness; and thus, the transportation progress across it can be greatly enhanced (Lai et al. 2001). The differences between the deposition of aerosol particles on a smooth and rough surface is illustrated in Fig. 8.4.

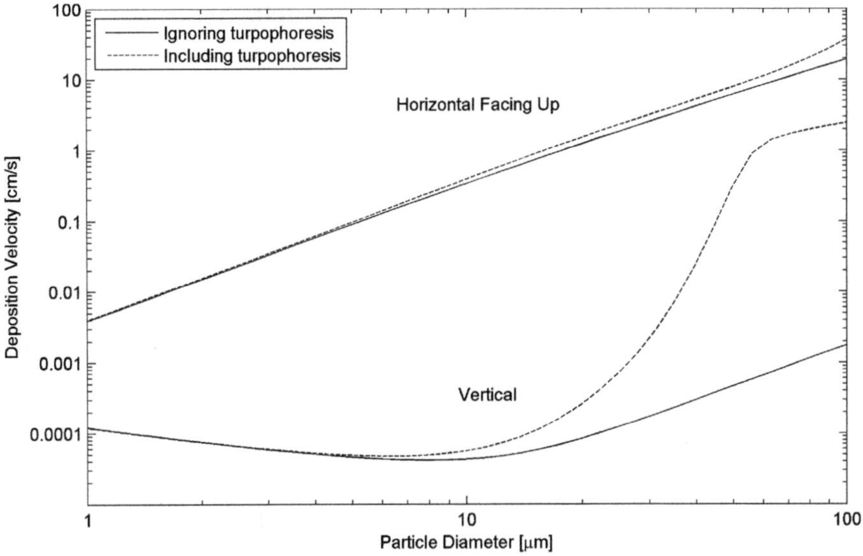

Fig. 8.3 Deposition velocity of indoor aerosol particles onto a vertical smooth surface or horizontally oriented facing up according to Lai and Nazaroff (2000) by ignoring turbophoresis and Zhao and Wu (2006a) by including turbophoresis. A friction velocity u* = 10 cm/s was used to calculate the deposition velocity

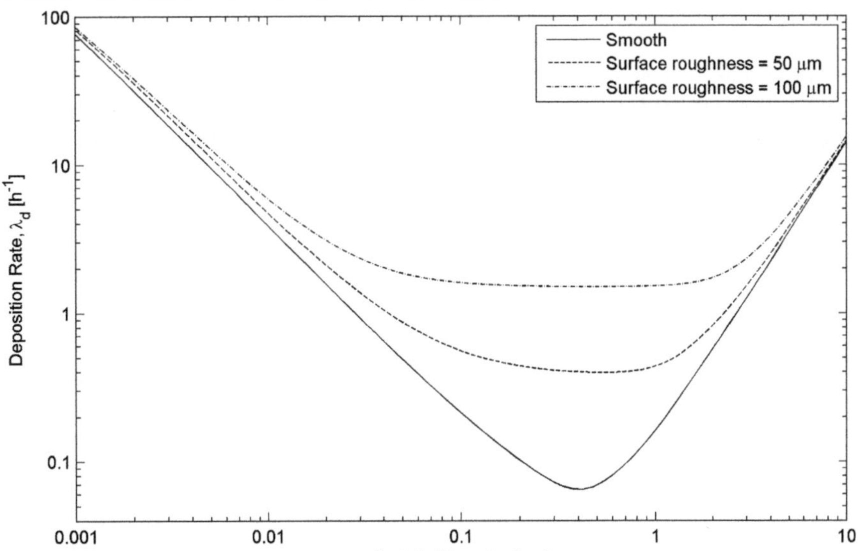

Fig. 8.4 Deposition rate of aerosol particles inside a 1-m^3 cubic box according to Zhao and Wu (2006a) with a friction velocity u* = 10 cm/s

8.3 Multiple-Component Indoor Air Models: Sectional Indoor Air Models

The simple indoor air model describes the dynamic behaviour of only one component of pollutants or assumes the same physical-chemical properties throughout the whole particle size-range, which is not realistic. We can easily extend this simple model to describe more than one pollutant or consider several size-ranges where the physical-chemical properties of each section are representative for that size-range. This approach is known as multiple-component indoor air model or sectional indoor air model. The change rate of each component concentration is described with a separate particle-number-balance equation similar to Eq. (8.1). In principle, indoor pollutants interact between each others and such interaction between two different pollutants or sections is handled with an additional term in the particle-number-balance equation. Interaction between components can be chemical reactions, disintegration of a radioactive gas such as radon, evolution of particle size distributions, etc.

A common example of multiple-component indoor air models is the sectional (size-resolved) indoor air models (SIAM) in which the dynamic behaviour of indoor particle size distributions is the main interest (e.g. Thatcher et al. 2002; Long et al. 2001; Abt et al. 2000; Thatcher and Layton 1995). An integrated aerosol dynamic model (e.g. nucleation, coagulation, condensation, etc.) is usually utilized in a SIAM (e.g. Asmi et al. 2004; Nazaroff 2004). A SIAM is superior to a simple indoor air model, such as the one presented in the previous section, because a SIAM provides more flexibility to describe the particle size distribution and it also provides better understanding on the particle size dependence of aerosol dynamic processes.

Let us make our SIAM more complete by introducing a combination of mechanical and natural ventilation. The particle-number-balance equation of each particle size-section can be then extended from Eq. (8.1) to include both mechanical and natural ventilations and aerosol dynamics

$$\frac{d}{dt}N_{in,i} = \frac{1}{V}(P_i Q_{0,in} + P_{\inf,i} Q_{\inf})N_{out,i} - \frac{1}{V}(Q_{0,ex} + Q_{exf})N_{in,i} - \frac{1}{V}\sum_{j}^{surfaces} A_j v_{d,j,i} N_{in,i} + [AD]_{in,i} + [RS]_{in,i} + S_{in,i} \tag{8.5}$$

where i denotes that the equation is valid for a certain size-section (i.e. size-range) of aerosol particles that have approximately similar dynamic behaviour and physical-chemical properties. N_{in} and N_{out} are again the indoor and outdoor aerosol particle concentrations, V is the volume of the indoor domain, P_i and $P_{inf,i}$ are respectively the penetration factors of aerosol particles through the standard filter installed in the mechanical ventilation system and the penetration factor across the building shell (natural ventilation), Q_{inf} and Q_{exf} are the air flow rates between the indoor and outdoor air due to the natural ventilation, and $Q_{0,in}$ and

$Q_{0,ex}$ are the air flow rates between the indoor and outdoor air due to the mechanical ventilation system. Note that Q_{inf} and Q_{exf} may not be equal, but $Q_{inf} + Q_{0,in} = Q_{exf} + Q_{0,ex}$ must be verified. v_d (m/s) is the deposition velocity of aerosol particles onto an indoor surface with an area A (m^2) and a certain orientation denoted by j $[AD]_{in,i}$ is the change rate of indoor particle concentration due to other aerosol dynamics, and S_{in} (m^{-3} s^{-1}) denotes the emission rates of aerosol particles from indoor sources.

In this model, we introduced the re-suspension process $[RS]_{in,i}$, that will be discussed in more details later in the integrated form of the multi-compartment and size-resolved indoor air model. Other aerosol dynamic processes such as nucleation, coagulation, condensation, etc. will be discussed in details in Section 3.1. Models of aerosol particle deposition onto indoor surfaces can be incorporated in a SIAM. Modelling of the deposition process was first introduced by Corner and Pendlebury (1951) and later on it has been well established after Nazaroff and colleagues (Nazaroff and Cass 1989; Lai and Nazaroff 2000) for Smooth surfaces.

Even though this model is more comprehensive than the simple one introduced in the previous section, it still has a limitation for one compartment and to the extent of indoor air mixing. In real life situations, dwellings consist of several rooms, floors, or even a mechanical ventilation system that consists of several sections (fresh and exhaust air sections and some cases a recycled air section). In long duct lines, aerosol particle concentration can be reduced due to deposition on the internal walls of the duct (e.g. Sippola and Nazaroff 2003). Therefore, it is sometimes recommended to consider an effective penetration factor that replaces the effects of both deposition and penetration processes of aerosol particles in duct lines until they emerge out from the fresh air inlet into the indoor air (Hussein 2005).

8.3.1 Interpretation of Particle Size-Sections in a SIAM

Based on the number of molecules or monomers that are contained in an aerosol particle, we can characterize the aerosol distribution by the number concentration (per m^3 of air) of all aerosol particles each distinguished by a certain number of molecules. However, a better way to interpret the aerosol distribution is the particle size instead of the number of molecules in that particle size, and thus, known as the particle number size distribution.

In the particle number size distribution, the whole particle size range is divided into smaller size-sections for which the number concentration in that size-section is recorded and the aerosol particles have similar physical-chemical properties. The number concentration in that size-section is then denoted by $dN(D_p)$, which is equivalent to N_i in Eq. (8.5). Usually, the particle number for each size-section is normalized with the width of that particle size-section, that is $dN(D_p)/dD_p$. In practice, aerosol particles are log-normally distributed, and therefore, their particle number distribution is normalized with $dlog(D_p)$ instead of dD_p (Seinfeld and Pandis 1998).

In general, particle number size distributions are recorded as $dN/dlog(D_p)$, and therefore, we have to transform them into an acceptable format for the SIAM. The acceptable format is the number of concentration in every size-section, N_i, and it is obtained from the particle number size distribution according to

$$N_i \leftarrow dN(D_p) = n(D_p) \cdot d\log(D_p), \tag{8.6}$$

where $n(D_p)$ is the particle number size distribution.

8.4 Multiple-Compartment and Multiple-Component Indoor Air Models: General Formulation

The previous indoor air models assume a well-mixed indoor air profile and therefore they are limited to a single-compartment approach. In real-life situations, a dwelling consists of several rooms or the indoor pollutant concentrations show spatial gradients where a SIAM is not applicable. Therefore, multiple-compartment and size-resolved approach indoor air models (MC-SIAM) are introduced to describe the dynamic behaviour of aerosol particles with better accuracy to consider the spatial distribution of aerosol particles. MC-SIAM is very complicated and it is numerically demanding. For simplicity, we will present the general formulation of the multiple-compartment and size-resolved indoor aerosol model (MC-SIAM) according to Hussein et al. (2005b, 2006). Several multiple-compartment and size-resolved indoor air models can be found in literature (e.g. Thornburg et al. 2001; Mosley et al. 2001; Schneider et al. 1999; Tung et al. 1999). MIAQ is another multiple-compartment model of aerosol particles indoors that also includes aerosol dynamics (Miller and Nazaroff 2001; Nazaroff and Cass 1986, 1989).

According to the MC-SIAM algorithm, the particle-number-balance equation for the particle number concentration in a compartment, k, and for a particle size-section, i, can be written in a compact form

$$\frac{d}{dt} N_{k,i} = \sum_l J_{k,l,i,} \tag{8.7}$$

where $N_{k,i}$ is the number concentration (m^{-3}) of particles in compartment k and size-section i. On the right-hand-side of the equation, $J_{k,l,i}$ denotes the change rate of particle number concentration ($m^{-3}s^{-1}$) due to a process l.

8.4.1 *Aerosol Dynamic Scheme UHMA*

Aerosol dynamics may include, and not limited to, coagulation, condensation, evaporation, and nucleation. These processes can be included in an indoor air model from available aerosol dynamics modules such as the MIAQ or the "University of Helsinki

Multi-component Aerosol model" – UHMA – (Korhonen et al. 2004). However, the compounds leading to new particle formation (e.g. nucleation processes) have remained unsolved in the indoor air. For example, pealing oranges produces a significant amounts of aerosol particles that grow from the very small sizes bellow 0.01 μm to the Aitken mode (0.025–0.1 μm) due to the release of certain terpens (e.g. Vartiainen et al. 2006). On the other hand, cooking may generate different kinds of vapours that stimulate secondary particle formation in the indoor air.

8.4.1.1 Coagulation

Coagulation refers to a process in which particles collide and coalesce with each other (Seinfeld and Pandis 1998). In indoor air models it is enough to include in the coagulation kernels the Brownian diffusion only; other processes seem to be negligible under normal indoor conditions. The change rate of the particle number concentration due to coagulation is

$$J_{k,coag,i} = \frac{1}{2}\sum_{j=1}^{i-1} K_{j,i-j} N_{k,j} N_{k,i-j} - N_{k,i} \sum_{j=1}^{\infty} K_{i,j} N_{k,j}, \tag{8.8}$$

where $K_{i,j}$ is the Coagulation kernel between particles in size sections i and j. For small particles, the correction for non-continuum effects is based on the work by Fuchs (1964).

The coagulation processes can be thought of a removal process, especially for small particles, and therefore coagulation of aerosol particles reduces their number concentration but their mass concentration remains unchanged. However, the geometric mean diameter of the particle size distribution is shifted towards bigger sizes as a result of coagulation. Coagulation is important when particle concentrations are relatively high and it can be ignored in indoor air models when the particle number concentration is below 10^4 cm^{-3} (e.g. Hussein et al. 2009).

8.4.1.2 Condensation of Vapours

The change rate of aerosol particle number concentration due to condensation vapours can be written as

$$J_{k,cond,i} = \frac{\dot{v}_{i-1}}{v_i - v_{i-1}} N_{i-1}(t) - \frac{\dot{v}_i}{v_{i+1} - v_i} N_i(t), \tag{8.9}$$

where v_i refers to the volume of particles in size section i. The change rate of particle volume is

$$\dot{v}_i = \sum_j 2\pi\beta_i v_{j,molec} \left(D_{p,i} + D_{j,molec}\right)\left(d_{p,i} + d_j\right)\left(C_j - C_{j,eq} Ke(D_{p,i})\right) \tag{8.10}$$

where the index j refers to the condensing vapour, $D_{j,molec}$ and $v_{j,molec}$ are the diameter and the volume of one vapour molecule and β_i is the transitional regime correction factor given by Fuchs and Sutugin (1971). The diffusion coefficients of the particles and vapour molecules are denoted with $d_{p,i}$ and d_j whereas C_j and $C_{j,eq}$ stand for the ambient and equilibrium concentrations of the vapor, respectively. The curvature effect of the particles is taken into account through the Kelvin factor $Ke(D_{p,i})$. According to UHMA, the condensation of water and ammonia is not calculated dynamically but their content in the particles is determined through thermodynamic equilibrium with the ambient air.

8.4.2 Indoor-Outdoor Air Exchange and the Penetration Process

The rate of incoming aerosols into an indoor compartment is

$$J_{k,P,i} = \frac{1}{V_k} \sum_m Q_{mk} P_{m,i} N_{out,i}, \tag{8.11}$$

where V_k is the volume (m^3) of compartment k and the summation is over the pathways that bring outdoor aerosol particles into the compartment, Q_{mk} denotes the air flow rate (m^{3}s^{-1}) that brings outdoor aerosol particles with a concentration $N_{out,i}$ (m^{-3}) via pathway m, and $P_{m,i}$ is the penetration factor of aerosol particles via that path.

Consequently, the rate of aerosol removal due to the outgoing air can be written as

$$J_{k,removal,i} = -\frac{1}{V_k} Q_{k,removed} N_{k,i}, \tag{8.12}$$

where V_k is again the volume (m^3) of compartment k, $Q_{k,removed}$ denotes the removed air flow rate (m^{3}s^{-1}), and $N_{k,i}$ is the aerosol particle number concentration (m^{-3}) in compartment k.

In the presence of mechanical ventilation system and assuming there is no air leakage across the building shell, $Q_{k,removed}$ and Q_{mk} are well controlled. However, in the presence of natural ventilation, $Q_{k,removed}$ and Q_{mk} are not well known and they are governed by several factors such as wind speed and direction and its turbulence, thermal buoyancy, the size and position of ventilation opening or leakage path, heat source and solar radiation, the conductance of the envelope, and etc. (e.g. Li and Delsante 2001; Dascalaki et al. 1996).

8.4.3 Internal Air Exchange Between Indoor Compartments

Internal air exchange between compartments can strongly influence indoor air pollutant concentrations. The driving forces of internal air exchange are pressure differences caused by winds, temperature differences, and fan operation. The internal air exchange of aerosol particle number concentration between indoor compartments can be written as

$$J_{k,exchange,i} = \frac{1}{V_k} \sum_j \left(Q_{jk} N_{j,i} - Q_{kj} N_{k,i} \right), \tag{8.13}$$

where V_k is again the volume (m^3) of compartment k and the summation is over all adjacent compartments. Q_{jk} is the air flow rate (m^3s^{-1}) from compartment j into compartment k, and $N_{j,i}$ and $N_{k,i}$ are the aerosol particle number concentrations (m^{-3}) in these compartments, respectively. In practice, it is not necessary that Q_{jk} and Q_{kj} are equal.

From the point of view of indoor air quality, internal air exchange has received too little attention. Specifically with respect to aerosol particles, a few studies have explored concentration variability among rooms and the factors that influence them (e.g. Hussein et al. 2006; Miller and Nazaroff 2001; Ju and Spengler 1981).

The internal air exchange between compartments can be very difficult to be measured. However, CFD modelling provides better understanding on the air flow inside a single-room dwelling or even inside a more complicated construction that consists of several rooms (e.g. Posner et al. 2003; Ziskind et al. 2002; Fan 1995; Gan 1995). Multiple-compartment air flow models have been widely used; some of these models are COMIS and CONTAM with a variety of applications (e.g. Ren and Stewart 2003; Haas et al. 2002; Borchielliini and Fürbringer 1999; Feustel 1999; Roulet et al. 1999; Walton 1997). There are also other methods presented in literature such as the lattice Boltzmann method (Crouse et al. 2002). Feustel (1999) presented a comprehensive review on different multiple-compartment air flow models and their applications.

8.4.4 Deposition and Re-suspension Processes

As mentioned before, the deposition rate of aerosol particles onto indoor surfaces can be described by

$$J_{k,d,i} = -\frac{1}{V_k} \sum_j A_{kj} v_{kj,d,i} N_{k,i}, \tag{8.14}$$

where A_{kj} is the total area (m^2) of the deposition surface j in compartment k. In the current indoor aerosol model, the deposition velocity, $v_{kj,d,i}$ (m s^{-1}), is estimated with the model developed by Lai and Nazaroff (2000). According to this deposition model, it is important to assume that the indoor air is well mixed within each compartment and the aerosol particle concentration gradient exists only very close to the deposition surfaces. This deposition model requires the friction velocity and particle size as input parameters. According to the deposition model, indoor surfaces are distinguished as upward oriented (ground), downward oriented (ceiling), and horizontal oriented (walls).

A fraction of aerosol particles deposited onto surfaces may undergo re-suspension. Ordinary indoor activities (including walking and housekeeping) can cause particle re-suspension and they may also generate new particles as a result of surface wear. The re-suspension process can be expressed mathematically by

$$J_{k,re,i} = \frac{1}{V_k} \sum_j f_{kj,i} A_{kj} \lambda_{kj,re,i} B_{kj,i}, \tag{8.15}$$

where $B_{kj,i}$ is the aerosol particle concentration (m^{-2}) accumulated on an indoor surface j of area A_{ki} (m^2). A fraction $f_{kj,i}$ of the accumulated particles are available for re-suspension from the surface with a re-suspension rate $\lambda_{kj,re,i}$ (s^{-1}). Consequently, the particle-number-balance equation that describes the change rate of aerosol particle number concentrations on an indoor surface is

$$A_{kj} \frac{d}{dt} B_{kj,i} = A_{kj} v_{kj,d,i} N_{k,i} - f_{kj,i} A_{ki} \lambda_{kj,re,i} B_{kj,i}, \tag{8.16}$$

The re-suspension process is not yet well understood. Only few studies considered this process (e.g. Ferro et al. 2004; Theerachaisupakij et al. 2003; Friess and Yadigaroglu 2002; Kulmala et al. 1999; Kildeso et al. 1999; Lazaridis and Drossinos 1998; Thatcher and Layton 1995), but there is not yet a generalized approach that can be utilized in the current version of MC-SIAM. However, the MC-SIAM with the semi-empirical method to estimate emission rates due to indoor sources of aerosol particles can be utilized to predict the re-suspension rate of deposited aerosol particles. Due to the complexity and the lack of experimental data, the re-suspension process is neglected in the current study, and we only mentioned the mathematical treatment for the completeness of the indoor aerosol model.

8.5 Mathematical Solutions and Applications of Indoor Air Models

The required input variables to simulate or predict the concentrations of indoor pollutants are: the pollutant concentrations in the outdoor air, penetration factor (or filtration efficiency), ventilation rate (or air exchange rate), deposition rate (or deposition velocity), dwelling geometries (volume, internal surface area including walls, ceilings, floors, and furniture), internal air-exchange rates between indoor compartments, and a diary of indoor activities. All these input parameters and variables are set-up for a numerical solution that solves for indoor particle concentrations of particle size-section in each compartment, $N_{k,i}$.

8.5.1 Analytical Solution for a Simple Indoor Air Model

Even though a numerical solution is always valid for all indoor air models, regardless to their degree of complexity, an analytical solution exists for a simple version of the models, which is described by Eq. (8.1), if we assume that the following:

- Penetration factor, ventilation rate, and deposition rate are all constant in time.
- The pollutant concentration is rather constant in the outdoor air.
- The indoor source has a constant emission rate.
- The re-suspension process is negligible.

The analytical solution would be

$$N_{in}(t) = N_{0,in} e^{-(\lambda+\lambda_d)t} + \frac{\lambda P N_{out} + S_{in}}{\lambda + \lambda_d}\left(1 - e^{-(\lambda+\lambda_d)t}\right), \qquad (8.17)$$

where $N_{0,in}$ is the initial concentration of the indoor pollutant at time $t = 0$. All other terms in Eq. (8.17) were introduced in Eq. (8.1).

In practice, the first two assumptions can be easily justified in chamber measurements or dwellings equipped with mechanical ventilation systems; that is well controlled indoor-to-outdoor relationship of aerosol particles. Consequently, the controlling parameters maintain rather constant values. In real life situations when the indoor-to-outdoor relationship of aerosol particles is not well controlled such as in dwellings with natural ventilation, the controlling parameters are not known and they vary with time. Therefore, the analytical solution is no longer valid and a numerical solution is needed. Figures 8.5 and 8.6 illustrate this analytical solution for total particle number concentrations inside a room.

Similarly, an analytical solution in the form of Eq. (8.17) can be obtained for each particle size-section described in a SIAM if we further assume that aerosol dynamics are negligible; that is non-interacting aerosol particles.

8.5.1.1 Steady-State Conditions

Usually, we are interested in a steady-state condition when the concentration of indoor pollutant maintain a constant value. In the absence of indoor sources; that is $S_{in} = 0$, the analytical solution presented in Eq. (8.17) can be reduced in the form

$$\text{I/O} = \left.\frac{N_{in}}{N_{out}}\right|_{Steady-State} \cong \frac{\lambda P}{\lambda + \lambda_d}, \qquad (8.18)$$

which defines the so called indoor-to-outdoor concentration (I/O) ratio of a pollutant in the absence of indoor sources. As illustrated in Fig. 8.5, the steady-state level for the I/O ratio is maintained after a certain time period, which is inversely proportional to the ventilation rate.

The difference between the temporal variations of the indoor and outdoor particle concentrations is known as the time-lag, which is illustrated in Fig. 8.7 for the response of the indoor particle concentration to the variation to the outdoor particle concentrations. Empirically, the steady-state of the I/O ratio can be evaluated from the measured indoor and outdoor particle concentrations after elimination of the

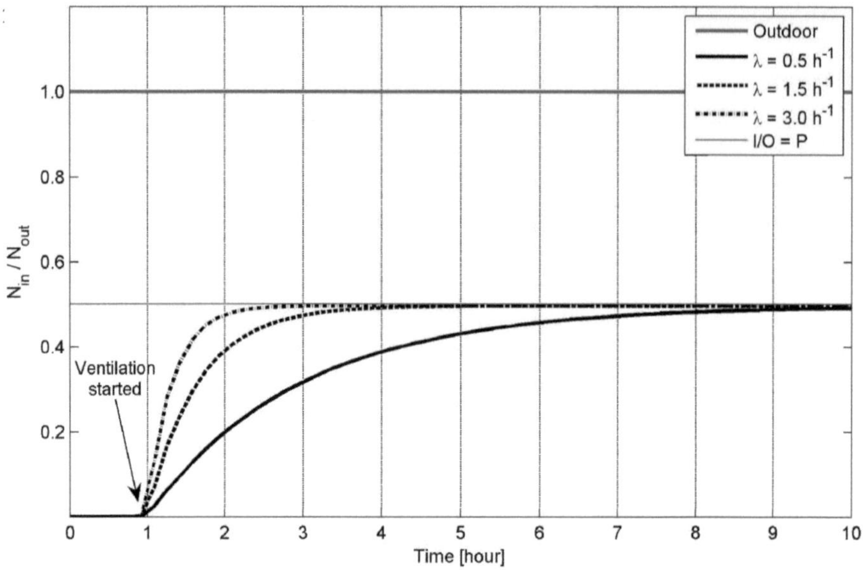

Fig. 8.5 Simple indoor air model simulations illustrate the effect of the ventilation factor (λ) on the total particle number concentrations inside an empty room (4 × 4 × 3 m³). Model parameters: outdoor particle concentrations are assumed constant N_{out} = 50,000 cm^{-3}, initial particle concentrations $N_{in,0}$ = 0 cm^{-3}, and P = 0.5. No indoor sources, re-suspension, or deposition. The ventilation process was turned on after 1 h. As result, the indoor particle number concentration increases to maintain a steady-state level. Because the deposition rate was neglected, the steady-state level for the indoor-to-outdoor concentration ratio is equivalent to the penetration factor. Clearly, the time needed to maintain this steady-state level is inversely proportional to the ventilation rate

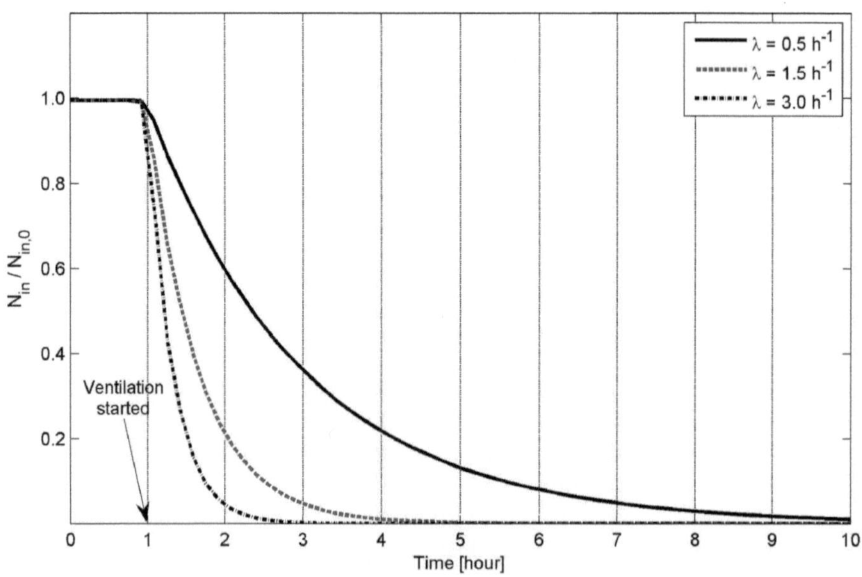

Fig. 8.6 Same as in Fig. 8.5 except that outdoor concentrations N_{out} = 0 cm^{-3} and initial particle concentrations $N_{in,0}$ = 50,000 cm^{-3}. When the mechanical ventilation is turned on, particle number concentrations are removed from the indoor air. The time needed to remove all indoor particles is proportional to the ventilation rate

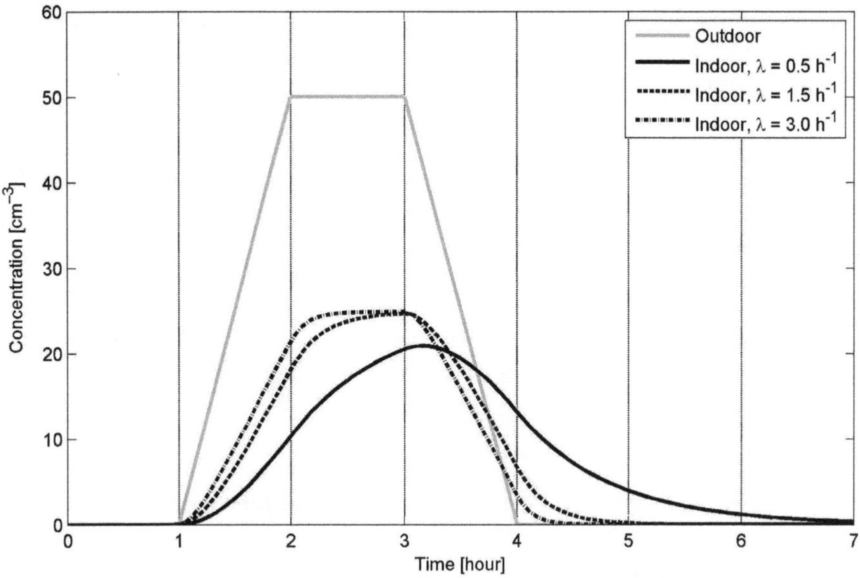

Fig. 8.7 Simple indoor air model simulations illustrate the time-lag between the indoor and outdoor particle number concentrations at different ventilation rates (λ) inside an empty room (4 × 4 × 3 m^3). Model parameters: $N_{in,0}$ = 0 cm^{-3}, and P = 0.5. No indoor sources, re-suspension, or deposition. The ventilation process is turned on all the time. As result, the indoor particle number concentration peaks at a later time; the higher the ventilation is the shorter the time-lag

time-lag between their temporal variations (e.g. Morawska et al. 2001). Note that in the example presented in Fig. 8.7, the steady-state of the I/O ratios were not maintained for the ventilation rates bellow 1.5 h^{-1}. On the other hand, at ventilation rate of 3 h^{-1} the steady-state was maintained with I/O ratio equal to the penetration factor as the deposition process was ignored.

8.5.1.2 The Infiltration Factor, INF

The I/O ratio must be distinguished from the infiltration factor (INF) and the penetration factor (*P*). As stated before, the penetration factor is the fraction of pollutants that pass across the building shell or via the ventilation system into the indoor air and it is less than unity. The INF is defined as the equilibrium fraction of ambient pollutants that penetrate indoors and remain suspended (Long et al. 2000); this is regardless to the air-exchange path. By definition, indoor sources of pollutants do not contribute to the INF, but they do contribute to the I/O ratio.

The I/O ratio and the INF are equivalent in the absence of indoor sources and they are both influenced by three main parameters: penetration factor, ventilation rate, and deposition rate (Kulmala et al. 1999). In other words, the value of the INF varies between 0 and 1 whereas the I/O ratio can have any value larger than zero.

If we further assume that the ventilation rate is much bigger than the deposition rate, then Eq. (8.18) indicates that both the INF and the I/O ratio are equivalent to the penetration factor

$$\left.\frac{N_{in}}{N_{out}}\right|_{Steady-State} \xrightarrow{\lambda >> \lambda_d} P, \tag{8.19}$$

which is already illustrated by Fig. 8.5.

In summary, the INF and the I/O ratios are the most important terms used to determine whether the indoor pollutants are from indoor or outdoor origin. On the other hand, the penetration factor is the most important parameter controlling the indoor air quality when the indoor pollutants are from outdoor origin.

8.5.2 Estimation of the Controlling Parameters with a SIAM

Numerically, the controlling parameters can be determined by iterating their values until a best-match is achieved between the measured and simulate indoor particle number size distributions. For example, the penetration factor can be iterated between 0 and 1 for all particle sizes. A good starting guess for the penetration factor is the empirical I/O ratio at steady-state conditions. Note that the I/O ratios may not maintain a steady-state condition when the outdoor concentrations are highly variable and the ventilation rate is low (Fig. 8.7).

To estimate the ventilation rate, we should pick-up a particle size-section between 0.1 and 1.0 μm (in diameter) while iterating values for the ventilation rate. Typically, aerosol particles in that size range have the lowest deposition rate because deposition processes (impaction, settling, and diffusion) are not efficient mechanisms for that particle size range; that is the deposition process term in the particle-number-balance equation is negligible. Once the penetration factor and the ventilation rate are estimated the deposition rate is easily obtained.

The ventilation rate and deposition rate of aerosol particles can be also estimated analytically if the analytical solution, which is in the form of Eq. (8.17),

$$N_{in,i}(t) = N_{0,in,i}e^{-(\lambda+\lambda_{d,i})t} + \frac{\lambda P N_{out,i} + S_{in,i}}{\lambda + \lambda_{d,i}}\left(1 - e^{-(\lambda+\lambda_{d,i})t}\right) \tag{8.20}$$

Assume an indoor source that generated huge amounts of aerosol particles that are significantly higher than the outdoor particle concentrations; that is $N_{in,i} >> N_{out,i}$. If the indoor source is turned off at time t_0, aerosol particle concentrations decay exponentially with time due to the removal processes (ventilation and deposition); as illustrated in Fig. 8.6. After a time period Δt Eq. (8.20) can be approximated to the form

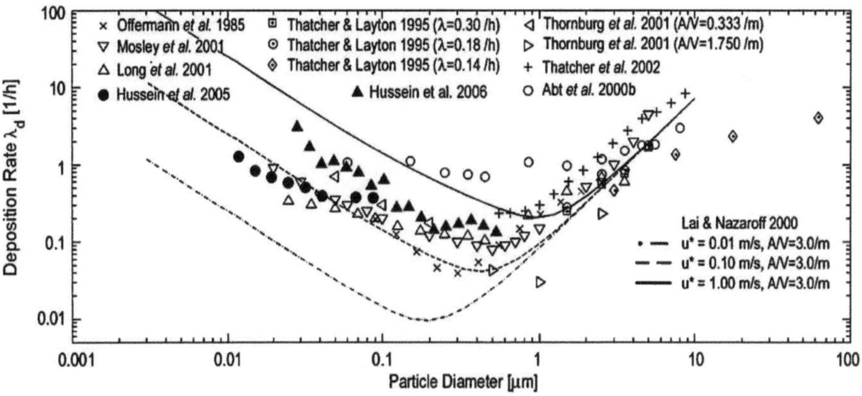

Fig. 8.8 Particle lose rates in different indoor environments. Note that each result was obtained with a different approach, measurement technique, and indoor conditions. This explains the differences between studies

$$\lambda_{d,i} + \lambda \cong \frac{1}{\Delta t} \ln\left(\frac{N_{in,i}(t_0)}{N_{in,i}(t_0 + \Delta t)}\right), \tag{8.21}$$

Again, this approximation also provides an estimate value for the ventilation rate if the ventilation rate is much bigger than the deposition rate of aerosol particles. For example, the deposition rate of particles between 0.1 and 1.0 μm in diameter is minimum and the particle loss rate of this particle size range provides an estimate for the ventilation rate. Typically, the deposition rate of aerosol particles in that diameter range varies between 0.02 and 0.1 h^{-1} (Fig. 8.8) and the ventilation rate is rarely lower than 0.2 h^{-1}.

8.5.3 Numerical Simulations

In our previous exercise, illustrated in Figs. 8.5 and 8.6, we assumed that all particles have similar physical-chemical properties, and therefore, we predicted the total particle number concentrations in the indoor air either by using a numerical or an analytical solution. However, the penetration and deposition processes depend on the particle size and we have to re-consider our simulation to group aerosol particles based on their physical-chemical properties; that is different particle size-sections.

In this exercise, we will assume typical urban traffic characteristics for the outdoor aerosol particles aiming to estimate the particle number concentrations inside a hypothetical room. Our room is rectangular (4 × 4 × 3 m^3) and it is not furnished. Assume the room is equipped with a mechanical ventilation ($\lambda = 3$ h^{-1}) with a standard class filter (F3, F6, or F7). We will assume a friction velocity

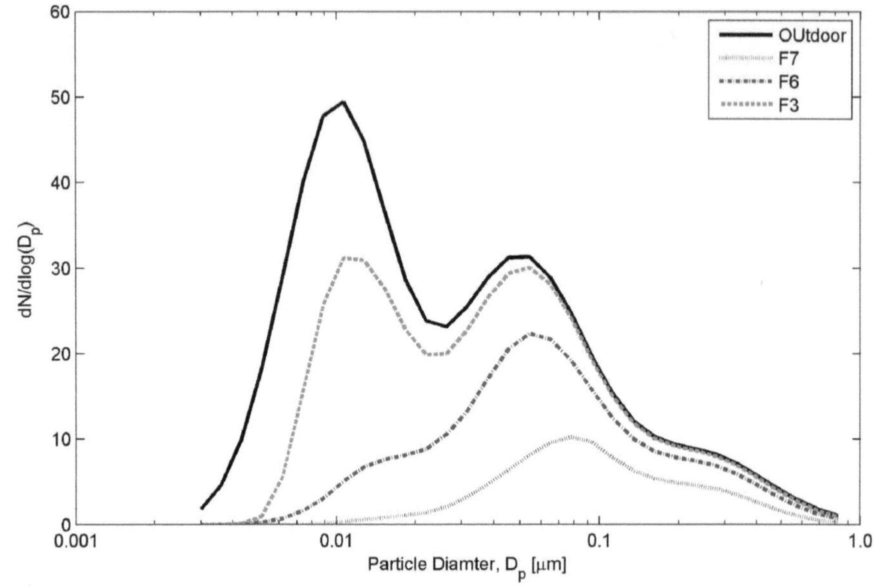

Fig. 8.9 Sectional indoor air model simulations illustrate the indoor particle number size distributions when using different class filters (F3, F6, and F7). The outdoor particle number size distribution is assumed constant (typical urban traffic aerosols) and the initial indoor concentrations is zero inside an empty room ($4 \times 4 \times 3$ m^3). The ventilation is mechanical ($\lambda = 3.0$ h^{-1}) and the friction velocity u* = 10 cm/s. It is assumed there are no indoor sources, re-suspension, or deposition

u* = 10 cm/s, which is not an actual value, for the deposition model by Lai and Nazaroff (2000). For the time being, we will neglect indoor sources, re-suspension, and other aerosol dynamics. As expected (Fig. 8.9), the F7 class filter provides the best indoor air quality by filtering a big fraction of ultrafine particles (UFP, D_p < 100 nm).

So far, our model simulations have been far from being complicated; only because we assumed ideal situations, which can be easily achieved in chamber measurements. However, model simulations are far from being simple in real-life situations. For example, the penetration factor, even for standard filters installed in mechanical ventilation systems, may deviate from its standard values and the ventilation rate may also fluctuate around the mean value. A major difficulty arises when estimating the deposition rate of aerosol particles onto indoor surfaces due to the lack of information about the indoor surfaces including their type, surface area, orientation, etc. When it comes to natural ventilation, these parameters are hardly estimated empirically with acceptable accuracy. Therefore, we utilize indoor air models to estimate the penetration factor, ventilation rate, and deposition rate (or fate) of indoor aerosols.

We mentioned these limitations for the penetration factor, ventilation rate, and deposition rate because they are the main factors that control the indoor-to-outdoor relationship of aerosol particles. These are not the only limitations when it comes to other processes such as re-suspension, indoor activities, new particle formation, chemical-physical properties, etc. When we gain good knowledge about these processes and related parameters, we are able to predict the indoor particle concentrations with acceptable accuracy. Here "acceptable accuracy" is the main key factor to introduce the user-influence on the model accuracy.

8.5.4 Indoor Sources and Emission Rate Estimation

Indoor sources of aerosol particles are usually produced during inhabitant's activities. Therefore, indoor sources can be classified based on the activity type. Indoor fine particles are generated by means of gas-to-particle conversion processes that are often associated with high temperature during combustion and cooking. For example, fine particles may consist of significant amounts of soot and metals. Coarse particles tend to be produced by mechanical means and they may contain common soil elements. Bio-aerosols including allergens, fungi, bacteria, and viruses can be either coarse of fine (e.g. Meklin et al. 2002; Lee et al. 2002; Otten and Burge 1999; Wanner 1993; Platts-Mills et al. 1991).

Even though many studies discussed indoor sources of aerosol particles (see for example, Hussein 2005 and Hussein et al. 2005, 2006 for detailed discussion and review of related studies) only a few focused on the particle number size distributions and even fewer presented quantitative determination of aerosol particle emissions from indoor sources (e.g. Hussein et al. 2005, 2006; Afshari et al. 2005; He et al. 2004; Fan and Zhang 2001). Quantitative determination of indoor aerosol particle emissions can be easily estimated based on semi-empirical methods. For example, the indoor aerosol particle concentration measurements quantify the suspended aerosol particles resulting from all possible sources and sinks. Similarly, simulations of indoor particle concentrations illustrate the amount of aerosol particles after being engaged in the processes included in the indoor aerosol model. The difference between the measured and simulated particle concentrations provides the initial step to determine the emission rate.

8.5.4.1 Semi-empirical Estimation of the Emission Rate

Assume there is an indoor source of aerosol particles and we know the best-fit values of the controlling parameters for the indoor-to-outdoor relationship of aerosol particles, then we can predict the number concentrations of indoor aerosol particles by ignoring the indoor sources. In this case, the difference between the measured particle concentrations and this simulation (with ignored indoor sources) represents the suspended particle concentrations due to that source. However, the difference between measurement and simulation should be corrected for particle removal and aerosol dynamic processes.

Based on that simple principle, the semi-empirical method is summarized as follows:

- Find a good estimate for the controlling parameters if they are not known before hand.
- Perform a model simulation to predict the indoor particle concentrations by ignoring the indoor source.
- Evaluate the difference between the measured (including indoor sources) and simulated aerosol concentrations (ignoring indoor sources) and let it represent the suspended aerosol particle concentrations $S_{k,Dp}$ (1/m^3) due to the emission processes of indoor activities.

$$S_{k,i} = N_{k,i,measured} - N_{k,i,simulated}. \tag{8.22}$$

For simplicity, we will assume that all indoor sources of aerosol particles are included in the term $S_{k,i}$.

- The change rate of aerosol particle number concentrations that remain suspended is

$$\frac{d}{dt}S_{k,i} = J_{k,emission,i} + J_{k,coag,i} + J_{k,cond,i} - \frac{1}{V_k}\sum_j A_{kj} v_{kj,d,i} S_{k,i} - \frac{1}{V_k}\sum_l Q_{k,removed} S_{k,i} \tag{8.23}$$

For which the emission rate terms $J_{k,emission,i}$ can be estimated numerically.

- Re-perform another model simulation with feed back of the emission rates obtained from the previous step and then adjust the emission rate according to Eqs. (8.22) and (8.23).
- The last step can be repeated as many times as needed until convergent solution is obtained for the emission rate term.

This principle can also provide quantification for aerosol particle losses due to un-identified sinks indoors, which are indicated by negative values in the emission rate term.

Recently, Ren and Stewart (2003) have presented an application of the international software COMIS to divide the indoor space into smaller virtual compartments to allow for better estimation of the temperature gradients indoors. Their method can be extended to estimate the emission rates of indoor pollutants.

References

Abadie, M., Limam, K., & Allard, F. (2001). Indoor particle pollution: Effect of wall textures on particle deposition. *Building and Environment, 36*, 821–827.

Abt, E., Suh, H. H., Catalano, P., & Koutrakis, P. (2000). Relative contribution of outdoor and indoor particle sources to indoor concentrations. *Environmental Science and Technology, 34*, 3579–3587.

Afshari, A., Matson, U., & Ekberg, L. E. (2005). Characterization of indoor sources of fine and ultrafine particles: A study conducted in a full-scale chamber. *Indoor Air, 15*, 141–150.

Alzona, J., Cohen, B. L., Rudolph, H., Jow, H. N., & Frohliger, J. O. (1979). Indoor-outdoor relationships for airborne particulate matter of outdoor origin. *Atmospheric Environment, 13*, 55–60.

Asmi, A. J., Pirjola, L. H., & Kulmala, M. A. (2004). Sectional aerosol model for submicron particles in indoor air. *Scandinavian Journal of Work, Environment and Health, 30*(Suppl 2), 63–72.

Borchiellini, R., & Fürbringer, J.-M. (1999). An evaluation exercise of a multizone air flow model. *Energy and Buildings, 30*, 35–51.

Corner, B. J., & Pendlebury, E. D. (1951). The coagulation and deposition of a stirred aerosol. *Proceedings of the Physical Society, B64*, 645–654.

Crouse, B., Krafczyk, M., Kühner, S., Rank, E., & van Treeck, C. (2002). Indoor air flow analysis based on lattice Boltzmann methods. *Energy and Buildings, 34*, 941–949.

Dascalaki, E., Santamouris, M., Argiriou, A., Helmis, C., Asimakopoulos, D. N., Papadopoulos, K., et al. (1996). On the combination of air velocity and flow measurements in single sided natural ventilation configurations. *Energy and Buildings, 24*, 155–165.

Fan, Y. (1995). CFD modelling of the air and contaminant distribution in rooms. *Energy and Buildings, 23*, 33–39.

Fan, C. W., & Zhang, J. J. (2001). Characterization of emissions from portable household combustion devices: Particle size distributions, emission rates and factors, and potential exposures. *Atmospheric Environment, 35*, 1281–1290.

Ferro, A. R., Kopperud, R. J., & Hildemann, L. M. (2004). Source strengths for indoor human activities that resuspend particulate matter. *Environmental Science and Technology, 38*, 1759–1764.

Feustel, H. E. (1999). COMIS–an international multizone air-flow and contaminant transport model. *Energy and Buildings, 30*, 3–18.

Fogh, C. L., Byrne, M. A., Roed, J., & Goddard, A. J. H. (1997). Size specific indoor aerosol deposition measurements and derived I/O concentrations ratios. *Atmospheric Environment, 31*, 2193–2203.

Friess, H., & Yadigaroglu, G. (2002). Modeling of the resuspension of particle clusters from multilayer aerosol deposits with variable porosity. *Journal of Aerosol Science, 33*, 883–906.

Fuchs, N. A. (1964). *The mechanics of aerosols*. New York: Dover.

Fuchs, N. A., & Sutugin, A. G. (1971). Highly dispersed aerosol. In G. M. Hidy & J. R. Brock (Eds.), *Topics in current aerosol research*. New York: Pergamon.

Gan, G. (1995). Evaluation of room air distributions systems using computational fluid dynamics. *Energy and Buildings, 23*, 83–93.

Goodfellow, H., & Tähti, E. (2001). *Industrial ventilation: Design guidebook* (p. 685). California: Academic (Gustavsson, J. [1996]. *Cabin air filters: Performance and requirements*. SAE Conference, Detroit, February 1996).

Guha, A. (1997). A unified Eulerian theory of turbulent deposition to smooth and rough surfaces. *Journal of Aerosol Science, 28*, 1517–1537.

Haas, A., Weber, A., Dorer, V., Keilholz, W., & Pelletret, R. (2002). COMIS v3.1 simulation environment for multizone air flow and pollutant transport modelling. *Energy and Buildings, 34*, 873–882.

Hanley, J. T., Ensor, D. S., Smith, D. D., & Sparks, L. E. (1994). Fractional aerosol filtration efficiency of in duct ventilation air cleaners. *Indoor Air, 4*, 169–178.

He, C., Morawska, L., Hitchins, J., & Gilbert, D. (2004). Contribution from indoor sources to particle number and mass concentrations in residential houses. *Atmospheric Environment, 38*, 3405–3415.

Hinds, W. C. (1999). *Aerosol technology* (2nd ed.). New York: Wiley.

Hussein, T. (2005). *Indoor and outdoor aerosol particle size characterization in Helsinki*. Report Series in Aerosol Science No: 74. Helsinki, Finland: Finnish Aerosol Association Research.

Hussein, T., Glytsos, T., Ondrá ek, J., Ždímal, V., Hämeri, K., Lazaridis, M., et al. (2006). Particle size characterization and emission rates during indoor activities in a house. *Atmospheric Environment, 40*, 4285–4307.

Hussein, T., Hruška, A., Dohányosová, P., Džumbová, L., Hemerka, J., & Kulmala, M., et al. (2009a). Evaluation of deposition rates of aerosol particles on smooth surfaces inside a test chamber. *Atmospheric Environment, 43*, 905–914.

Hussein, T., Korhonen, H., Herrmann, E., Hämeri, K., Lehtinen, K., & Kulmala, M. (2005b). Emission rates due to indoor activities: Indoor aerosol model development, evaluation, and applications. *Aerosol Science and Technology, 39*(11), 1111–1127.

Hussein, T., Kubincová, L., Dohányosová, P., Hruška, A., Džumbová, L., Hemerka, J., et al. (2009b). Deposition of aerosol particles on rough surfaces inside a test chamber. *Buildings and Environment, 44*, 2056–2063.

Jamriska, M., Morawska, L., & Ensor, D. S. (2003). Control strategies for sub-micrometer particles indoors: Model study of air filtration and ventilation. *Indoor Air, 13*, 96–105.

Ju, C., & Spengler, J. D. (1981). Room to room variations in concentration of respirable particles in residences. *Environmental Science and Technology, 15*, 592–596.

Kildeso, J., Vinzents, P., Schneider, T., & Kloch, N. P. (1999). A simple method for measuring the potential resuspension of dust from carpets in the indoor environment. *Textile Research Journal, 69*, 169–175.

Korhonen, H., Lehtinen, K. E. J., & Kulmala, M. (2004). Aerosol dynamic model UHMA: Model development and validation. *Atmospheric Chemistry and Physics, 4*, 757–771.

Kulmala, M., Asmi, A., & Pirjola, L. (1999). Indoor air aerosol model: The effect of outdoor air, filtration and ventilation on indoor concentrations. *Atmospheric Environment, 33*, 2133–2144.

Lai, A. C. K. (2006). Investigation of electrostatic forces on particle deposition in a test chamber. *Indoor Built Environment, 15*, 179–186.

Lai, A. C. K., & Nazaroff, W. W. (2000). Modeling indoor particle deposition from turbulent flow onto smooth surfaces. *Journal of Aerosol Science, 31*, 463–476.

Lai, A. C. K., Byrne, M. A., & Goddard, A. J. H. (2001). Aerosol deposition in turbulent channel flow on a regular array of three-dimensional roughness elements. *Aerosol Science, 32*, 121–137.

Lai, A. C. K., Byrne, M. A., & Goddard, A. J. H. (2002). Experimental studies of the effect of rough surfaces and air speed on aerosol deposition in a test chamber. *Aerosol Science and Technology, 36*, 973–982.

Lazaridis, M., & Drossinos, Y. (1998). Multilayer resuspension of small identical particles by turbulent flow. *Aerosol Science and Technology, 28*(6), 548–560.

Lee, S.-C., Guo, H., Li, W.-M., & Chan, L.-Y. (2002). Inter-comparison of air pollutant concentrations in different indoor environments in Hong Kong. *Atmospheric Environment, 36*, 1929–1940.

Li, Y., & Delsante, A. (2001). Natural ventilation induced by combined wind and thermal forces. *Buildings and Environment, 36*, 59–71.

Liu, D.-L., & Nazaroff, W. W. (2001). Modeling pollutant penetration across building envelopes. *Atmospheric Environment, 35*, 4451–4462.

Long, C. H., Suh, H. H., & Koutrakis, P. (2000). Characterization of indoor particle sources using continuous mass and size monitors. *Journal of the Air & Waste Management Association, 50*, 1236–1250.

Long, C. M., Suh, H. H., Catalano, P. J., & Koutrakis, P. (2001). Using time- and size-resolved particulate data to quantify indoor penetration and deposition behavior. *Environmental Science and Technology 35*, 2089–2099.

Lum, R. M., & Graedel, T. E. (1973). Measurements and models of indoor aerosol size spectra. *Atmospheric Environment, 7*, 827–842.

McMurry, P. H., & Radar, D. J. (1985). Aerosol wall losses in electrically charged chambers. *Aerosol Science and Technology, 4*, 249–268.

Meklin, T., Reponen, T., Toivola, M., Koponen, V., Husman, T., Hyvärinen, A., et al. (2002). Size distributions of airborne microbes in moisture-damaged and reference school buildings of two construction types. *Atmospheric Environment, 36*, 6031–6039.

Miller, S. L., & Nazaroff, W. W. (2001). Environmental tobacco smoke particles in multizone indoor environments. *Atmospheric Environment, 35*, 2053–2067.
Morawska, L., He, C., Hitchins, J., Gilbert, D., & Parappukkaran, S. (2001). The relationship between indoor and outdoor airborne particles in the residential environment. *Atmospheric Environment, 35*, 3463–3473.
Mosley, R. B., Greenwell, D. J., Sparks, L. E., Guom, Z., Tucker, W. G., Fortmann, R., et al. (2001). Penetration of ambient fine particles into the indoor environment. *Aerosol Science and Technology, 34*, 127–136.
Nazaroff, W. W., & Cass, G. R. (1986). Mathematical modeling of chemically reactive pollutants in indoor air. *Environmental Science and Technology, 20*, 924–934.
Nazaroff, W. W., & Cass, G. R. (1989). Mathematical modeling of indoor aerosol dynamics. *Environmental Science and Technology, 23*, 157–166.
Nazaroff, W. W. (2004). Indoor particle dynamics. *Indoor Air, 14*(Suppl. 7), 175–183.
Otten, J. A., & Burge, H. A. (1999). Bacteria. In J. Macher (Ed.), *Bioaerosols, assessment and control* (pp. 183–1810). Cincinnati, OH: American Conference of Governmental Industrial Hygienists.
Platts-Mills, T. A. E., Ward, G. W., Sporik, R., Gelber, L. E., Champman, M. D., & Heymann, P. W. (1991). Epidemiology of the relationship between exposure to indoor allergins and asthma. *International Archives of Allergy and Applied Immunology, 87*(2), 505–510.
Porstendörfer, J., & Reineking, A. (1992). Indoor behavior and characteristics of radon progeny. *Radiation Protection Dosimetry, 45*, 303–311.
Posner, J. D., Buchanan, C. R., & Dunn-Rankin, D. (2003). Measurement and prediction of indoor air flow in a model room. *Energy and Buildings, 35*, 515–526.
Raunemaa, T., Kulmala, M., Saari, H., Olin, M., & Kulmala, M. H. (1989). Indoor air aerosol model: Transport indoors and deposition of fine and coarse particles. *Aerosol Science and Technology, 11*, 11–25.
Ren, Z., & Stewart, J. (2003). Simulating air flow and temperature distribution inside buildings using a modified version of COMIS with sub-zonal divisions. *Energy and Buildings, 35*, 257–271.
Riley, W. J., Mckone, T. E., Lai, A. C. K., & Nazaroff, W. W. (2002). Indoor particulate matter of outdoor origin: Importance of size-dependent removal mechanisms. *Environmental Science and Technology, 36*, 200–207.
Roulet, C.-A., Fürbringer, J.-M., & Creton, P. (1999). The influence of the user on the results of multizone air flow simulations with COMIS. *Energy and Buildings, 30*, 73–86.
Schneider, T., Kildeso, J., & Breum, N. O. (1999). A two-compartment model for determining the contribution of sources, surface deposition and resuspension to air and surface dust concentration levels in occupied rooms. *Building and Environment, 34*, 583–595.
Seinfeld, H. S., & Pandis, S. N. (1998). *Atmospheric chemistry and physics: From air pollution to climate change* (2nd ed.). New York: Wiley.
Shimada, M., Okuyama, K., Kousaka, Y., Okuyama, Y., & Seinfeld, J. H. (1989). Enhancement of Brownian and turbulent diffusive deposition of charged particles in the presence of an electric field. *Journal of Colloid and Interface Science, 128*, 157–168.
Sippola, R., & Nazaroff, W. W. (2003). Modeling particle loss in ventilation ducts. *Atmospheric Environment, 37*, 5597–5609.
Thatcher, T. L., Lai, A. C. K., Moreno-Jackson, R., Sextro, R. G., & Nazaroff, W. W. (2002). Effects of room furnishings and air speed on particle deposition rates indoors. *Atmospheric Environment, 36*, 1811–1819.
Thatcher, T. L., & Layton, D. W. (1995). Deposition, resuspension, and penetration of particles within a residence. *Atmospheric Environment, 29*, 1487–1497.
Theerachaisupakij, W., Matsusaka, S., Akashi, Y., & Masuda, H. (2003). Reentrainment of deposited particles by drag and aerosol collision. *Journal of Aerosol Science, 34*, 261–274.
Thornburg, J., Ensor, D. S., Rodos, C. E., Lawless, P. A., Sparks, L. E., & Mosley, R. B. (2001). Penetration of particles into buildings and associated physical factors – Part I: Model development and computer simulations. *Aerosol Science and Technology, 34*, 284–296.

Tung, T. C. W., Chao, C. Y. H., & Burnett, J. (1999). A methodology to investigate the particulate penetration coefficient through building shell. *Atmospheric Environment, 33*, 881–893.

Vanmarcke, H., Landsheere, C., Van Dingenen, R., & Poffijn, A. (1991). Influence of turbulence on the deposition rate constant of the unattached radon decay products. *Aerosol Science and Technology, 14*, 257–265.

Vartiainen, E., Kulmala, M., Ruuskanen, T. M., Taipale, R., Rinne, J., & Vehkamäki, H. (2006). Formation and growth of indoor air aerosol particles as a result of D-limonene oxidation. *Atmospheric Environment* (in press), corrected proof.

Walton, G. N. (September 1997). *CONTAM96 User manual*. Report NSITIR 6056. Gaithersburg: US Department of Commerce, National Institute of Standards and Technology.

Wanner, H. U. (1993). Sources of pollutants in indoor air. *IARC Scientific Publications, 109*, 19–30.

Zhao, B., & Wu, J. (2006a). Modeling particle deposition from fully developed turbulent flow in ventilation duct. *Atmospheric Environment, 40*, 457–466.

Zhao, B., & Wu, J. (2006b). Modeling particle deposition onto rough walls in ventilation duct. *Atmospheric Environment, 40*, 6918–6927.

Ziskind, G., Dubovsky, V., & Letan, R. (2002). Ventilation by natural convection of a one-story building. *Energy and Buildings, 34*, 91–102.

Further Reading

Abadie, M., Limam, K., Bouilly, J., & Génin, D. (2004). Particle pollution in the French high-speed train (TGV) smoker cars: Measurement and prediction of passengers exposure. *Atmospheric Environment, 38*, 2017–2027.

Cheng, Y. S., Bechtold, W. E., Yu, C. C., & Hung, I. F. (1995). Incense smoke: Characterization and dynamics in indoor environments. *Aerosological Science and Technology, 23*, 271–281.

Chen, F., Yu, S. C. M., & Lai, A. C. K. (2006). Modeling particle distribution and deposition in indoor environments with a new drift-flux model. *Atmospheric Environment, 40*, 357–367.

Cole, C. (1998). Candle Soot deposition and its impacts on restorers, USA, Sentry Construction Company.

Dennekamp, M., Howarth, S., Dick, C. A. J., Cherrie, J. W., Donaldson, K., & Seaton, A. (2001). Ultrafine particles and nitrogen oxides generated by gas and electric cooking. *Occupational and Environmental Medicine, 58*, 511–516.

Fine, P. M., Cass, G. R., & Simoneit, B. R. T. (1999). Characterization of fine particle emissions from burning church candles. *Environmental Science and Technolology, 33*, 2352–2362.

Flückiger, B., Seifert, M., Koller, T., & Monn, C. (2000). Air quality measurements in a model kitchen using gas and electric stoves. *Proceedings of Healthy Buildings 2000, 1*, 567–572.

Helsper, C., Moltr, W., Loffler, F., Wadenpohl, C., Kaufmann, S., & Wenninger, G. (1993). Investigation of a non-aerosol generator for the production of carbon aggregate particles. *Atmospheric Environment, 27A*, 1271–1279.

Howard-Reed, C., Wallace, L. A., & Emmerich, S. J. (2003). Effect of ventilation system and air filters on decay rates of particles produced by indoor sources in an occupied townhouse. *Atmospheric Environment, 37*, 5295–5306.

Hussein, T., Hämeri, K., Heikkinen, M. S. A., & Kulmala, M. (2005a). Indoor and outdoor particle size characterization at a family house in Espoo – Finland. *Atmospheric Environment, 39*, 3697–3709.

Jones, A. P. (1999). Indoor air quality and health. *Atmospheric Environment, 33*, 4535–4564.

Kemens, R., Lee, C.-T., Wiener, R., & Leith, D. (1991). A study to characterize indoor particles in three non-smoking homes. *Atmospheric Environment, 25A*, 939–948.

Kleeman, M. J., Schauer, J. J., & Cass, G. R. (2000). Size and composition of fine particulate matter emitted from motor vehicles. *Environmental Science and Technology, 34*, 1132–1142.

Klepeis, N. E., Apte, M. G., Gundel, L. A., Sextro, R. G., & Nazaroff, W. W. (2003). Determining size-specific emission factors for environmental tobacco smoke particles. *Aerosological Science and Technology, 37*, 780–790.

Lai, A. C. K. (2004). Modeling of airborne particle exposure and effectiveness of engineering control strategies. *Building and Environment, 39*, 599–610.

Li, W., & Hopke, P. K. (1993). Initial size distributions and hygroscopicity of indoor combustion aerosol particles. *Aerosological Science and Technology, 19*, 305–316.

Li, C. S., Lin, W. H., & Jenq, F. T. (1993). Size distributions of submicrometer aerosols from cooking. *Environment International, 19*, 147–154.

Lioy, P. J., Wainman, T., & Zhang, J. J. (1999). Typical household vacuum cleaners, the collection efficiency and emissions characteristics for fine particles. *Journal of the Air and Waste Management Association, 49*, 200–206.

Luoma, M., & Batterman, S. A. (2001). Characterization of particulate emissions from occupant activities in offices. *Indoor Air, 11*, 35–48.

Morawska, L., He, C., Hitchins, J., Mengersen, K., & Gilbert, D. (2003). Characteristics of particle number and mass concentrations in residential houses in Brisbane, Australia. *Atmospheric Environment, 37*, 4195–4203.

Schauer, J. J., Kleeman, M. J., Cass, G. R., & Simoneit, B. R. T. (1999). Measurement of emissions from air pollution sources. 1. C_1 through C_{29} organic compounds from meat charbroiling. *Environmental Science and Technology, 33*, 1566–1577.

Siegmann, K., & Sattler, K. (1996). Aerosol from hot cooking oil, a possible health hazard. *Journal of Aerosol Science, 27*, 493–494.

Sohn, M. D., Lai, A., Smith, B. V., Sextro, R. G., Feustel, H. E., & Nazaroff, W. W. (1999). Modeling aerosol behavior in multizone indoor environments. *Proceedings of Indoor Air'99 Edinburgh, 4*, 785–790.

Schneider, T., Jensen, K. A., Clausen, P. A., Afshari, A., Gunnarsen, L., Wåhlin, P., et al. (2004). Prediction of indoor concentration of 0.5–4 mm particles of outdoor origin in an uninhabited apartment. *Atmospheric Environment, 38*, 6349–6359.

Wallace, L. (2000). Real-time monitoring of particles, PAH, and CO in an occupied townhouse. *Applied Occupational and Environmental Hygiene, 15*, 39–47.

Wallace, L., & Howard-Reed, C. (2002). Continuous monitoring of ultrafine, fine and coarse particles in a residence for 18 months in 1999–2000. *Journal of the Air and Waste Management Association, 52*, 828–844.

Chapter 9
Air Quality Management and Personal Exposure

Trond Bøhler

Abstract Atmospheric pollution occurs on different spatial and temporal scales. On the macro scale, global problems of importance are the depletion of the stratospheric ozone layer and global warming, which is caused by the emission and accumulation of greenhouse gases in the atmosphere. On a regional scale, the transboundary transport of pollutants can be responsible for acid deposition or the formation of photochemical smog. Air quality on a local and urban scale is closely related to emissions arising from almost all human activities, and to local characteristics such as topography, climate and also economy. Aspects of air quality investigated on a micro-scale include studies on emissions from a single point source, indoor pollution from oil or wood burning stoves, or the dispersion of pollutants in a street canyon.

9.1 Introduction

Many epidemiological investigations have demonstrated that the exposure of the population to air pollution causes adverse health effects, which can be divided into acute or chronic effects. The level of impact will depend on the duration and level of exposure (Dockery et al. 1993; Neuberger et al. 2004).

Pollutants enter the human body in three main different ways: by inhalation, ingestion or skin absorption. The amount of any given pollutant that is received is often termed the dose. The dose will be dependent on the duration and intensity of the exposure (ICRP 2002; Lazaridis et al. 2001).

In many cases, exposure may occur simultaneously from many sources and through multiple routes. Exposure assessment therefore typically requires detailed

T. Bøhler (✉)
Norwegian Institute for Air Research, PO Box 100, N-2007 Kjeller, Norway
e-mail: trond.bohler@nilu.no

M. Lazaridis and I. Colbeck (eds.), *Human Exposure to Pollutants via Dermal Absorption and Inhalation*, Environmental Pollution 17, DOI 10.1007/978-90-481-8663-1_9,

knowledge about the geographical distribution of the pollutants of concern, the temporal variations in pollution levels, and the processes of exposure.

People are often exposed to different pollutants simultaneously. Exposure to these may occur at different locations like in the workplace and at home at different times of the day and day of the week. The full range of factors that therefore may need to be examined is potentially large. It may include many different environmental pollutants from many different sources released either continuously or sporadically, and either under controlled conditions or accidentally.

In order to implement an appropriate strategy for improving local air quality, local authorities should establish the necessary air quality management system combining air quality monitoring network with atmospheric dispersion modeling to determine the levels of pollutants and identify the sources.

Monitoring enables local authorities to confirm previous assumptions that pollution levels are unacceptably high; it also enables them to set priorities and for improvement according to the resources available and the cost and benefits of particular courses of action. Long-term monitoring programmes show the efficiency of the policies put in place thus enabling the re-allocation of resources if this is thought appropriate.

9.2 Air Quality Management and Assessment

Before an effective and efficient air quality strategy can be developed, the current situation has to be assessed in detail, and all influential aspects and available resources have to be taken into account. The fundamental tool in air quality assessment is monitoring of pollutant concentrations in different locations in combination with determination of emissions, meteorological data and societal factors. It is important to take seasonal variances, trends and economic aspects, as well as existing and planned legislation, into account. However, the role of monitoring does not end after one air quality problem has been correctly identified and assessed; continuous provision of reliable data is necessary for any successful longer-term environmental management.

In Europe, as elsewhere, a variety of air quality management systems have been developed by cities in response to their perceived needs. These management strategies are at different stages of implementation and enforcement, and this gives rise to considerable differences in the capacities of cities to control air pollution. The assessment and comparison of air quality management capabilities is a complex task which needs to take a wide range of aspects into account. It can be assessed by answering the key questions (WHO 2006):

- Were appropriate air quality objectives set?
- Is representative and dependable data available and used in a suitable manner?
- Does the administrative and legislative framework allow the implementation and enforcement of control strategies?

Finding the answers to these three questions will indicate how the use of existing management capabilities and available human, technical and financial resources can be optimised. However, due to its diversity, the subject does not lend itself to finding easy and straightforward solutions. Answering these questions is an exercise that requires the application of a technique which facilitates an assessment on a qualitatively and quantitatively appropriate level.

Measurement of the concentrations of individual chemical species or a mixture of these in the atmosphere over a period of time, are the basic steps in air quality monitoring. Statistical analysis and evaluation of the collected data and the application of quality assurance and quality control (QA/QC) systems are essential to affirm the validity and reliability of the monitored data. The generation of decision relevant and reliable, with a sufficient temporal and spatial resolution that allows the identification of problems, is fundamental for successful air quality management.

9.3 Management Systems

An Environmental Management System is a set of processes and practices that enables an organisation to reduce its environmental impacts and increase the efficiency with which environmental policies are implemented. This is done by integrating fully the environmental objectives into the overall operation of the authority. A management system therefore helps ensure that strategies or plans (e.g., environmental management plan, sustainable urban transport plan) are implemented.

An air quality management system should provide the basis for air quality management through an integrated tool for emission inventorying, air quality modelling and assessment, enabling forecasting of future air quality and development of cost-effective abatement strategies. A brief technical description of such an air quality management system is given in the following chapters.

A modern air quality management system (AQMS) consists of several components and makes use of integrated menus and GIS to obtain a user friendly system (Fig. 9.1). As a minimum, the AQMS system should consists of:

1. An ambient air quality network
2. A measurement data base for meteorology and air quality
3. A modern emission inventory data base with emission models
4. Numerical models for transport and dispersion in air of pollutants
5. A module for exposure estimates and population exposure assessment
6. Statistical evaluation of measurements and modelling results

9.3.1 Design of Air Quality Monitoring Networks

The design and subsequent of a new air quality monitoring network, or the review and adaptation of an existing system, is a complex task. Finding the most appropriate and cost-effective solution under the special circumstances, which govern the

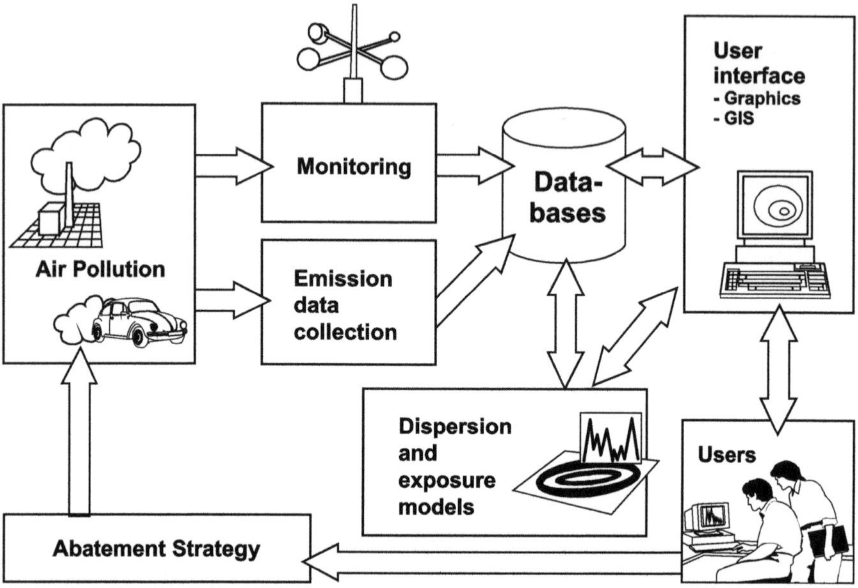

Fig. 9.1 The modules of an integrated air quality monitoring and management system

process of establishing and running a specific network, starts with setting the monitoring objectives. These will help to determine the sampling equipment, site selection n and the overall network design.

Typical monitoring objectives are (WHO 2002):

- Establishing a sound scientific basis for policy development
- Determining compliance with statutory criteria
- Assessment of population/ecosystem exposure
- Public information
- Identification of pollution sources or risks
- Evaluation of long term trends

Definition of priority pollutants, necessary data accuracy and time resolution, and consideration of available resources(capital, manpower, space) will restrict the choice of employable measurement methodologies and thus aid the selection of monitoring equipment. Air monitoring methods can be divided into five groups:

- Passive samplers
- Active samplers
- Automatic monitors for on-line information
- Remote sensors for multi component measurements and
- Bioindicators

Samples taken by both active and passive samplers, as well as many bioindicators require subsequent laboratory analysis, which can be labour intensive and expensive.

The density and resolution of the monitoring network and the location of the individual sites are the next point to be detected. Requirements which have to be taken into account are (WHO 2002):

- Monitoring objectives
- Pollutant sources and emissions in the area concerned
- Meteorology and topography
- Existing air quality data
- Model simulations
- Demographic/health/land use data
- Access and security of locations
- Site sheltering and
- Infrastructure (power supply, telephone etc.)

Once the implementation of the monitoring network is completed and is in operation, quality assurance and control programmes have to be applied to the generated data and the measurement equipment, using regular maintenance and calibration procedures. Only the use of set data quality standards will guarantee the production of accurate, representative, inter-comparable and reproducible information on a local and international level. The production of consistent, representative data of known quality and accuracy can be achieved by using either reference or equivalent techniques. The equivalence of procedures has to be established by inter-comparisons of different monitoring and QA/QC methods.

9.3.2 Emission Inventories

Emission estimates are collected together into inventories or databases which usually also contain supporting data on, for example: the locations of the sources of emissions; emission measurements where available; emission factors; capacity, production or activity rates in the various source sectors; operating conditions; methods of measurement or estimation, etc.

The sources of air pollution are divided in three categories:

- Emissions from single activities of some size, like industries, energy production etc., that are linked to single stacks, are treated as point sources.
- Emissions from road traffic are treated as line sources in the emission database.
- Emissions from home heating and cooling, public and private services, agricultural activities etc., which cannot be represented as a point or line, are treated as area sources. These may also be converted to grid.

9.3.2.1 Point Sources

The point sources or stationary sources are normally linked to individual well-defined stacks. Emission estimates are provided on an individual plant or emission outlet (usually large) usually in conjunction with data on location, capacity or throughput, operating conditions etc.

Examples of point sources include industrial and commercial boilers, electric utility boilers, turbine engines, wood and pulp processors, paper mills, industrial surface coating facilities, refinery and chemical processing operations, and petroleum storage tanks. Area sources that may fall under the point source definition are piping leaks, industrial wastewater treatment ponds, rock and quarry operations, and tank farms. Insignificant point sources are included by category in the area source inventory.

One must pay special attention where both combustion and fuels and processing of materials may have effects on emissions. Fuel mixture as well as specific energy demands may change over time. As a consequence, both fuel input and product output need to be accounted.

Whenever point sources are estimated individually, the estimated sum of the activity represented by these sources should be subtracted from the estimated collective activity. This is to avoid double-counting the individually considered point sources when estimating the rest of the source activity emissions (the collective approach).

As in the case of point sources treated individually in the accounting for processes with combustion, attention should be paid to avoid double counting of energy consumption statistics. Reference activity data may be available from public and private statistics, institutions or research projects. Information on fuels should include non-commercial fuels and wastes used for energy generation.

9.3.2.2 Line Sources

In the majority of European cities, and indeed around the world, mobile emitters, particularly road traffic, are the predominant source of atmospheric emissions. Several contaminants, such as atmospheric lead (Pb), nitrogen oxides (NOx) and carbon monoxide (CO) originate predominantly from road vehicle emissions.

Line sources or mobile sources are normally composed from those emitters that are moving along roads or tracks. In some inventories vehicle emissions from road transport, railways, inland navigation, shipping or aviation etc are provided for sections along the line of the road, railway-track, canal or sea-lane.

The emissions of CO and NO_x from traffic are calculated by multiplying the traffic work with an emission factor. The emission for a given road is a function of:

- Speed
- Road gradient
- Year of calculation (this determines the technology level of the vehicle)
- Number of cars in each vehicle class

The emissions increase with the age of the car. There are also increased emissions from cars in cold start mode. Both of these factors can be accounted for in a model.

The total emission from the road network (t/year) is estimated from the mean daily traffic parameters. The peak emission calculations utilise rush-hour parameters.

The calculation of emissions/generation of PM_{10} (road dust) is usually based on a different method than for the other components. The reason for this is that PM_{10} refers to a 24-h average, whereas CO and NO_2 are 1 h averages.

9.3.2.3 Area Sources

Area sources are used to describe sources where geographical distribution is not exactly known and where emissions are small but in large numbers so that they have a significant impact on concentrations. These kinds of emissions can be from house heating, traffic or various type of land use. The emissions are normally connected to the use of different fuels in an area that is distributed according to population distribution.

Area source models are also used to estimate emissions of ammonium from agriculture. The area sources in a city have local influence, they are linked to consumption and emission factors are needed.

The area sources are most often a collection of smaller or more diffuse sources of pollution. The emission from these large number of scattered small sources are provided on an area basis either for administrative areas, such as counties, regions etc., or for regular grids.

Area source inventories generally report emissions by categories rather than by individual source; a common method in reporting point source emissions. Area source emissions are calculated by various methods and depend on the type of data available for each category.

Major categories of area sources are:

- Stationary source fuel combustion such as residential fuel combustion
- Solvent use (e.g., small surface coating operations)
- Product storage and transport distribution (e.g., gasoline)
- Light industrial/commercial sources
- Agriculture (e.g., feedlots, crop burning)
- Waste management (e.g., landfills)

Miscellaneous area sources (e.g., forest fires, wind erosion, unpaved roads)

9.3.3 Meteorological Models

9.3.3.1 Meteorological Measurements

Meteorological data are important as part of a complete air quality management system. As important input to the dispersion models. Continuous measurement of meteorology should include sensors for the most important parameters such as:

1. Wind speeds
2. Wind directions
3. Relative humidity

4. Temperatures or vertical temperature gradients
5. Net radiation
6. Wind fluctuations or turbulence
7. Precipitation and
8. Atmospheric pressure

The dispersion capacity of the atmosphere may also have to be estimated using meteorological pre-processors in combination with meteorological measurements. Input data for these meteorology model estimates are hourly measurements of wind speed, wind direction, temperature, vertical temperature gradient or net radiation. Measurements of wind fluctuations or turbulence may also be used to estimate the dispersion.

9.3.3.2 Wind Field Modelling

The numerically simulating air pollutant transport and transformation were made in the local-to-regional scale, which, broadly speaking corresponds to the mesoscale. In this context, it has been recognised for a long time that urban scale problems can only be treated successfully by the aid of mesoscale air pollution models where either a large enough domain is considered or accurate boundary conditions are established. The former is in conflict with the limited hardware resources – an aspect of paramount importance for practical applications. For the latter, models with nesting capabilities are required – and those have only recently become available.

Mesoscale air pollution models require at input considerable meteorological information. In the last years, two different approaches were followed in this respect:

- Diagnostic wind field calculation, in conjunction with an empirical parameterisation for turbulence quantities.
- Prognostic calculation of both wind fields and turbulence quantities.

In view of the above, a mesoscale air pollution model usually represents a model system consisting of:

- A wind model (either a diagnostic or a prognostic one) and
- A dispersion model

9.3.4 Atmospheric Dispersion Models

On a spatial scale from about 1–100 km there are several types of numerical models available; both Lagrangian type and Eulerian type models. The Lagrangian type models follow puffs of air pollutants estimating in each puff the turbulent diffusion, chemical reactions and deposition processes. The turbulence description and the diffusion processes may be treated in different ways.

The air quality model requests on local scale are normally:

- *Near-field models* with on-line modelling capabilities, for areas near roads and industrial installations
- *Larger scale models* for prediction of air quality in urban centres
- *Models for accidental releases* of gases (heavy gases or not), with on-line operation capabilities

The needed functionality of the models in relation to other parts of an air quality assessment system is that input data, such as emission data and meteorological data, will be automatically introduced into the models, to enable on-line operation of the models for short-term forecasting. This will be especially relevant in the industrial emission cases related to near-field and accidental releases.

The coupling of the models and the presentation of their results within a GIS system is also a requirement. This type of integration of air quality models into an Air Quality Monitoring and Management System is described below.

For standard air quality objectives the model needs can be divided into the following categories:

- Gaussian type multiple source models
- Puff trajectory type models for local and regional impacts
- Numerical models, local scale dispersion
- Numerical regional scale dispersion models (with chemistry)
- Accidental release models

9.3.5 *Population Exposure*

Exposure of air pollution may occur simultaneously from many sources and through multiple routes. Valid exposure assessment therefore typically requires detailed knowledge about the geographical distribution of the pollutants of concern, the temporal variations in pollution levels, and the processes of exposure.

An air quality management system should contain different methods for estimating the population's exposure to air pollution. Air pollution impact on health can be estimated by combining calculated concentrations, either in grid or building addresses, and the population distribution. Exposure estimates can be used to describe how many people that are exposed to air pollution above air quality guidelines.

Based on dispersion calculation, exposure calculation might be performed by combining the concentrations by population data either in field or in building points.

9.3.5.1 Grid Square Exposure

A first simplified method for estimating population exposure calculations is to calculate only in grid squares. Traffic is a major contributor to the air pollutant peak concentrations in urban areas, and by using this method the peak exposure values will not be calculated and the location not identified.

9.3.5.2 Sub Grid Exposure

Exposure estimates from traffic pollution for a population can be calculated based on information of population distribution by addresses or in buildings, using the sub grid dispersion model to calculate the concentration contribution from traffic to these buildings.

9.3.5.3 Personal Exposure

The two methods above gridded or by addresses, gives both fixed values in given positions or areas. A third method is to introduce personal exposure, meaning that persons are followed in their daily activities(indoor/outdoor, home or work) to give a more realistic description of their exposure to air pollution.

This method is described in the following chapters followed with typical examples in how the people are exposed to air pollution during different activities.

9.4 Urban Exposure Management Tool

9.4.1 Introduction

The exposure management tool has been developed for calculating personal exposure to air pollution. It can be used to study human exposure from compounds that account for up to two important pathways of exposure, inhalation and dermal absorption, and to quantify exposure specifically for particulate matter and chloroform in urban areas. In this way, the assessment of human exposure from indoor and outdoor air to particulate matter and human exposure from house water and ambient air to drinking water disinfection by-products, has become available as a support mechanism for urban management decisions.

The various environmental models described previously have been integrated in the exposure module for calculation of human exposure in urban environments. The module is implemented as part of the Air Quality Management System AirQUIS (Bøhler 1998).

9.4.2 Particulate Matter

Particle concentrations of PM_{10}, $PM_{2.5}$ and PM_1, for both outdoor and indoor environments and the resulting deposition in the respiratory system is calculated. The calculations are made for a person along a predefined daily route and based on

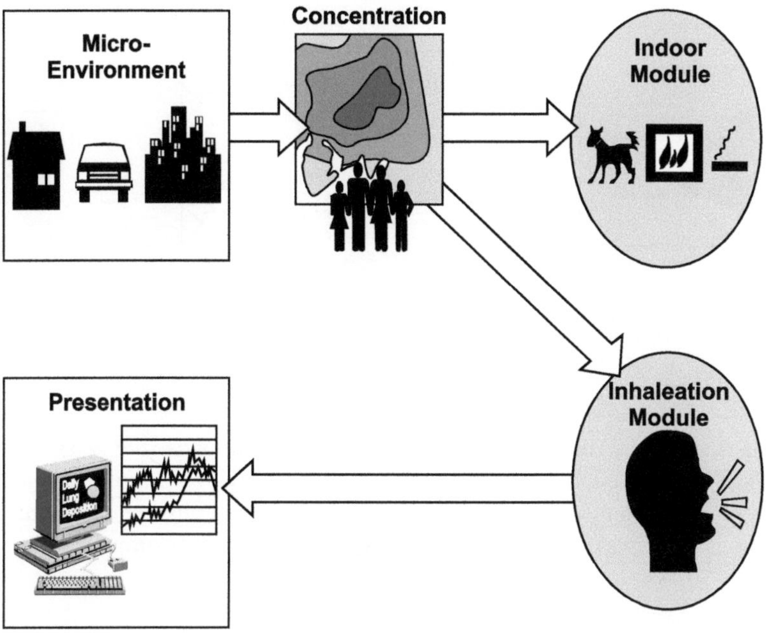

Fig. 9.2 The schematic description of the calculations of particulate matter in urban exposure

a user defined activity pattern. Selected indoor sources are activated over certain periods of the day in the various microenvironments.

For estimating the individual's exposure and respiratory deposition, the dispersion model, the indoor model and the respiratory model are applied (Fig. 9.2). The dispersion model EPISODE in AirQUIS calculates the outdoor concentrations in the individual microenvironments. For a selected period, both PM_{10} and $PM_{2.5}$ are calculated. The particle concentrations from the dispersion model are divided into 48 logarithmically equidistant size bins from 10 nm to 100 μm.

The calculated outdoor concentration is used as input parameter for both the indoor module and the respiratory deposition module (Fig. 9.2). If the person is indoors for a given hour, the indoor module is activated and the calculated outdoor concentration becomes an additional source for the indoor concentration. The calculated indoor concentration is then used as input for the inhalation module. If the person is outdoors during the specified calculation hour, the outdoor concentration is used as input for the inhalation module directly.

9.4.3 *Multi Pathway Gas Uptake*

Uptake of chloroform through inhalation and through skin is calculated as an example of multi pathway gas uptake. Chloroform is a by-product of chlorine disinfection of both tap water and swimming pools.

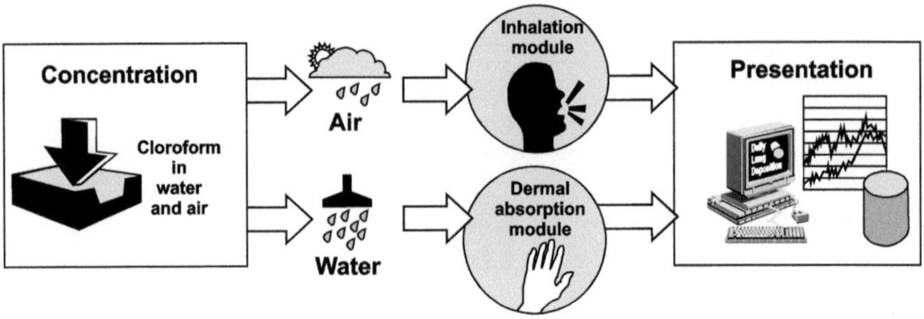

Fig. 9.3 The schematic description of the calculations of multi pathway gas uptake in urban exposure

The dose of chloroform is calculated in two different parts (Fig. 9.3). Based on the water concentration and exposure time, the amount of chloroform accumulated in the blood (systemic dose) is calculated by the dermal absorption module. The respiratory gas uptake is calculated using the respiratory deposition module.

9.4.4 The Tool for Calculating Personal Exposure

The Urban Exposure user interface is shown in Fig. 9.4. The interface has three sub forms:

- Person characteristics and daily routine
- Indoor sources
- Multi pathway gas uptake

The main functionalities of the tool are accessed through the toolbar at the top of the interface.

9.4.4.1 Personal Characteristics and Daily Routine

As mentioned, the outdoor air concentrations determine the outdoor exposure and contribute to the indoor exposure. Therefore the outdoor concentration must be calculated for all the person's microenvironments. Before running the dispersion model for calculating the outdoor PM_{10} and $PM_{2.5}$ concentrations, the geographical positions for all the microenvironments must be defined. For each hour throughout the day, the microenvironments where the person stays are defined. The microenvironments defined are:

- Home
- Work
- Travel to

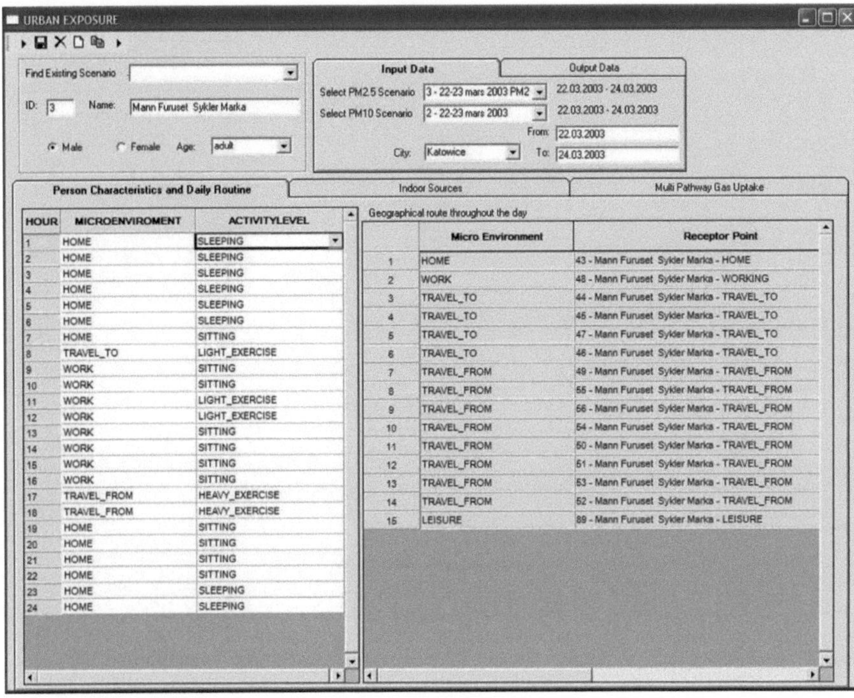

Fig. 9.4 The urban exposure management tool user interface, featuring the form for defining person characteristics and daily routine

- Travel from
- Leisure

The geographical coordinates for a scenario's microenvironment are defined by using a GIS interface as described in Fig. 9.5.

Activity levels are required input for calculating the respiratory deposition of particulate matter. For each hour throughout the day the user must define the scenario person's activity level. The different activity level options are :

- Sleeping
- Sitting
- Light exercise
- Heavy exercise

9.4.4.2 Indoor Sources

The indoor module is activated for the calculation hours the person spends indoors. Various indoor sources can be activated and thereby contribute to the indoor concentrations for specified hours. The indoor module uses discrete 1-h source contributions.

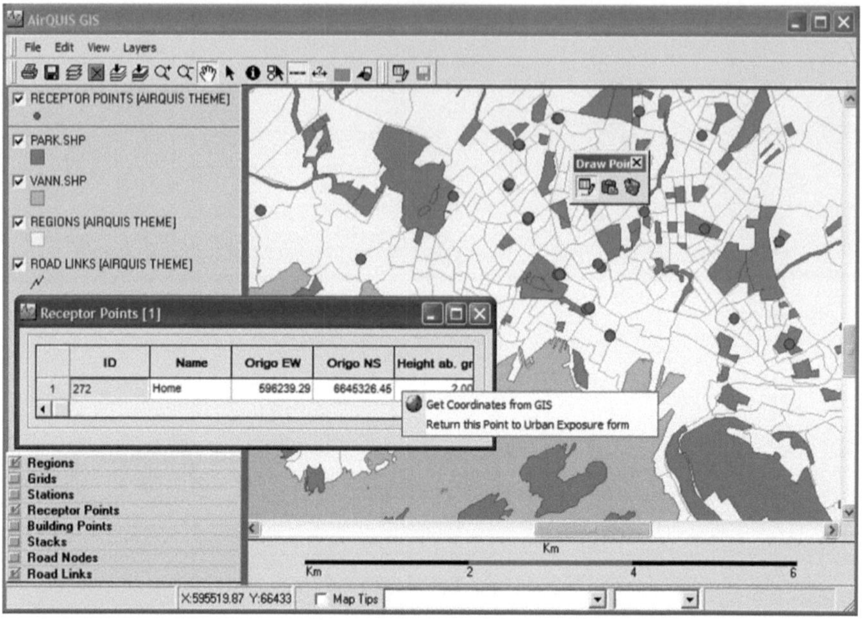

Fig. 9.5 Defining the coordinates for the various microenvironments (exemplified by Home), using the GIS functionality

9.4.4.3 Features

Smoking	Fine fraction, passive smoking
Gas stove	Mostly fine fraction
Indoor heating	Defined as coal fire heating in the current data set
Vacuuming	Coarse fraction
Pets	Mostly Coarse fraction
Filter cleaning	As a sink for particles

9.4.4.4 Multi Pathway Gas Uptake

The Urban Exposure management tool includes an option for calculating the multi pathway gas uptake of chloroform through inhalation and dermal absorption. In addition to specifying personal information, microenvironment and activity level, the exposure time has to be specified and the water source have to be selected between Swimming pool, Bathtub and Shower.

9.5 Case Studies

The Urban Exposure Management Tool has been developed with the needs of the city authorities and national governments in mind and aims at establishing a framework for addresses complexities of interactions between the various meteorological and emission components and actors/inhabitants in the urban areas. Two comprehensive case studies in Katowice and Oslo have been carried out in order to validate and test the Urban Exposure module in the framework of an existing air quality management tool. Only results from Oslo are presented in this chapter. The tool was developed during the Urban-Exposure project (Colbeck et al. 2005).

Katowice (PL) and Oslo (NO) were selected as end user cities in the project. Both urban areas were already using or started using the air quality management systems, AirQUIS. The two cities have somewhat different pollution situations with respect to both chloroform and particulate matter. The use of chlorine in treatment of tap water is for example more common in Katowice than in Oslo. Oslo has rather low chloroform concentrations in tap water since the use of chlorine for disinfection is limited. However, chloroform concentrations in swimming pools can vary in both cities.

In terms of particulate matter, the emission sources for outdoor air are different in the two cities. In Oslo, the main sources are traffic, including dust resuspension and dust production from studded tires, and wood burning. In Katowice, on the other hand, industry inside and around the city is an important contributor to emissions. In addition heating with coal is very common. The different sources to particulate matter also result in different size distributions of particles in the two cities. Oslo has a larger fraction of coarse particles, which is taken into account in the calculations. Data for the available indoor sources to particulate matter is the same for both cities. This is reasonable for smoking, pets, vacuuming, filter cleaning and gas stove. These sources and source strengths are mainly based on data from Germany (Hollander and Windt 2002). Home heating with coal or wood is, of course, also a source to indoor particle pollution. Unfortunately, it was not possible to find sufficient data for wood burning, so that the source information for home heating, for the time being, is based on coal burning data from Katowice.

Two different approaches in defining scenarios were used for the two cities. In Katowice time activity data from questionnaires was combined with data from measurement campaigns and this study formed the basis for scenarios. This makes it possible to compare monitored and modelled concentrations and exposure, which in turn can contribute to improve modelling or to calculate long-term exposure for specific realistic cases. Another possibility, as applied in Oslo, is to use statistical data on population distribution, transport patterns and pollution levels to define several relevant or statistically representative example cases. A more comprehensive version of such an approach would make it possible to assess a statistical distribution of exposure for a whole population in a city.

9.5.1 Oslo Case Study Results

Oslo, the capital of Norway, is the largest city in Norway with ca. 520,000 inhabitants and an area of 450 km^2, whereof only one third is developed.

The city is located in a pot-formation bordering to the fjord in the south and hills up to 450 m high in the north, see Fig. 9.6. The topography of Oslo contributes to

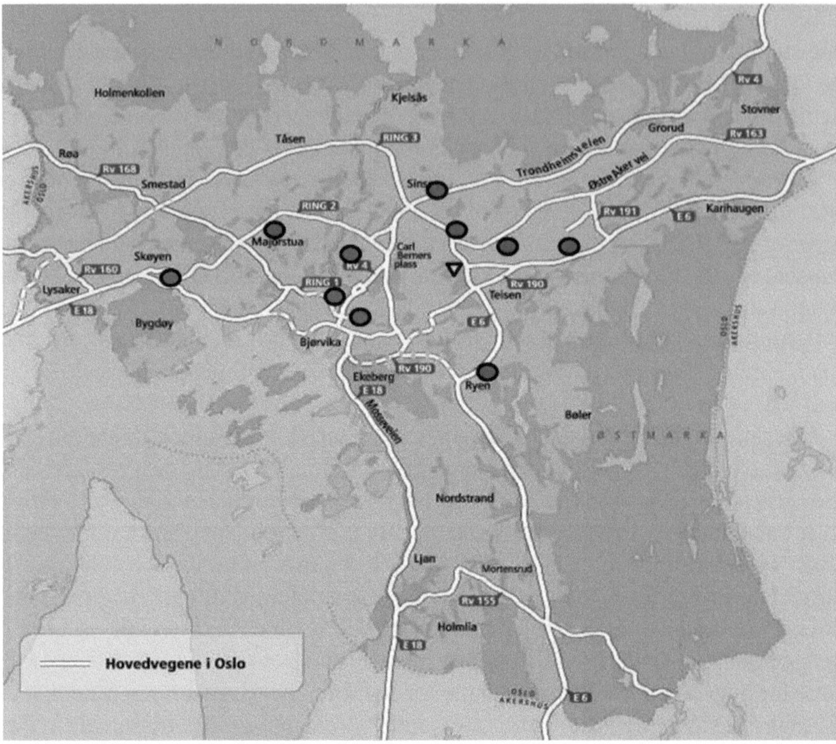

Fig. 9.6 Map of Oslo with monitoring stations (*dots*): developed city area in *grey*, main roads in *white*

inversion episodes in winter and spring, which in turn result in high air pollution levels. The main sources of particles in Oslo are traffic and wood burning in households. Traffic generates particles partly through exhaust, but resuspension of road dust including the use of studded tires is the main pollution source from traffic. The sources of particulate matter outdoors in Oslo influence the choice of scenarios, which study distance to main roads and consider home heating in particular. The number of exceedences for PM10 is given in Fig. 9.7.

All the cases for studying particulate matter have been calculated for 1 day each during the winter, spring and summer season of 2003. In winter the outdoor air is mainly dominated by $PM_{2.5}$, in spring by PM_{10}, while in summer the air quality is generally good and the PM concentrations are low. In December, it is evident that ca. 75% of PM_{10} consist of $PM_{2.5}$. In March, however, the $PM_{2.5}$ fraction of PM_{10} only accounts for ca. 35–40%. Within the framework of the project the following case studies have been calculated and evaluated.

1. Man living in the outskirts and working downtown
   - Indoor transport along a major road
   - Outdoor transport/exercise along the same road
   - Outdoor transport/exercise in areas with little traffic
   - Indoor sources: smoking, pets or filter cleaning at work

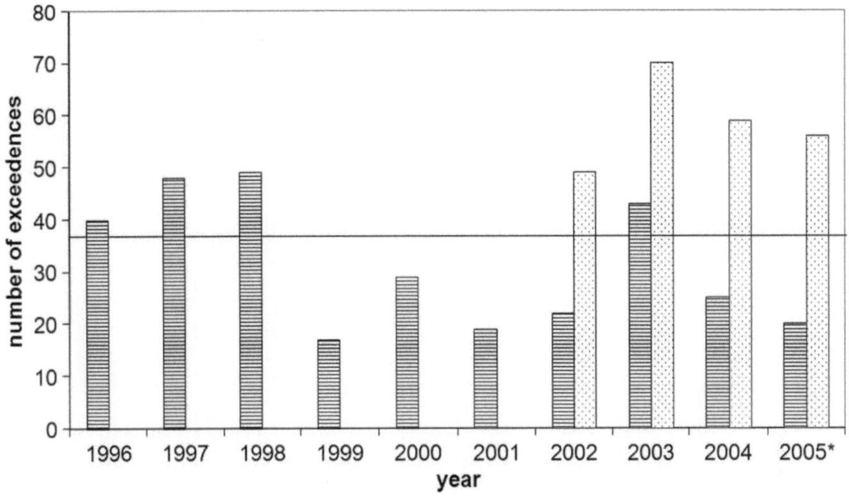

Fig. 9.7 Number of exceedences of the threshold value for PM_{10} for one representative (*dark grey*) and one hot spot (*dotted light gray*) station (*black line* indicates threshold)

2. Living in the same area – comparing
   - Adult and child with work/day-care near major road
   - Child with day-care near major road and child with day-care near the forest
   - Run for all sources – individual and all (adult, sources on all day)
3. Woman working downtown
   - Exercising downtown with and without indoor heating at home
   - Exercising in the forest
   - Filter cleaning at work

9.5.1.1 Case 1: Travelling Alternatives

Figure 9.8 presents results from Case 1 without any indoor sources, to determine outdoor and indoor PM_{10} concentrations, while Fig. 9.9 shows the PM outdoor/indoor size distribution. Even though concentrations of particulate matter (PM_{10}) are clearly higher in outdoor air than in an indoor environment (for the case of no indoor sources), most of the particles that penetrate through the building shell and windows are fine particles. Figure 9.9 shows that the coarse fraction of particles is considerably reduced from outdoor to indoor, while the smaller particles are much less reduced inside. In indoor air almost all particles are $PM_{2.5}$ and approximately half of them are PM_1. Outdoor air pollution thus has an impact on indoor air quality with mainly fine particles penetrating to the inside.

Figure 9.10 present hourly and daily PM_{10} concentrations caused by different indoor emission sources. It is evident that the impact of indoor sources on the concentration is growing with decreasing room size. This is mainly due to a low mixing rate. As of now, the tool uses discrete on-hour indoor source contributions. This means

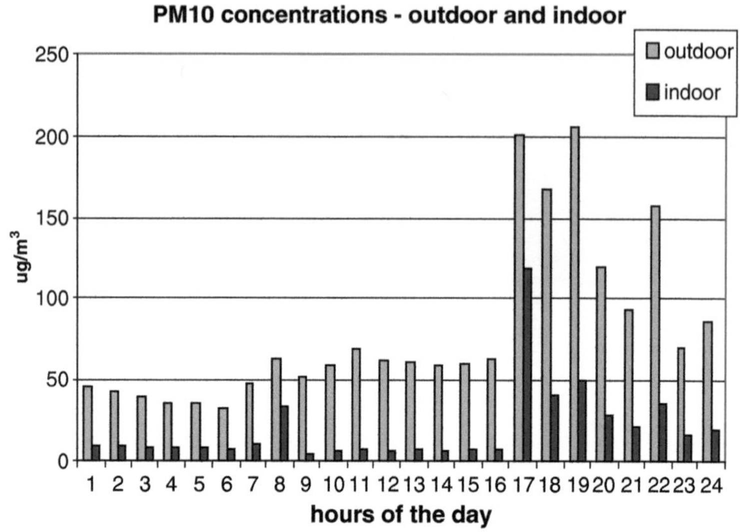

Fig. 9.8 Hourly PM_{10} outdoor and indoor concentrations without indoor sources

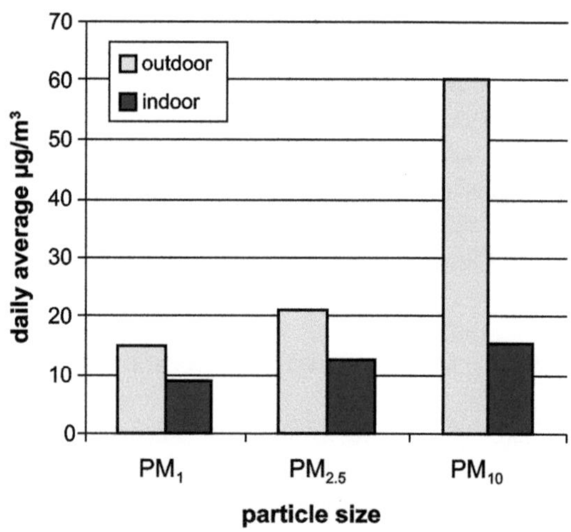

Fig. 9.9 Outdoor and indoor PM size distribution

that indoor sources are either on or off for a given hour and an after-effect is not taken into account. Therefore, when smoking stops in hour 23, for example, the indoor concentrations are reduced to indoor concentration without indoor sources. Furthermore, the source strengths of the indoor sources are calculated as a statistical average of possible source strengths (e.g., variations in smoking frequency). This can cause small variations of indoor particle concentrations, for example for smoking in the evening. Such variations in indoor concentrations can also be due to variations in the outdoor concentrations and/or in ventilation.

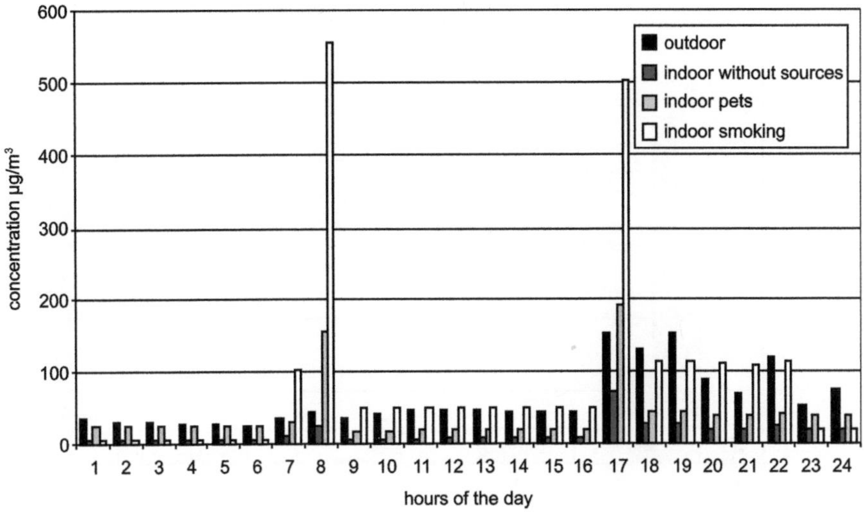

Fig. 9.10 Hourly PM_{10} concentrations caused by different indoor emission sources

Without any indoor sources, the outdoor concentrations are generally much higher than those indoors (Fig. 9.8). However, when considering indoor sources, indoor concentrations of particulate matter are similar to or even higher than outdoor concentrations (Fig. 9.10). These findings correspond to indoor and outdoor PM_{10} measurements carried out in Oslo in 2002 as part of the Urban Aerosol project (Lazaridis et al. 2006).

Another feature of the Urban Exposure management tool is the calculation of inhalation and deposition. Deposition in the respiratory tract is mainly dependent on concentration, activity level and physiological parameters. In the management tool, respiratory deposition is calculated as mass in µg for PM_{10}, $PM_{2.5}$ and PM_1. Particle size is an important factor in deposition. Smaller particles, for example, spread throughout the entire respiratory system and certain types of particles will remain in the body for months up to years. The Urban Exposure management tool displays deposition for the whole respiratory tract. However, deposition in the four regions of the respiratory tract is calculated and can be extracted if desired for all three particle sizes.

Figure 9.11 shows the aggregated dose received in the various microenvironments for Case 1. The dose received in each microenvironment depends, of course, on concentration and activity level, but also on the duration of the stay in the particular microenvironment. In the example displayed in Fig. 9.11, the scenario with smoking gives the highest uptake of all in all microenvironments except for the "travel from"-environment for the scenario with heavy exercise along the major road in a period with high pollution. The high pollution level in combination with heavy exercise results in a very high uptake. Even though exposure is limited to only one microenvironment and to 2 h that day, this extremely high dose dominates the total daily respiratory deposition also compared to the other scenarios. When exercising heavily in less polluted areas, the uptake in that microenvironment is

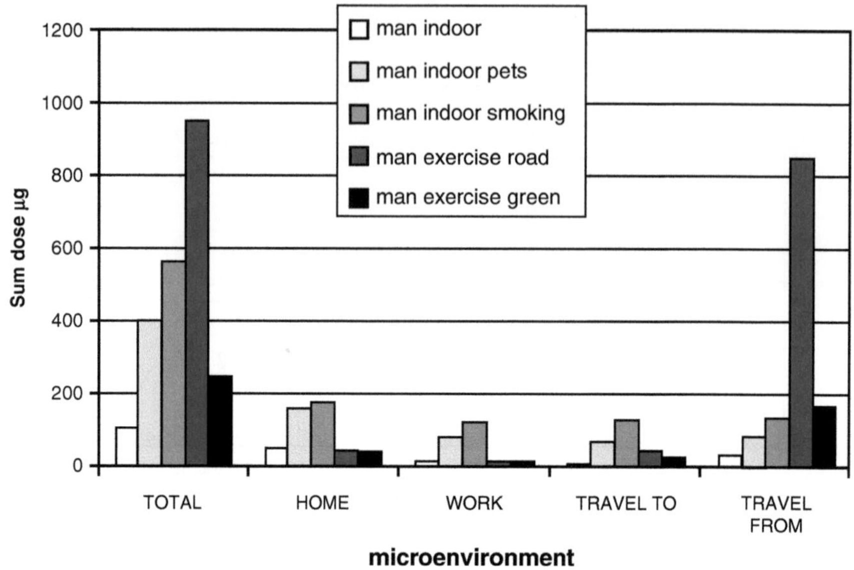

Fig. 9.11 Daily PM total deposition and the doses inhaled from different microenvironments

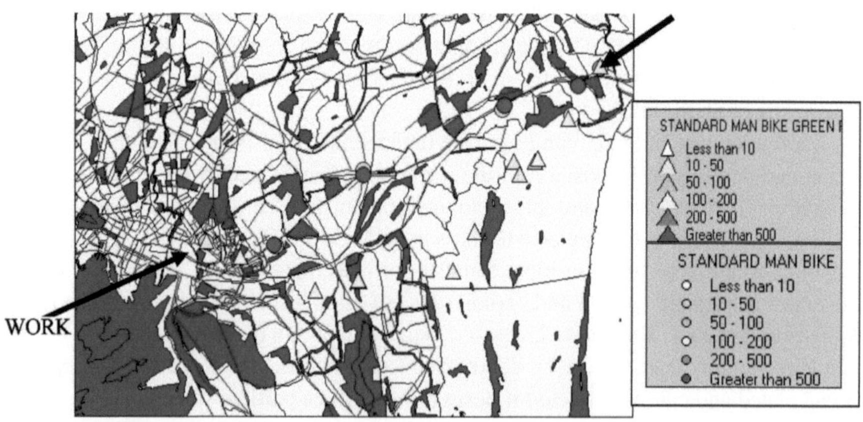

Fig. 9.12 Aggregated respiratory deposition for various microenvironments for the person cycling along a major road or through a less polluted area

increased compared to the other scenarios ("for travel from"). Nevertheless, when considering total daily respiratory deposition this special exposure does not play such a predominant role as when exercising in high pollution.

Respiratory deposition in the different microenvironments can be displayed on a map as in Fig. 9.12. The presentation shows the difference in uptake when exercising along a major road (circle) and in areas with less traffic (triangle).

9.5.2 *Cases 2 and 3*

For cases 2 and 3 an overview of total daily respiratory deposition is displayed in Tables 9.1 and 9.2. For both cases, the uptake for the different scenarios is generally highest for the day in winter. Basically, this can be explained by a high percentage of smaller particles and their deposition as well as the general pollution distribution in Oslo on the selected days.

For the multi-pathway calculations, more general cases were defined comparing four persons of different age and size who are exposed to the same chloroform concentrations presented in Table 9.3.

The hourly respiratory deposition and dermal absorption for a male adult in the course of a day are displayed in Fig. 9.13. Uptake increases with increased concentration

Table 9.1 Total daily doses in microgram for case 2 for three seasons

	Road		Forest
	Adult	Child (5 years)	Child (5 years)
Winter	410.03	247.21	39.34
Spring	230.80	131.08	114.31
Summer	35.46	19.54	15.97

Table 9.2 Total daily doses in microgram for case 3 for three seasons

Woman downtown	Exercise		
	Town		Forest
	With indoor heating	Without indoor heating	Without indoor heating
Winter	1,035.35	582.62	279.47
Spring	627.34	203.87	172.51
Summer	n. a.	24.32	23.93

Table 9.3 Chloroform concentrations in and above water and exposure time applied for all chloroform cases

Water source	$CHCl_3$ in water (μg/l)	$CHCl_3$ in air (μg/m^3)	Exposure times (min)
Shower	9.15	27.46	5, 10, 20
	14	42	5, 10, 20
	15	45	10, 20
	30	90	10
Bath tub	9.15	9.15	20
	14	14	20
Swimming pool	14	39	20, 40, 60
	69	206	20, 40, 60
	25	75	30, 60
	50	150	30

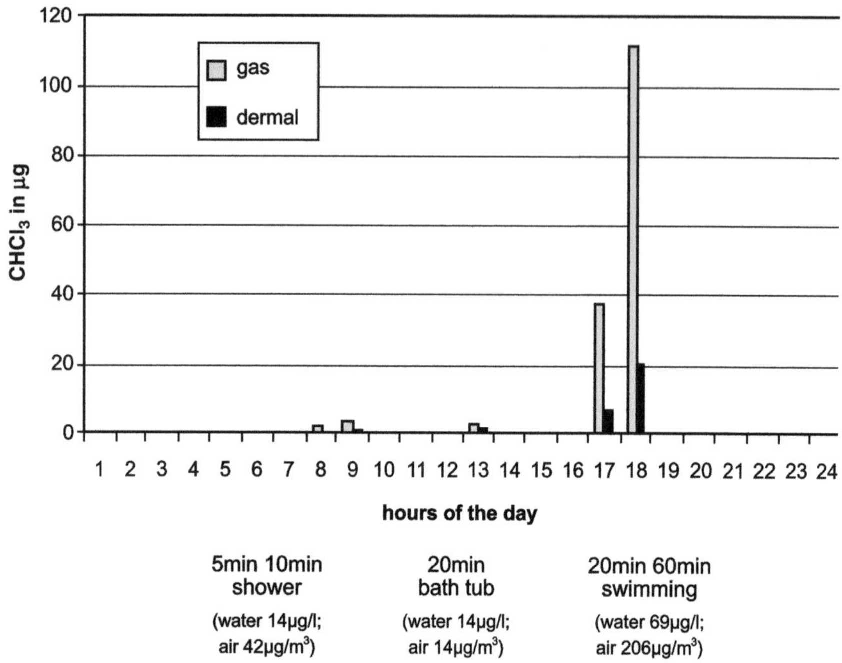

Fig. 9.13 Chloroform dermal and inhaled (gas) dose for an example cases (Male adult, CHCl3 concentration in and above water are given in parenthesis below the graph)

or exposure time. In addition the respiratory deposition is obviously dominant to dermal absorption. This is true for all water sources but is clearly more pronounced for shower and swimming pool where air concentrations are considerably higher than water concentrations.

Comparing total uptake for different exposure times for the cases, it is evident that at the same chloroform concentration levels, uptake increases linearly with increasing exposure time, see Fig. 9.14. In addition, the dose is, as expected, higher the greater the height, weight and lung volume are. However, it is important to bear in mind that children generally are more vulnerable and that high chloroform concentrations, therefore can be more damaging for children than for adults.

Moreover, the water source is important when considering exposure. Figure 9.15 shows the total chloroform dose for the cases for different water sources with the same chloroform concentration in the water. The chloroform concentrations in the surrounding air vary, that is they are highest for a shower environment, a little lower over a swimming pool and lowest over the bathtub. This in turn results in different inhaled doses and the gaseous part of the uptake is, as mention before, predominant. Therefore, total uptake is highest in the shower, a little lower in the swimming pool and less than half in the bathtub for all cases.

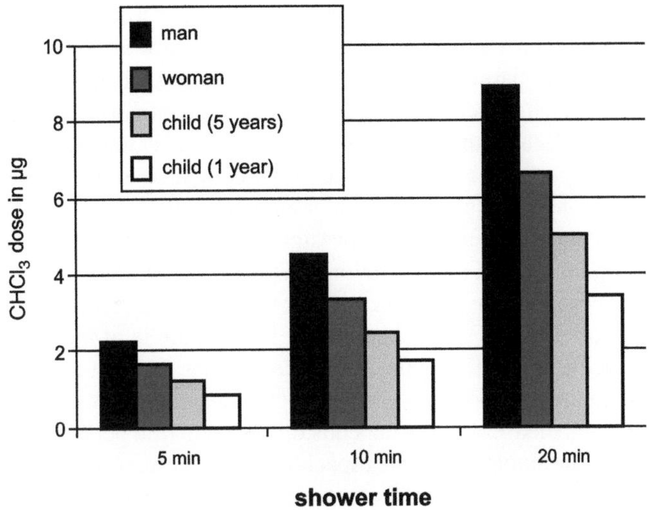

Fig. 9.14 Total $CHCl_3$ dose for different exposure times depending on gender and age (water concentration – 14 μg/l, air concentration– 42 μg/m^3

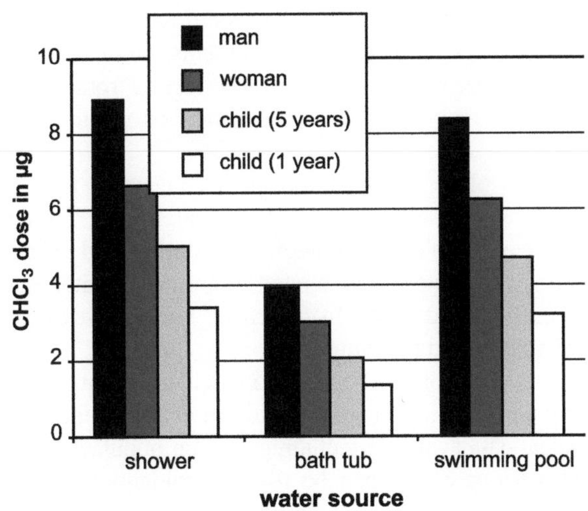

Fig. 9.15 Total $CHCl_3$ dose for different water sources with the same water concentration of $CHCl_3$ – 14 μg/l and time exposure – 20 min

9.6 Summary of Case Studies

The results from these case studies show that migration of particulate matter from the outdoor environment is the main source of indoor PM air pollution. However, occurrence of additional emission sources for example heating sources, cigarette smoking results in an increase of the PM concentration to the level significantly exceeding outdoor concentrations. The indoor PM10 fraction is mainly of outdoor origin. However, the particle size distribution is altered when air penetrates into indoor spaces, which is reflected in the higher concentrations of particles with smaller diameter. Hence, the results show that the concentration of fine particles relative to coarse particles is higher in indoor air compared to outdoor air. This tendency is even stronger in the case of indoor environments with cigarette smokers.

Using the management tool it was possible to assess an impact of indoor emission sources on indoor PM concentrations. Without indoor sources, PM concentrations in apartments can be lower compared to outdoor in the range of 40–70%. Including indoor emission sources can result in an increase of indoor PM concentration by a factor of 10. Based on the defined scenarios, the main indoor source reflecting on the inhaled dose in Oslo was cigarette smoking.

The level of inhaled PM doses depends on the style of a person's life. A higher level of activity in ambient air and especially close to main roads causes higher exposure and increase of respiratory dose of particulate matter. The results show, however, that the home environment makes a substantial contribution to the accumulated daily dose, mostly because a significant part of a day is spent there.

The developed Urban Exposure management tool provides the possibility to calculate more realistic personal exposure than usually provided in air quality management tools. Both following a person on a specific route and calculating concentration in different indoor and outdoor environments, multi-pathway exposure and deposition in the body are new possibilities. The results from calculations provide more detailed and more realistic information on what people in urban environments are exposed to.

The functionalities of the Urban Exposure management tool provide a lot of information, which can be useful in assessing and attempting to reduce exposure of citizens in urban areas. Even though guidelines and threshold values for air quality are presently defined for outdoor air related to certain monitoring and modelling methods, the possibilities and results from Urban Exposure are of particular interest for health related studies and planning.

On the one hand, the tool can be used to calculating statistical distribution of personal exposure and dose in a city. For Oslo, for example, there is a considerable amount of statistical data available including data on population and transport patterns. These calculations could then be used in epidemiological studies, that is comparing exposure, deposition and dose from the tool to data from health studies on hospitalization and occurrences of certain diseases and medical conditions.

On the other hand, the tool could be used to attempt to identify and present the most important sources and locations of exposure both outdoors and indoors. This type of information may in turn be used to set a focus on the most significant factors contributing to health problems from air pollution, possibly in the form of some kind of health strategy. Presumably such strategies would basically be information campaigns such as carried out in the past against smoking and related to abatement measures for reducing outdoor air pollution.

In addition, the Urban Exposure management tool provides information on concentration levels, activity and dose. From a health perspective, the illustration of this interrelation and of the high dose when in activity in polluted areas should be included more actively in planning of outdoor recreational areas, day-care centres for children, bike paths, sports fields and the like. Through this, better air quality in public places in urban areas can be ensured which in turn can contribute to reduce health effects from air pollution.

References

Bøhler, T. (1998). *AirQUIS – A modern air quality management system.* Presented at 5th International Conference on Harmonization, Rhodes, May 1998. NILU F 12/98.

Colbeck, I., Bøhler, T., Broday, D. M., Coulson, G., Fløisand, I., Fudala, J., et al. (2005). *Exposure risks from pollutants in domestic environments: The urban exposure project.* Romanian International Conference on Chemistry and Chemical Engineering, Bucharest, 22–24 September 2005.

Dockery, D. W., Pope, C. A., Xu, X., Spengler, J. D., Ware, J. H., Fay, M. E., et al. (1993). An association between air pollution and mortality in six US cities. *New England Journal of Medicine, 329*, 1753–1808.

EMEP. (2001). *Transboundary particulate matter in Europe: Status report 2001.* (K. Tørseth, L. Tarrason, K. Markus Amman, Eds., EMEP Report 5/2002). Kjeller, Norway: Norwegian Institute for Air Research.

Hollander, W., & Windt, H. (2002). *A simple model for simulating the distribution of indoor particle concentrations and measurements for providing the input data required.* Proceedings from the Sixth International Aerosol Conference Taipei, Taiwan, 8–13 September 2002.

International Commission on Radiological Protection (ICRP). (2002). Guide for the practical application of the ICRP human respiratory tract model. *Annals of the ICRP, 32*(1–2), 1–312. Oxford: Pergamon.

Lazaridis, M., Aleksandropoulou, V., Smolík, J., Hansen, J. E., Glytsos, T., Kalogerakis, N., et al. (2006). Characterization of indoor/outdoor particulate matter physico-chemical characteristics in two residential houses in Oslo, Norway: measurements overview and physical characteristics of the particulate matter – URBAN-AEROSOL project. *Indoor Air, 16*(4), 282–295.

Lazaridis, M., Broday, D. M., Hov, Ø., & Georgopoulos, P. G. (2001). Integrated exposure and dose modeling and analysis system. 3. Deposition of inhaled particles in the human respiratory tract. *Environmental Science Technology, 35*, 3727–3734.

Neuberger, M., Schimekc, M. G., Horak, F., Jr., Moshammerb, H., Kundib, M., Frischerd, T., et al. (2004). Acute effects of particulate matter on respiratory diseases, symptoms and functions: epidemiological results of the Austrian Project on Health Effects of Particulate Matter (AUPHEP). *Atmospheric Environment, 38*, 3971–3981.

Seinfeld, J. H., & Pandis, S. N. (2006). *Atmospheric chemistry and physics, from air pollution to climate change* (2nd ed.). New York: Wiley.

WHO. (2002). *World Health Organisation. Role of human exposure assessment in air quality management* (24 pp., E79501 Report EUR/03/5039760). Joint workshop report, Bonn.

WHO. (2006). *World Health Organisation air quality guidelines, global update 2005* (484 pp.). Copenhagen: WHO. Retrieved June 14, 2007, from http://www.euro.who.int/Document/E90038.pdf

Index

MIX
Papier aus verantwortungsvollen Quellen
Paper from responsible sources
FSC® C105338
FSC
www.fsc.org

If you have any concerns about our products,
you can contact us on
ProductSafety@springernature.com

In case Publisher is established outside the EU,
the EU authorized representative is:
Springer Nature Customer Service Center GmbH
Europaplatz 3, 69115 Heidelberg, Germany

Printed by Libri Plureos GmbH
in Hamburg, Germany